Design Acoustics

Gh. Reza Sinambari

Design Acoustics

Primary and Secondary Noise Mitigation

Springer

Gh. Reza Sinambari
Weisenheim am Berg, Germany

ISBN 978-3-658-40182-5 ISBN 978-3-658-40183-2 (eBook)
https://doi.org/10.1007/978-3-658-40183-2

This book is a translation of the original German edition „Konstruktionsakustik“ by Sinambari, Gh. Reza, published by Springer Fachmedien Wiesbaden GmbH in 2017. The translation was done with the help of artificial intelligence (machine translation by the service DeepL.com). A subsequent human revision was done primarily in terms of content, so that the book will read stylistically differently from a conventional translation. Springer Nature works continuously to further the development of tools for the production of books and on the related technologies to support the authors.

This Springer imprint is published by the registered company Springer Fachmedien Wiesbaden GmbH, part of Springer Nature.
The registered company address is: Abraham-Lincoln-Str. 46, 65189 Wiesbaden, Germany

Preface

The present book is based on my professional experience of more than 40 years in the fields of technical acoustics, sound and vibration cause analyses, design acoustics, and primary and secondary noise reduction. Basic research in the mentioned fields, 7 years of industrial research, and 26 years of university activity predominantly shaped my professional experience. In almost 27 years as managing director of the company IBS-Frankenthal, of which approximately 20 years were part-time, parallel to my employment as a professor at the FH Bingen, I was responsible, among other things, for the low-noise design of numerous machines and systems in practice.

After the introduction in Chap. 1, the physical basics necessary for the understanding of design acoustics are given in Chap. 2. Chapter 3 deals with the sound generation mechanisms, which are explained by examples. Chapter 4 gives the essential measurement techniques necessary for a sound weak point analysis. Chapter 5 describes, in addition to the systematic approach, the principal noise reduction techniques for primary noise reduction in sound generation, transmission, and radiation. Furthermore, the mechanisms of action for secondary measures, especially acoustic hoods and silencers, are given and supported by examples. Finally, Chap. 6 explains the procedure for design acoustics and primary and secondary noise reduction, especially on existing machines and systems, based on practical examples.

I would like to express my sincere thanks to Prof. Dr.-Ing. Stefan Sentpali and Dipl.-Ing. Udo Thorn for their careful review of the manuscript and valuable suggestions. Thanks are also due to Mrs. Veronika Novotny for the typing work. I would like to thank Dipl.-Ing. Thomas Zipsner, editorial office, for his thorough proofreading and useful comments. I would also like to thank my dear wife Ursel for her unlimited understanding and support.

Neustadt an der Weinstraße
January 2017

Gh. Reza Sinambari

Preface to the English Edition

This book is an English translation of the German edition of 2017. The content of the English translation has not changed. Only several errors have been eliminated and small corrections have been made.

The translation was done by Springer with the help of artificial intelligence. A subsequent human revision became necessary to correct many specific technical terms which had not been translated correctly or have not reproduced the idea or didactic flow of the book material appropriately.

By thorough proofreading by experts who are also familiar with the English language, the non-correct expressions have been corrected as far as possible. Despite all care, it is possible that some translations are not fully correct or lead to confusion.

I would like to apologize in advance for any mistranslations and encourage feedback and assistance of readers so that the errors can be corrected in the next edition.

I would like to thank Prof. Dr. Josef Kallrath (Weisenheim am Berg, Germany) for his thorough proofreading of the English edition. Special thanks are also directed to the staff of the company IBS Ingenieurbüro für Schall- und Schwingungstechnik GmbH (Frankenthal, Germany), Dipl.-Ing. Udo Thorn, Dipl.-Ing. Eric Tschöp, and Dipl.-Ing. Andreas Sinambari for checking and proofreading.

Weisenheim am Berg, Germany
November 2022

Gh. Reza Sinambari

List of Formula Sizes

Symbol	Meaning	Units
A	Equivalent absorption area	m^2
$\overline{A}_{\text{total}}$	Average total absorption of a room	m^2
a,b,c	Dimensions of the measuring or enveloping surface	m
a_0	Reference acceleration (freely selectable)	m/s^2
a_e	Vibration acceleration at the point of discharge	m/s^2
a_i	Longest edge of a rectangular duct	m
B	Auxiliary variable	–
B	Silencer width	m
B'	Longitudinal rigidity of the plate	N m
c	Speed of sound	m/s
c_B	Flexible shaft speed	m/s
c_{De}	Strain wave velocity	m/s
c_F	Speed of sound in a fluid	m/s
c_L	Sound velocity of the surrounding air	m/s
C_{met}	Meteorological correction	dB
d	Damping coefficient	N s/m, kg/s
d	Nozzle diameter, width, insulation thickness	m
d	Half-width	m
d_i	Inner diameter of a pipeline	m
d_{gap}	Gap width	m
D	Attenuation of an absorption silencer	dB
D	Sound level difference	dB
$D(f)$	Insertion insulation measure	dB
D_2	Impeller diameter	m
D_h	Standardized damping	dB
E	Modulus of elasticity	N/m^2

(continued)

Symbol	Meaning	Units
e	Thickness of the damping material	m
$E_0 = 1$	Reference modulus of elasticity	N/m^2
E_{dyn}	Modulus of elasticity of the damping material	N/m^2
E_F	Volume elasticity of the fluid	N/m^2
F	Force, excitation force	N
f	Frequency	Hz
$F(f)$	Spectral function of the force	N s
f_0	Natural frequency, fundamental frequency of pulsation	Hz
$f_{0,\ \mathrm{P}}$	Natural frequency of the plate	Hz
f_d	Natural frequency of the damped oscillation	Hz
f_d	Pass frequency	Hz
f_g	Cut-off frequency, coincidence frequency	Hz
$f_{\mathrm{vortex\ frequency}}$	frequency of vortex shedding	Hz
f_k	Harmonics	Hz
F_k	Alternating force at the k-th discharge point	N
$f_{m,\mathrm{oct}}$	Octave center frequency	Hz
$f_{m,\mathrm{third\text{-}oct}}$	Third-octave center frequency	Hz
$F_n(f)$	Partial amplitudes of the line spectrum of intermittent force pulses	N
f_{Os}	Oscillation frequency	Hz
$f_{\mathrm{Os},\ k}$	Harmonic of the oscillation frequency	Hz
f_p	Harmonics of the pulsation frequency	Hz
F_{pl}	Pressure intensity index	dB
f_R	Ring strain frequency of the pipeline	Hz
f_{gap}	Splitting frequency, splitting tones, fundamental frequency of vortex shedding	Hz
f_U	Unbalance frequency	Hz
h	Plate thickness, wall thickness	m
h	Admittance or the mobility	s/kg
h	Half slit width	m
H	Silencer height	m
$h_{\ddot{\mathrm{u}}_{\mathrm{k,j}}}$	Transfer function between the *kth* discharge point and the *jth* subarea or component	–
I	Sound intensity	W/m^2
I_0	Reference sound intensity	W/m^2
I_α	Absorbed sound intensity	W/m^2
I_e	Incident sound intensity	W/m^2
I_H	Sound intensity in the diffuse field	W/m^2
I_r	Reflected sound intensity	W/m^2
$\mathrm{j} = \sqrt{-1}$	Imaginary unit	–
J_S	Integral area below that of the curve $v(t)$ in the time domain Δt	m

(continued)

Symbol	Meaning	Units
J_S	Pulse strength	N s
k	Wave number	1/m
k	Natural numbers (1, 2, 3, ...)	–
k	Spring stiffness of the insulation	N/m
K	Correction level	dB
K	Systematic error measure	dB
K_0	Sound characteristic impedance	dB
K_1	Correction level for extraneous noise	dB
K_2	Ambient correction for room reflections	dB
K_A	Proportionality factor of the intake noise	–
K_α	Correction level to take into account the sound power reduction within the pipeline	dB
K_D	Proportionality factor of the dipole sources	–
K_D	Correction level to take account of sound reflections in the pipeline	dB
K_I	Surcharge for impulsiveness	dB
K_m	Correction level due to the mass of the accelerometer	dB
K_n	Proportionality factor at exponent n	–
K_p	Factor for the partial amplitude of the sound velocity in the frequency spectrum	–
K_p	Parameter for pulsation frequencies	–
K_Q	Proportionality factor of quadrupole sources or free-radiated noise	–
K_R	Surcharge for rest periods	dB
K_T	Appreciation of tones	dB
K_v	Correction level due to interference or extraneous vibrations	dB
K_Z	Parameter	–
L	Level	dB
l	Silencer length, length	m
L_A	A-weighted level	dB(A)
L_a	Acceleration level	dB
L_{Aeq}	Continuous energy equivalent sound pressure level	dB(A)
L_d	Workspace	dB
$L_{\mathrm{EX,\ 8\,h}}$	Daily noise exposure level	dB(A)
L_F	Noise or extraneous level	dB
L_G	Total measured level	dB
L_I	Sound intensity level	dB
L_M	Level of the machine without interference level	dB
L_p	Sound pressure level	dB
L''_p	Sound pressure level with the machine switched off, noise level	dB
$L_{p,\,\mathrm{i}}$	Internal sound pressure level	dB

(continued)

Symbol	Meaning	Units
L'_p	Sound pressure level in the presence of the noise level	dB
L_{pA}	A-sound pressure level	dB(A)
L_r	Rating level	dB(A)
L_S	Square measure	dB
L_v	Speed level	dB
L_W	Sound power level	dB
$L_{W,\text{ total}}$	Total sound power level	dB
$L_{W,\text{ K}}$	Total body sound power level	dB
$L_{W,\text{ L}}$	Total airborne sound power level	dB
L_{WA}	A-sound power level	dB(A)
L_{Wa}	External sound power level	dB
$L_{Wi,\text{ Dp}}$	Internal sound power level caused by the pressure equalization processes	dB
$L_{Wi,\text{ DV}}$	Internal sound power level, caused by the temporal fluctuations of the flow rate	dB
$L_{Wi,\text{ total}}$	Total internal sound level	dB
$L_{W,\text{ spec}}$	Specific sound power level	dB
m	Mass	kg
M	Total number of sub-areas	–
m''	Mass occupancy	kg/m^2
Ma	Mach number	–
m_A	Mass of the accelerometer	kg
m_b	Moving resp. dynamic mass	kg
m_F	Dynamic mass of the spring element	kg
m_z	Additional mass	kg
n	Number of levels	–
n	Speed	1/min
n	Exponent	–
N	Number of measuring points	–
p	Sound pressure	N/m^2
P	Sound power	W
p_0	Atmospheric pressure (resting pressure)	N/m^2
P_0	Reference sound power	W
p_0	Reference sound pressure	N/m^2
P_D	Sound power of dipole sources	W
P_{DV}	Internal sound power due to temporal fluctuations of the flow rate	W
P_{total}	Total sound power	W
$P_{i,\text{Dp}}$	Internal sound power due to the pressure equalization processes	W
P_K	Structure-borne sound power	W
$P_{\text{K}}(f)$ $P_{\text{K},\,j}(f)$	Total structure-borne sound power or structure-borne sound power of the *i-th* component at frequency *f*	W
P_L	Airborne sound power	W

(continued)

Symbol	Meaning	Units
$P_L(f)$ $P_{L,i}(f)$	Total airborne sound power or airborne sound power of the *i-th* opening at frequency f	W
P_M	Mechanical performance of the machine	W
P_Q	Sound power of the source	W
P_{S_i}	Partial sound power of the surface S_i	W
P_S	Flow capacity of a turbomachine	W, kW
p_{sta}	Static pressure	N/m^2
$p_{sta,0}$	Static reference pressure	N/m^2
p_W	Effective amplitude of the sound pressure	N/m^2
$\widetilde{p}$	RMS value of sound pressure	N/m^2
r	Distance, radius	m
r	Reflection factor	–
R	Sound reduction index	dB
R	Specific gas constant	N m/ kg K
R_0	Reference gas constant	N m/ kg K
Re	Reynolds number	–
R_f	Sound reduction index of the pipe wall at frequency f	dB
r_H	Reverberation radius	m
R_i	Sound reduction index of the components or partial areas	dB
R_m	Average sound insulation value	dB
r_N	Near-field radius	m
R_{res}	Resultant sound reduction index	dB
R_S	Flow resistance	$N\,s/m^3$
r_S	Length-related flow resistance	$N\,s/m^4$
S	Area, measuring area, enveloping area, radiating area, flow cross-section	m^2
S_0	Reference area	m^2
S_{total}	Total area	m^2
S_K	Capsule surface	m^2
$S_{Ö}$	Opening area	m^2
St	Strouhal number	–
St_{oct}	Strouhal number at the octave center frequency	–
$St_{third\text{-}oct}$	Strouhal number at the one-third octave center frequency	–
T	Measuring time	s
T	Reverberation time	s
T	Absolute temperature	K
t	Temperature	°C
t	Time	s
T	Time intervals of intermittent force pulses	s
T_0	Period of an oscillation, period of oscillation	s

(continued)

Symbol	Meaning	Units
T_0	Reference temperature	K
T_0	Assessment time for working noise	h
T_d	Period of oscillation of the damped oscillation	s
T_e	Effective exposure time	h
T_r	Assessment time	h
T_S	Sabine's reverberation time	s
t_{St}	Application time of the force pulse	s
u	Flow velocity	m/s
U	Circumference of the plate supported moment-free on all sides	m
U	Absorbent lined perimeter	m
U_0	Incoming flow velocity	m/s
$u_0 = 1$	Reference speed	m/s
u_2	Peripheral speed of the impeller	m/s
u_s	Splitting speed, speed between the baffles	m/s
v	Sonic	m/s
V	Volume, room volume	m^3
$\dot{V}$	Volume flow	m^3/s m^3/h
v_0	Reference sound fast	m/s
$v_{e,\ k}$	Vibration velocity (velocity) at the discharge point *k*	m/s
$v(f)$	Fourier spectrum of the sound velocity	m
$v(f_p)$	Partial amplitude of the sound velocity in the spectrum	m/s
v_i, v_j	Component of the velocity fluctuations	m/s
$v_{j,\mathrm{k}}$	Mean velocity of the *j-th* partial surface caused by the *k-th* alternating force	m/s
$\widetilde{v}$	RMS value of sound velocity	m/s
W_r	Energy recovered per vibration	N m
W_v	Energy lost or converted into heat per oscillation	N m
x	Vibration path	m
x_i, x_j	Component *i* and *j of* the coordinate directions	m
$\dot{x}$	Vibration velocity	m/s
$\ddot{x}$	Vibration acceleration	m/s^2
Z	Acoustic impedance	$N\ s/m^3$
z	Number of delivery volumes per revolution, number of pulsations per revolution	–
z	Number of blades, number of teeth, number of vanes	–
Z_1, Z_2	Impedance of the medium 1 or 2	$N\ s/m^3$
Z_D	Damper impedance	kg/s

(continued)

Symbol	Meaning	Units
Z_e	Mechanical input impedance	N s/m, kg/s
$Z_{e,\ k}$	Mechanical input impedance of the *k-th* introduction point	kg/s
Z_F	Spring impedance	kg/s
Z_m	Mass impedance	kg/s

Symbol	Meaning	Units
$\|\alpha\|$	Magnitude of the amplitude frequency response of the floor force during force excitation	–
∂	Symbol partial differential	–
α	Angle, phase shift	grad, °
α	Sound absorption coefficient	–
α	Frequency-dependent absorption coefficient	–
δ	Decay constant	1/s
Δ	Laplace operator	$1/m^2$
Δf	Bandwidth	Hz
Δf_{oct}	Bandwidth of octaves	Hz
$\Delta f_{\text{Third-oct}}$	Bandwidth of Third-octave	Hz
Δ_φ	Phase difference	°
ΔL	Correction level	dB
$\Delta L_{\text{A, fm}}$	Attenuation level of the A-weighting at the center frequency f_{m}	dB
ΔL_{oct}	Attenuation level of the A-weighting for octaves	dB
$\Delta L_{\text{third-oct}}$	Attenuation level of the A-weighting for Third-octav	dB
$\Delta L_{W,\ \text{oct}}$	Relative octave spectrum of sound power	dB
$\Delta L_{W,\ \text{third-oct}}$	Relative third octave spectrum of sound power	dB
$\Delta L_{W\text{d}}$	Blade frequency cone surcharge	dB
ΔL_z	Level surcharge for reflections in the pressure line	dB
Δp	Static pressure difference	N/m^2
$\Delta p_0 = 1$	Reference of pressure difference	N/m^2
δ_{plo}	Cross-field suppression	dB
Δp_{t}	Total pressure increase of the fan	N/m^2
Δp_{th}	Theoretical pressure increase of the fan	N/m^2
Δp_{v}	Total pressure drop of the fan	N/m^2
Δr	Distance	m
Δt	Duration of compression or pulse duration during pressure equalization	s
ΔV	Volume compression	m^3
Δz	Level surcharge for reflections in the pressure line	dB
ε	Stretch	–

(continued)

Symbol	Meaning	Units
ϕ	Speed potential	m^2/s
ϕ'	Auxiliary variable	m^3/s
η	Fan efficiency	–
η	Loss factor	–
η	Frequency ratio	–
η	Normal frequency	–
η_A	Acoustic efficiency of the machine	–
η_u	Auxiliary variable	–
φ	Phase angle	grad, °
φ	Number of deliveries	–
κ	Adiabatic exponent	–
κ_0	Reference adiabatic exponent	–
λ	Wavelength	m
Λ	Logarithmic decrement	–
Λ	Standard lining depth	–
λ_B	Bending wavelength	m
λ_d	transmission frequency wavelength	m
λ_L	Air wavelength	m
μ	Transverse contraction coefficient	–
μ	Relative temporal change of the volume flow	–
Ψ	Print number	–
ρ	Density	kg/m^3
ρ_0	Reference density	kg/m^3
$\rho_0 \cdot c_0$	Reference impedance	$N\,s/m^3$
ρ_D	Density of the damping material	kg/m^3
ρ_W	Density of the pipe wall	kg/m^3
σ	Radiation factor	–
σ	Voltage	N/m^2
σ	High-speed number	–
σ'	Logarithm of the radiation factor	dB
σ'^*	Auxiliary quantity related to the logarithm of the radiation factor	dB
σ_R	Standard deviation	dB
ϑ	Degree of damping	–
ω	Circular frequency	1/s
ω_d	Natural angular frequency of the damped oscillation	1/s
ζ	Pressure loss coefficient of the nozzle	–

Contents

Introduction and Motivation

Noise reduction is always associated with the reduction of people's exposure to noise. This is due to the fact that the term "noise" is understood to mean the unpleasant part of the subjective perception of sound events by the human ear.

The task of design acoustics is to avoid or reduce the noise development of machines responsible for disturbing sound events by suitable measures, preferably design measures.

Noise reduction measures are divided into **primary** and **secondary** measures. The task of primary measures is to avoid or reduce the generation of sound at the source, i.e. to reduce the **sound emission**. Secondary measures are intended to avoid or reduce the exposure of people to noise if the noise generation remains unchanged, i.e. to reduce the **noise immission**.

The prerequisite for the development of suitable noise reduction measures is the knowledge of the sound generation mechanisms and the localisation or identification of level-determining sources, components or frequencies. In this context, a distinction is made between two cases.

I.	Low-noise and low-vibration design of a newly developed product, machine or system
II.	Low-noise and low-vibration optimization of existing products, machines and systems

In case I, it is possible to **roughly** describe the air-borne and structure-borne sound generation, transmission and radiation only when design drawings have been submitted and all operating data have been determined

In this case, further investigations, such as FEM or CFD calculations, may be necessary. The problem here is the reliability of calculation data, since some influencing parameters, such as clamping conditions, damping, backlash, etc., cannot always be modelled accurately. Also, the necessary dynamic material parameters are often not available, especially at higher frequencies. However, it is possible to avoid at least the rough acoustic weak

G. R. Sinambari, *Design Acoustics*,
https://doi.org/10.1007/978-3-658-40183-2_1

points in the planning phase. Very helpful here are the experiences from previous products or constructions, which can be integrated into the new design. After constructing the first prototype, the exact data should then be objectively determined by suitable acoustical and vibration-related investigations, and the effectiveness of the measures planned in the planning phase should be checked.[1, 2]

In case II, all necessary data for sound and vibration optimization can be determined through measurements on existing machines and systems. By means of a sound and vibration weak point analysis, it is possible to localise the main noise sources, to determine airborne and structure-borne noise development as well as level-determining frequency components, and to determine the necessary noise reduction measures. Based on the data determined in this way, suitable primary and secondary noise reduction measures can then be developed.

With the help of further investigations, e.g. problem-oriented time analyses, it is possible to determine the exact causes of decisive work processes. Figure 1.1 shows the time analysis of a production machine.

Through this analysis, it was possible to prove, among other things, that the largest forces occurring over time are not responsible for the maximum A-sound pressure level. The highest A-sound pressure level occurs at an earlier point in the course of time of the force progression, when smaller forces are present. The reason for this can be seen in the fact that in the case of sound generation it is not the maximum force that is responsible, but above all the gradient or slope, i.e. dF/dt, of the force-time curve. With the help of these investigations, which can of course only be carried out on existing machines, it was possible to localise the decisive work process for the generation of sound. The noise reduction of approx. 8 dB(A) was achieved, among other things, by modifying the machine control, in which the critical force-time curve was stretched, i.e. the gradient (dF/dt) was reduced. The exact mechanisms of action for the force-time curve are discussed in Chaps. 3 and 4.

With the help of special measuring techniques, e.g. laser scanning vibrometry, it is possible to visualise the structural vibrations on existing machines, which are responsible for the structure-borne noise radiation or noise development of machines. This makes it possible to clarify complicated relationships. Figure 1.2 shows the vibration modes of a brake disc and a brake caliper of a passenger car measured with the aid of a laser scanning vibrometer. From this it can be seen that, depending on the braking condition, different natural frequencies of the brake disc, unbraked 982 Hz (Fig. 1.2a) and braked 932 Hz (Fig. 1.2b), are excited or become effective.

[1] ISO 11688-2: Acoustics – Guidelines for the design of low-noise machinery and equipment – Part 2: Introduction to the physics of noise reduction by design measures, 2001.

[2] VDI-3720, Sheet 1: Design of low-noise machinery and equipment, design tasks and methodology, 2014.

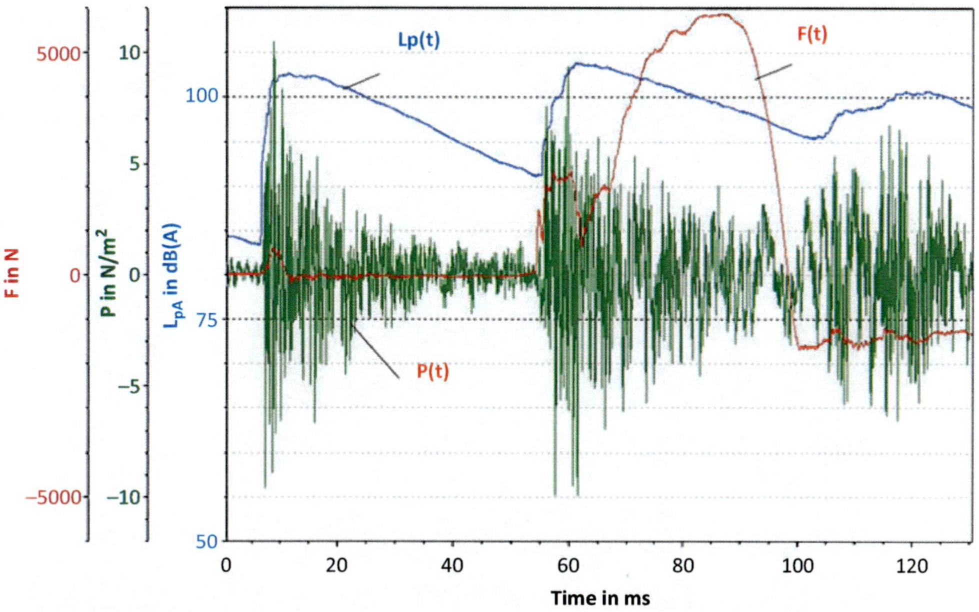

Fig. 1.1 Measurement of the force-time curve or level-time curve of a production machine (IBS-Frankenthal-Germany)

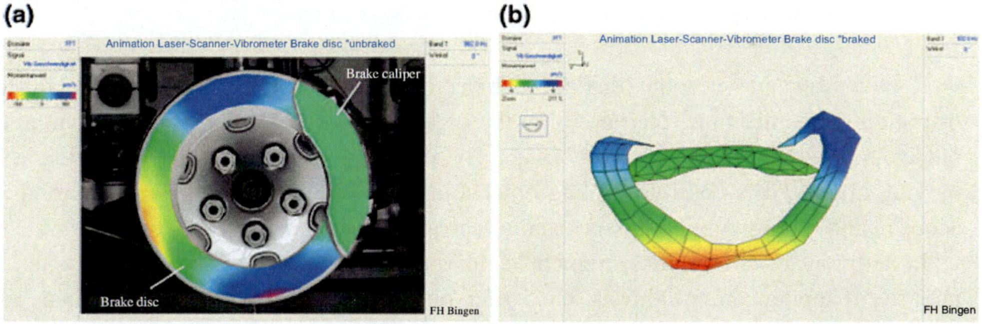

Fig. 1.2 Vibration modes of a brake disc and a brake caliper of a passenger car. (**a**) 2D representation, unbraked (982 Hz), (**b**) 3D representation, braked (932 Hz)

Furthermore, Fig. 1.2b shows that when the brakes are applied, the brake disc and the brake caliper oscillate in phase opposition at 932 Hz, i.e. the two components oscillate at the same frequency and amplitude but in exactly the opposite direction. Such investigations are helpful, for example, in clarifying squeaking or creaking noises[3] in the vehicle, which are primarily caused by friction between surfaces.

[3] F. Klein: Psychoacoustic evaluation of brake creaking noises, Dipl. thesis FH Bingen, 2006.

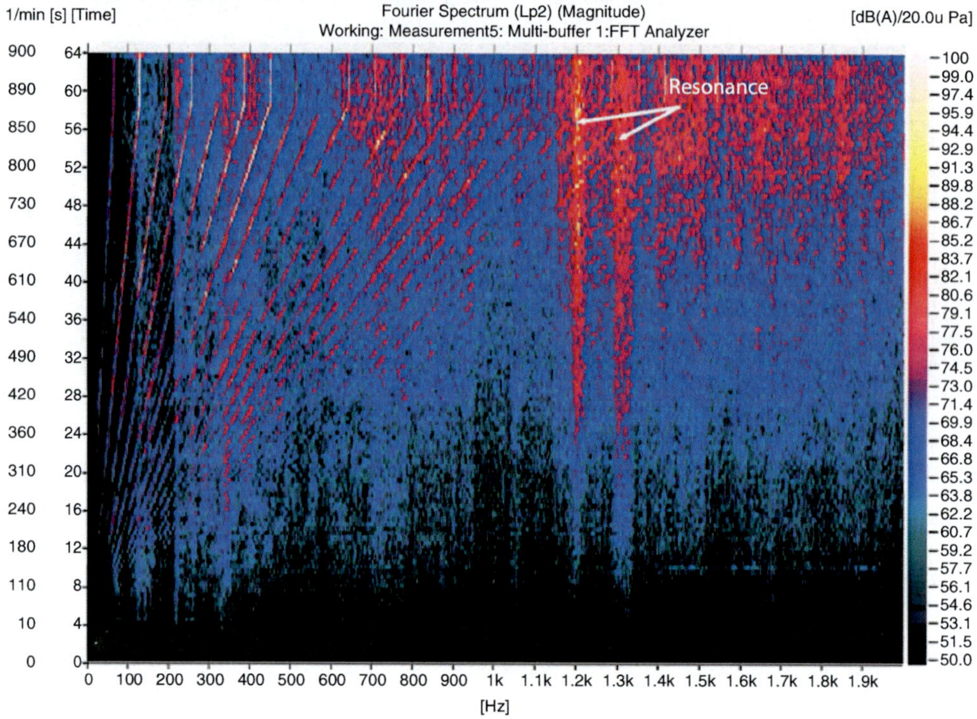

Fig. 1.3 Order analysis of a rotating machine (IBS-Frankenthal-Germany)

If one takes into account that these investigations can also be carried out during operation of the machine, then not only the eigenmodes but also the operating vibration modes can be determined or represented. By storing object photos in connection with moving images, it is possible, among other things, to better clarify the understanding of acoustic and vibration-related issues and to carry out more precise cause analyses.

In addition to various measurement techniques, special analysis techniques, such as intensity mapping, order analysis, etc., offer further possibilities for cause analysis and description of sound generation. Figure 1.3 shows the order analysis of a rotating machine (Campbell diagram).

Among other things, this can be used to determine the critical resonances of rotating machines, which are often responsible for excessive noise development.

In the context of this book, in addition to physical fundamentals, above all the methodical approach to the development of primary and secondary noise reduction measures on existing machinery and equipment (Case II) is presented.

A main objective of this book is to reduce the noise produced by existing machines and/or plants, based on experimentally determined critical acoustic points, preferably by primary measures. Therefore, the identification of critical sources of noise of a machine and/or plant for the development of suitable measures is an essential part of this book.

In addition to the general principles, Chap. 2, knowledge of the sound generation mechanisms, Chap. 3, is decisive for this. In Chap. 4, the measurement techniques necessary for the determination of sound and vibration parameters and the performance of an acoustic weak point analysis are presented. The development of suitable primary and secondary measures, based on these findings and characteristic values, is given in Chap. 5. Chapter 6 deals with the implementation and the systematic approach to noise and vibration reduction by means of design acoustic measures on the basis of selected application examples from practice.

An essential task in design acoustics is to know, for each planned noise reduction measure, the reason for which this measure is intended and what it is intended to achieve. If a certain measure, which is specifically used to solve an acoustic problem, does not achieve the planned noise reduction, it is absolutely important to know the reasons for this. This supports planning further steps more accurately, avoids unnecessary work and, of course, reduces costs.[4, 5]

Finally, it should be noted that the objective of this book is not to design a low-noise machine on the drawing board by means of "low-noise design". However, the special knowledge, especially about the generation of sound and primary noise reduction, is also very helpful with regard to "low-noise designs" for new machine developments.

[4] Sinambari, Gh. R.; Kunz, F.: Primary noise reduction by acoustic weak point analysis. VDI Reports No. 1491 (1999).

[5] Sinambari, Gh. R.: Low noise design using acoustical weak point analysis. Construction **10** (2001).

Physical Basics

The physical foundations are the prerequisite for understanding the interrelationships in design acoustics and/or noise reduction. The theoretical foundations of the sound field have been dealt with in detail in Engineering Acoustics [1]. In the following, only the essential basics and terms are reproduced, which should help to understand the interrelationships in Chaps. 3, 4, 5 and 6 as they are treated in this book.

2.1 Terms

Sound

The transport of energy in an elastic medium in the form of mechanical vibrations (waves) is referred to as sound. Sound is an **objective quantity** and can be measured. The frequency range below 16 Hz is called infrasound, that above 16,000 Hz is called ultrasound.

In the acoustically audible frequency range, approx. 16–16,000 Hz, various terms are used depending on the medium:

Airborne sound	in air or gases
Fluid or liquid sound	in liquid
Structure-borne sound	in solid structures

Noise

Noise is a **subjective quantity** and characterises the perception of sound processes by the human ear and the evaluation in the brain. Noise cannot be measured directly. When a sound event becomes noise or noise pollution depends on many factors and situations and

G. R. Sinambari, *Design Acoustics*,
https://doi.org/10.1007/978-3-658-40183-2_2

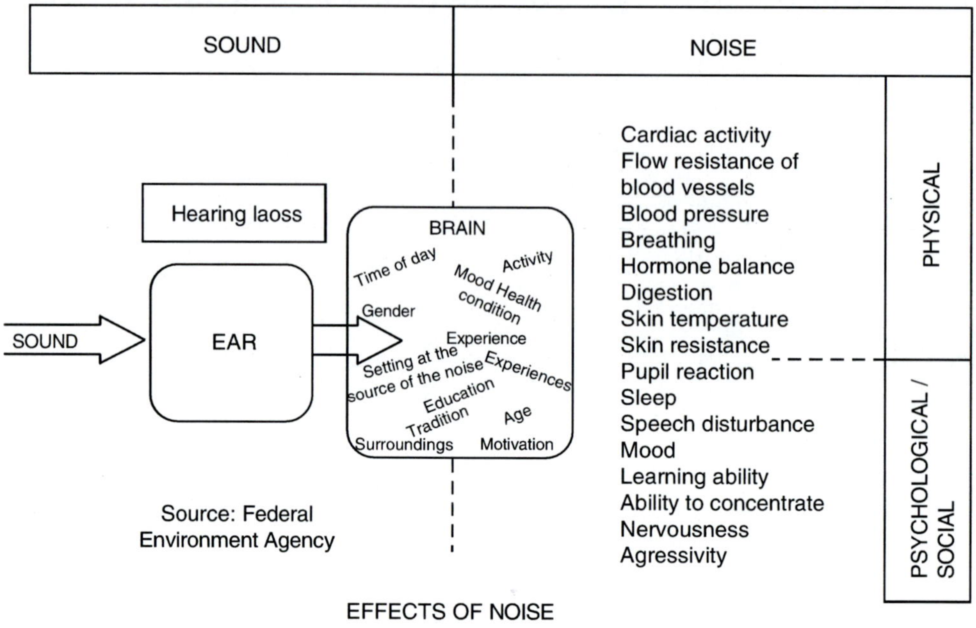

Fig. 2.1 Relationship between sound, noise and defensive reactions

is being decided by the affected person in the brain. The following example should clarify the difference:

A ticking clock with approx. 25 dB(A) can be perceived as disturbing noise pollution when falling asleep, whereas the same person perceives 115 dB(A) at an open-air concert as music and not as noise.

Figure 2.1 schematically illustrates the relationship between sound and noise and the human defensive reactions.

The hearing process is initiated when the human ear is exposed to the physical quantity of **sound pressure**.

Sound Pressure *p*

The pressure fluctuations superimposed on the atmospheric pressure (resting pressure), caused by the compression and relaxation of air particles during a sound event, are referred to as sound pressure (Fig. 2.2).

Sound pressure is a scalar quantity and is not directional.

Acoustically audible pressure variations range from approximately 10^{-5} to 10^2 N/m^2.

Sound pressure is the physical quantity used to characterise sound **immission**.

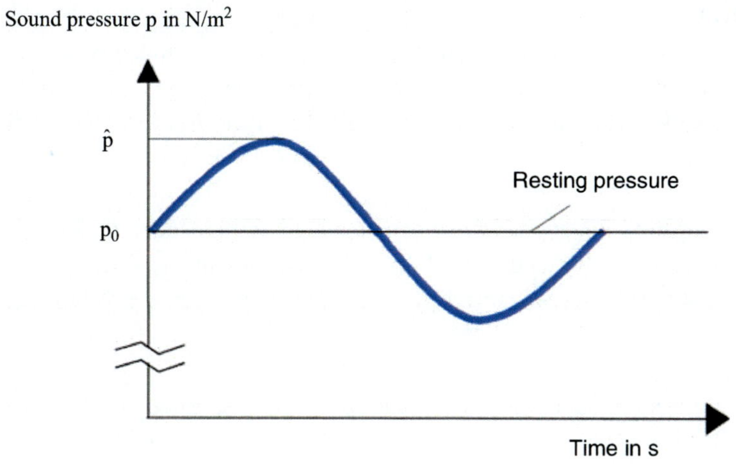

Fig. 2.2 Sound pressure *p*

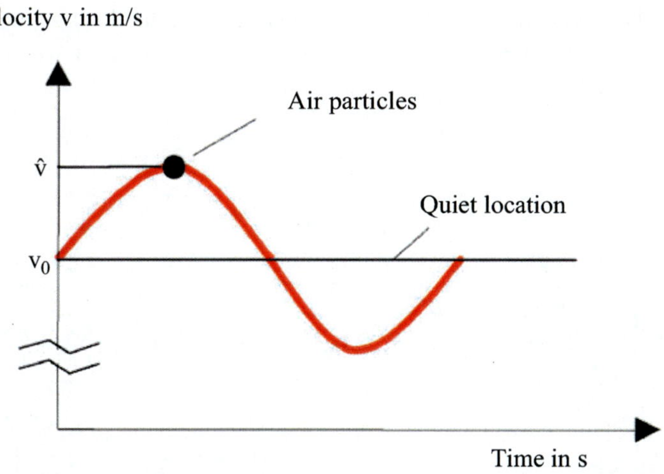

Fig. 2.3 Sound velocity *v*

Sound Velocity *v*

The velocity fluctuations with which air particles oscillate around the rest position during a sound event are referred to as sound velocity (Fig. 2.3).

Acoustically audible sound velocities are between approx. $5 \cdot 10^{-8}$ and 0.25 m/s.

The sound velocity has a vector character as a kinematic quantity and indicates the direction of sound propagation.

Sound Field

External disturbances – sound events – such as clapping hands, excite air particles in a room to pressure and velocity fluctuations, sound pressure and sound velocity. The resulting sound distribution in the room is called a **sound field**. A sound field is unambiguously defined if the sound pressure and the sound velocity are known at each point in time and in the room.

The propagation of sound waves in a sound field occurs at the speed of sound. Propagation occurs spherically in all spatial directions in the open air, similar to wave propagation in a fluid at rest caused by external disturbances, e.g. by throwing in a stone.

Free Field

If the sound generated by a source can propagate unhindered, then the sound field generated by the source is called a free field. In the free field, the sound propagates spherically. The sound intensity decreases with the square and the sound pressure decreases linearly with the distance. In practice, the free field is realised by so-called anechoic chambers (Fig. 2.4), in which the reflection component can be neglected above a certain frequency – the lower cut-off frequency (see ISO 26101). In this case, it is possible to carry out acoustic measurements and investigations independently of weather conditions and external disturbances.

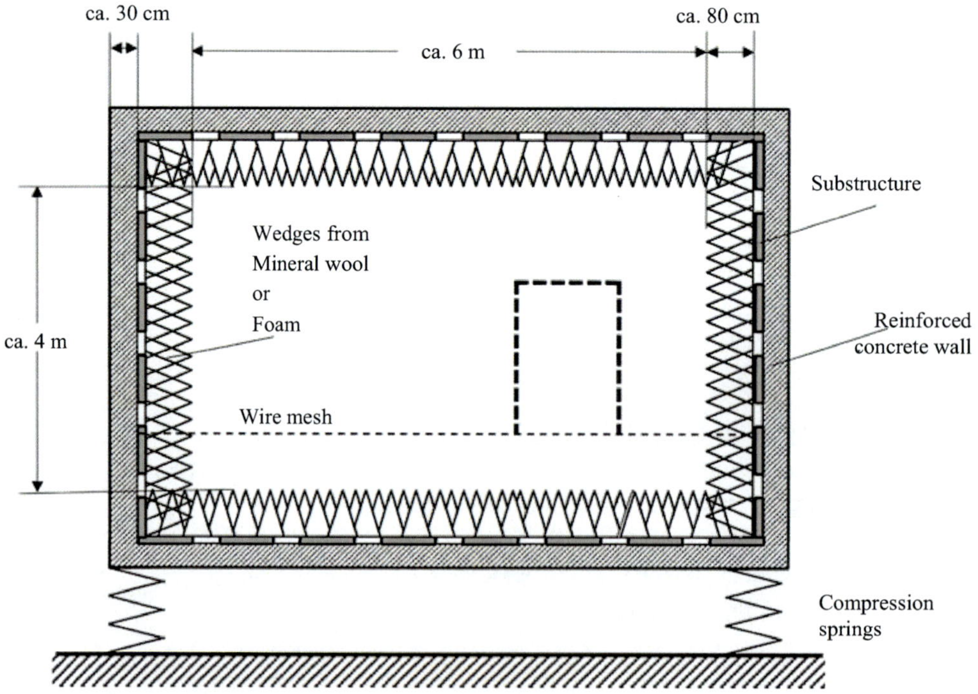

Fig. 2.4 Schematic representation of an anechoic chamber

Diffuse Field

If the sound generated by a source can be reflected at room boundary surfaces, then a constant sound field is built up in the room with a constantly emitting sound source, which contains direct components of the source (free field) as well as reflection components. If the reflection component is so large that the same sound pressure prevails at all points in the room, then the sound field corresponds to a diffuse field. The sound pressure decrease in a diffuse field is almost zero. In practice, the diffuse field is generated in so-called reverberation chambers, Fig. 2.5. Reverberation chambers are particularly suitable for the investigation of stationary broadband noises.

Real Sound Field

The sound field in ordinary living or working spaces lies between the free and diffuse fields. This is referred to as a semi-diffuse or real sound field. Figure 2.6 shows the sound field of a real room schematically. The sound field in the vicinity of the machine, which is not clearly defined, is referred to as the near field. Outside the near field up to a radius r_H the sound field can be regarded approximately as a free field. From the so-called reverberation radius r_H, the sound field is largely diffuse.

Wavelength λ

The wavelength is the smallest distance between two points of a sine wave that have the same amplitude and phase (Fig. 2.7). In air (gases), the wavelength is inversely proportional to the frequency f.

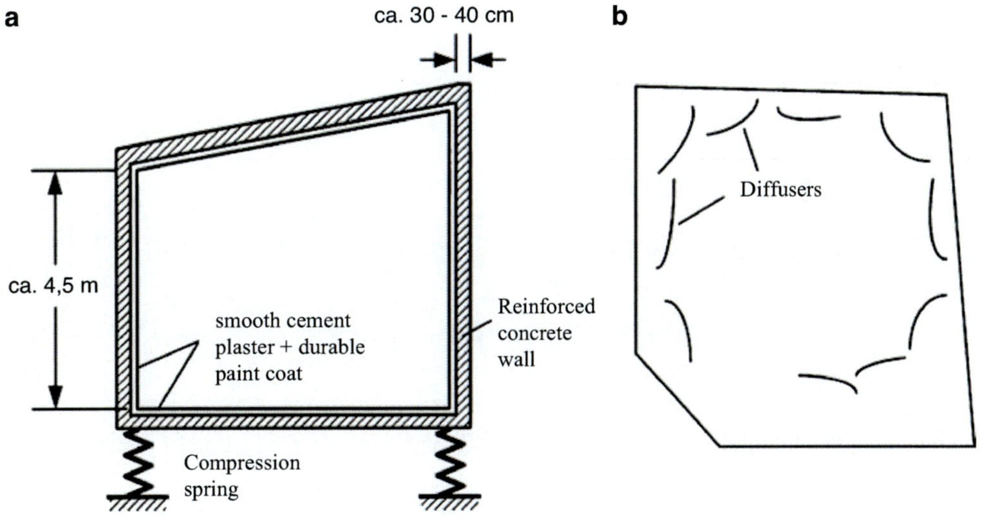

Fig. 2.5 Schematic representation of a reverberation chamber, (**a**) longitudinal section, (**b**) plan view

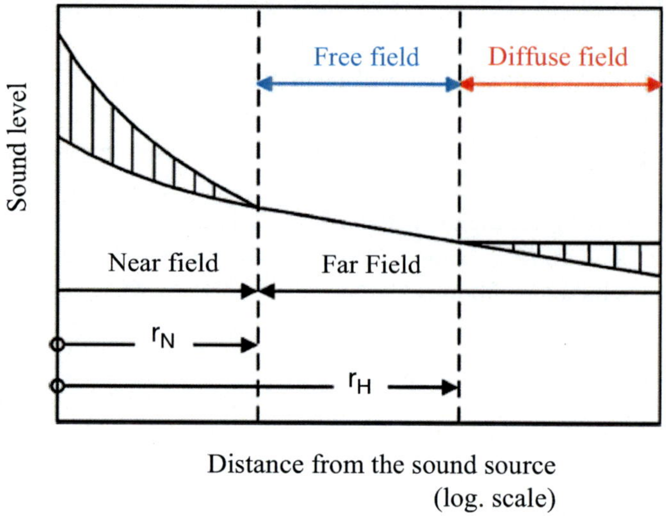

Fig. 2.6 Schematic representation of a real sound field

Fig. 2.7 Definition of wavelength

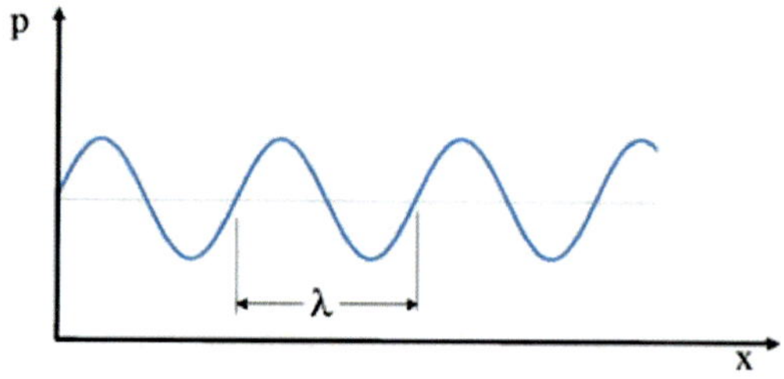

$$\lambda = \frac{c}{f} \text{ m} \tag{2.1}$$

c	Speed of sound in m/s
f	Frequency in Hz

For air at room temperature (15 °C), $c \approx 340$ m/s, and $\lambda_{100\,\text{Hz}} \approx 3.4$ m $\lambda_{1000\,\text{Hz}} \approx 0.34$ m.

Frequency

The number of oscillations per second is called the frequency f. The unit of measurement of frequency is Hertz (Hz). The frequency indicates the pitch.

The acoustically audible frequency range is between 16 and 16,000 Hz.

Speed of Sound *c*

The propagation speed of a sound wave is called the speed of sound. It depends on the elastic properties (density, compression modulus or modulus of elasticity) of the medium.

$$c_{\text{air}} \approx 340\frac{\text{m}}{\text{s}} \quad \text{at} \quad t \approx 15\,^\circ\text{C}$$

$$c_{\text{water}} \approx 1450\ \frac{\text{m}}{\text{s}} \quad \text{at} \quad t \approx 10\,^\circ\text{C}$$

$$c_{\text{steel}} \approx 5200\frac{\text{m}}{\text{s}}$$

2.2 Sound Field Quantities, Wave Equation

The relationship between the sound field quantities is best described by the wave equation of the velocity potential ϕ [1]:

$$\Delta\phi = \frac{1}{c^2}\frac{\partial^2\phi}{\partial t^2} \tag{2.2}$$

Here L is the so-called Laplace operator:

$$\Delta = \frac{\partial^2}{\partial r^2} + \frac{2}{r}\frac{\partial}{\partial r}\,(\text{spherical coordinates}) \tag{2.3}$$

Note: Due to the isotropic propagation of sound in our applications its suffices to consider only the radial component of the Laplace operator.

$$\Delta = \frac{\partial^2}{\partial x^2} + \frac{\partial^2}{\partial y^2} + \frac{\partial^2}{\partial z^2}\,(\text{Cartesian coordinates}) \tag{2.4}$$

The sound field quantities p and v can be specified using the velocity potential as follows:

$$p = -\rho\frac{\partial\phi}{\partial t} \quad \text{N/m}^2 \tag{2.5}$$

$$v = \frac{\partial\phi}{\partial r} \quad \text{m/s} \tag{2.6}$$

The solution approach for the wave equation (2.2) is:

$$\phi = -\frac{\phi'}{r} \cdot e^{-jkr} \cdot e^{j\omega t} \quad m^2/s \tag{2.7}$$

Here r is the distance from the source and is ϕ' the constant auxiliary quantity with dimension m^3/s. k is the so-called wave number:

$$k = \frac{\omega}{c} = \frac{2\pi}{\lambda} \quad 1/m \tag{2.8}$$

ω is the angular frequency:

$$\omega = 2\pi f \quad 1/s \tag{2.9}$$

The complex e-functions represent general approaches to periodic solutions:

$$e^{\pm j\alpha} = \cos(\alpha) \pm j \cdot \sin(\alpha) \tag{2.10}$$

With the imaginary unit

$$j = \sqrt{-1}$$

In acoustics, the factor j means a phase shift of $\pi/2 = 90°$.

$$e^{\pm j\frac{\pi}{2}} = \cos\left(\frac{\pi}{2}\right) \pm j \cdot \sin\left(\frac{\pi}{2}\right) = \pm j$$

Using the wave equation (2.2) and the solution approach (2.7), the sound field quantities p and v, (2.5) and (2.6), can be calculated:

$$p = -\rho\frac{\partial\phi}{\partial t} = j\phi' \cdot \frac{\omega\rho}{r} \cdot e^{-jkr} \cdot e^{j\omega t} \quad N/m^2 \tag{2.11}$$

$$v = \frac{\partial\phi}{\partial r} = \frac{\phi'}{r^2}(1 + jkr) \cdot e^{-jkr} \cdot e^{j\omega t} \quad m/s \tag{2.12}$$

2.2.1 Acoustic Impedance

The ratio between sound pressure p and sound velocity v is called acoustic impedance or acoustic resistance. With (2.11) and (2.12) the acoustic impedance can be calculated [1, 2]:

$$Z = \frac{p}{v} = (\rho \cdot c) \cdot \frac{\mathrm{j}kr}{1 + \mathrm{j}kr} = \frac{\widehat{p}}{\widehat{v}} \cdot \frac{\mathrm{j}}{1 + \mathrm{j}kr} = \frac{\widehat{p}}{\widehat{v}} \cdot \frac{1}{\sqrt{1 + k^2 r^2}} \cdot \mathrm{e}^{\mathrm{j}\varphi} \quad \mathrm{N\,s/m^3} \tag{2.13}$$

The phase angle between φ sound pressure and sound velocity is defined as follows:

$$\phi = \arctan \frac{\mathrm{Im}(p/v)}{\mathrm{Re}(p/v)} = \frac{\pi}{2} - \arctan(k \cdot r) = \frac{\pi}{2} - \arctan\left(2\pi \frac{r}{\lambda}\right) \tag{2.14}$$

It follows from this that the transition from $\varphi = 90°$ to $\varphi \approx 0$ (approximate phase equality) takes place at a relatively short distance.

$$\begin{aligned} r &= 0 \quad \varphi = 90\,° \\ r &= \lambda \quad \varphi \approx 9\,° \\ r &= 3\lambda \quad \varphi \approx 3\,° \end{aligned}$$

For practical applications, this means that at a distance of a wavelength $r \approx \lambda$, the sound field can be regarded as quasi-planar, i.e. sound pressure and sound velocity are approximately in phase. For industrial noise and noises whose dominant frequencies lie in the range of $f > 300$ Hz, phase equality already results at a distance $r \approx 1$ m, which is also prescribed or recommended in many measurement regulations and standards, e.g. ISO 3744 [7], as the measurement distance for sound pressure measurements.

At a large distance r from the source, ($r > \lambda$) the acoustic impedance is a real quantity (p and v are in phase) and equal to:

$$Z \approx \rho \cdot c \quad \mathrm{N\,s/m^3} \tag{2.15}$$

In practice, a real impedance is usually assumed for a distance of $r \approx \lambda$. The sound pressure and the sound threshold can then be determined from (2.11) and (2.12) as follows:

$$p = \mathrm{j}\phi' \cdot \frac{\omega\rho}{r} \cdot \mathrm{e}^{-\mathrm{j}kr} \cdot \mathrm{e}^{\mathrm{j}\omega t} = \mathrm{j} \cdot \widehat{p} \cdot \mathrm{e}^{-\mathrm{j}kr} \cdot \mathrm{e}^{\mathrm{j}\omega t} \quad \mathrm{N/m^2} \tag{2.16}$$

$$v \approx \mathrm{j} \cdot \frac{\phi' \cdot \omega}{r \cdot c} \cdot \mathrm{e}^{-\mathrm{j}kr} \cdot \mathrm{e}^{\mathrm{j}\omega t} = \mathrm{j} \cdot \widehat{v} \cdot \mathrm{e}^{-\mathrm{j}kr} \cdot \mathrm{e}^{\mathrm{j}\omega t} \quad \mathrm{m/s} \tag{2.17}$$

For air at room temperature ($c \approx 340$ m/s; $\rho = 1.21$ kg/m^3) is:

$$Z_{\mathrm{air}} \approx \rho \cdot c \approx 410 \quad \mathrm{N\,s/m^3}$$

The acoustic impedance is responsible for the structure of the sound field. With real impedance, the sound velocity can be determined from the sound pressure, which can be measured relatively easily.

$$v = \frac{p}{Z} = \frac{p}{\rho \cdot c} \quad \text{m/s} \tag{2.18}$$

Equation (2.18) states that a large impedance is detrimental to the establishment of a sound field if excitation is via sound pressure. A value of about 400 N s/m^3 represents the lowest impedance for practical applications. Therefore, it is relatively easy to build up a perceptible sound field with speech, for example. In contrast, it is very difficult to build up a perceptible sound field in water because the impedance of water at room temperature ($c \approx 1450$ m/s; $\rho = 1000$ kg/m^3) is much greater than in air:

$$Z_{\text{Wasser}} \approx \rho \cdot c \approx 1.45 \cdot 10^6 \ \text{N s/m}^3$$

2.2.2 Sound Intensity

The sound energy transported per unit area and time caused by an excitation source is called **sound intensity** with the unit [N · m/m^2 · s) = W/m^2].

The sound intensity is defined as the product of the sound pressure and the sound velocity in $p(t) \cdot v(t)$ W/m^2. The mean intensity is obtained by averaging (integration over the period or measurement time T):

$$\boldsymbol{I}(r,t) = \frac{1}{T}\int_0^T p(r,t) \cdot \boldsymbol{v}(r,t)\mathrm{d}t \quad \text{W/m}^2 \tag{2.19}$$

The intensity is a function of the location and has vector character and points in the direction of the associated velocity.

Acoustic intensities of interest range from 10^{-12} to 10 W/m^2.

For large r or $r/\lambda > 1$, the alternating pressure and velocity are defined according to (2.16) and (2.17). The sound intensity averaged $T = \frac{2\cdot\pi}{\omega}$ over the period is then obtained [1]:

$$I(r) = \frac{1}{2}\widehat{p}(r)\widehat{v}(r) = \frac{\widehat{\Phi}^{2} \cdot \omega^2 \cdot \rho}{2 \cdot r^2 \cdot c} = \widetilde{p}(r) \cdot \widetilde{v}(r) \quad \text{W/m}^2 \tag{2.20}$$

Here, $\widetilde{p}$ and $\widetilde{v}$ are the rms values of the sound pressure and the sound velocity:

$$\widetilde{p}(r) = \sqrt{\frac{1}{T}\int_0^T \widehat{p}^2(r,t) \cdot \mathrm{d}t} = \frac{\widehat{p}(r)}{\sqrt{2}} \quad \text{N/m}^2 \tag{2.21}$$

$$\widetilde{v}(r) = \sqrt{\frac{1}{T}\int_0^T \widehat{v}^2(r,t)\cdot \mathrm{d}t} = \frac{\widehat{v}(r)}{\sqrt{2}} \quad \mathrm{m/s} \tag{2.22}$$

For large r or r/λ, the intensity of the spherical wave field decreases along with $1/r^2$, but is constant at radius r itself. With (2.18) it follows from (2.20):

$$I(r) = \widetilde{p}(r)\cdot\widetilde{v}(r) = \frac{\widetilde{p}(r)^2}{Z} = \frac{\widetilde{p}(r)^2}{\rho\cdot c} = \widetilde{v}(r)^2\cdot Z \quad \mathrm{W/m^2} \tag{2.23}$$

2.2.3 Sound Power

The sound power P passing through a surface S_i, at a distance r from the source, is the product of the sound intensity with the surface perpendicular to the direction of energy transport (Fig. 2.8).

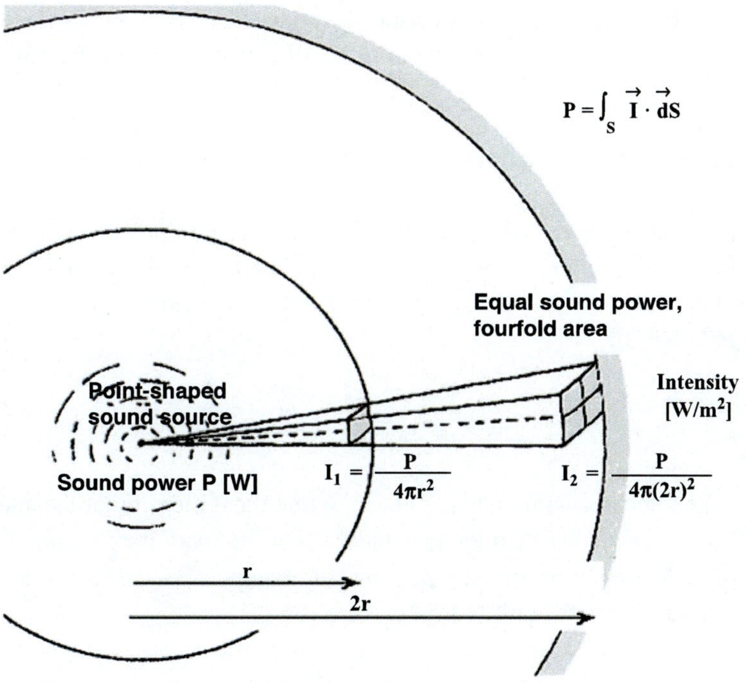

Fig. 2.8 Relationship between sound intensity and sound power

$$P_{S_i} = I \cdot S_i = \frac{\widetilde{p}^2(r)}{\rho \cdot c} \cdot S_i \quad \text{W} \tag{2.24}$$

The total sound power of a source is obtained by summing up (integration of) the partial sound powers over a closed area around the sound source (enveloping surface):

$$P_Q = \oint_S \frac{\widetilde{p}(r)^2}{Z} \cdot \mathrm{d}S = \sum_{i=1}^{n} P_{S_i} = \frac{\widetilde{p}^2(r)}{\rho \cdot c} \cdot S \quad \text{W} \tag{2.25}$$

$\widetilde{p}^2(r)$	RMS value of the mean square of sound pressure on the closed surface S at distance r *from* the source
S	Closed measurement area around a sound source. In the free field, the measurement surface corresponds to a spherical surface $S = 4\pi \cdot r^2$

Remark
The relation given in (2.25) is only valid for a real impedance, free sound propagation and without losses between source and measuring surface.

The sound power is the physical quantity for the identification of the **sound emission**.

2.2.4 Mechanical Impedance

The mechanical impedance is the resistance which an elastic structure opposes to the acting forces. If the excitatory force (periodic excitation) and the system velocity are at the same point and if both are rectified, then the force related to the velocity represents the so-called **input impedance** Z_e:

$$Z_e = \frac{F}{v} \quad \text{kg/s} \tag{2.26}$$

The point-like force excitation means (Fig. 2.9) that the dimension of the attack surface is considerably ($< \frac{1}{6}$) smaller than the generated wavelengths of the structure-borne sound.

The mechanical input impedance Z_e is usually complex, i.e. the force and the velocity are out of phase. Z_e has the unit N s/m, or kg/s.

Physically, Z_e indicates the strength of structure-borne sound generated in the structure when a force is applied. A large input impedance therefore leads to a low level of structure-borne sound in the structure when a force is applied. In the following, the mechanical input impedances are given for some special cases.

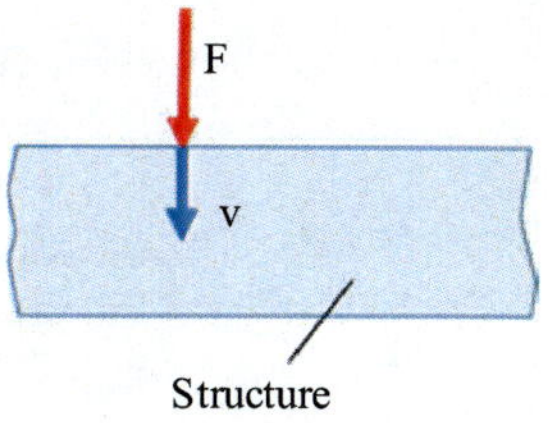

Fig. 2.9 Point force excitation of a structure

2.2.4.1 Mechanical Impedances of Idealised Components

The idealised input impedance can be illustrated for many components, e.g. the coupling points of machines to the structures, using the example of a damped single-mass oscillator (EMS). The equation of motion of the damped single-mass oscillator given in Fig. 2.10 can be stated using Newton's basic equation ($\sum F = m \cdot a$ sum of all external forces equals mass times acceleration) as follows:

$$\left.\begin{array}{c} F - x \cdot k_\mathrm{F} - d \cdot \dot{x} = m \cdot \ddot{x} \\ \text{respectively} \\ m \cdot \ddot{x} + x \cdot k_\mathrm{F} + d \cdot \dot{x} = F \end{array}\right\} \quad \mathrm{N} \tag{2.27}$$

with:

$x \cdot k_\mathrm{F}$ N	Spring force (k_F = spring stiffness in N/m)
$d \cdot \dot{x}$ N	Damping force proportional to speed (d = damping coefficient in Ns/m = kg/s)
$m \cdot \ddot{x} = m \cdot a$ N	Mass or acceleration force
$\dot{x} = \frac{\mathrm{d}x}{\mathrm{d}t} = v = \frac{\ddot{x}}{\mathrm{j}\omega}$ m/s	Speed, vibration velocity
$a = \ddot{x} = \frac{\mathrm{d}^2x}{\mathrm{d}t^2} = \dot{x} \cdot \mathrm{j}\omega$ m/s^2	Acceleration, vibration acceleration
$F = \widehat{F} \cdot \mathrm{e}^{\mathrm{j}\omega t}$ N	Periodic excitation force at the angular frequency $\omega = 2\pi f$

For forced oscillations, force excitation with angular frequency ω, the structure also oscillates with angular frequency ω. The solution approach for the oscillation path x is:

$$x = \widehat{x} \cdot \mathrm{e}^{\mathrm{j}\omega t} = \frac{\dot{x}}{\mathrm{j}\omega} \quad \mathrm{m} \tag{2.28}$$

By (2.26) and (2.28), it follows from (2.27):

$$Z_\mathrm{e} = \frac{F}{v} = \frac{F}{\dot{x}} = \mathrm{j}m\omega + \mathrm{d} + \frac{k_\mathrm{F}}{\mathrm{j}\omega} \quad \mathrm{kg/s} \tag{2.29}$$

It follows from this that the input impedance is made up of three components:

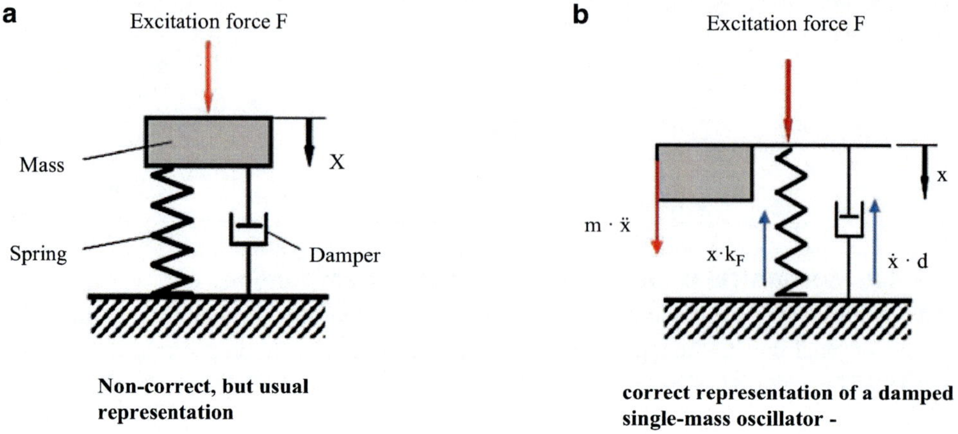

Fig. 2.10 Damped single-mass oscillator

1. **Mass impedance Z_m**

$$Z_m = jm\omega \quad \text{kg/s} \tag{2.30}$$

The impedance of a point mass is a purely imaginary quantity, i.e. the oscillating speed (velocity) of the mass is phase-shifted by 90° with respect to the exciting force. The magnitude of the impedance is proportional to the mass m and the exciting angular frequency ω.

2. **Damper impedance Z_D**

$$Z_\text{D} = d \quad \text{kg/s} \tag{2.31}$$

The impedance of a velocity-proportional damper is a real quantity, i.e. the oscillating velocity of the damper is in phase with the exciting force. The magnitude of the impedance is equal to the damping constant *d*.

3. **Spring impedance Z_F**

$$Z_\text{F} = \frac{k_\text{F}}{j\omega} \quad \text{kg/s} \tag{2.32}$$

The impedance of a spring is again purely imaginary, i.e. the oscillating velocity is phase-shifted by with respect to −90° the exciting force. Its magnitude is proportional to the spring stiffness k_F and inversely proportional to the exciting angular frequency ω.

The input impedance of the damped single-mass oscillator, which is composed of the three basic impedances according to (2.30), (2.31) and (2.32), can assume very different values depending on the properties of the individual elements mass, damper and spring.

Graphical representation of the idealised impedance curves
To illustrate the physical meaning of mechanical impedance, the ratio is shown below as a function of $\frac{|Z|}{\omega}$ frequency. This ratio has the dimension of a mass and is usefully referred to as the moving or **dynamic mass:**

$$m_{\mathrm{b}} = \frac{|Z|}{\omega} = \frac{\text{Amount of the mech.impedance}}{\text{Circular frequency}} \quad \mathrm{kg} \tag{2.33}$$

m_{b} co-moving or dynamic mass

The dynamic mass m_{b} characterises the resistance of a structure to external excitation forces. In contrast to the static mass of structures, which is a constant, the dynamic mass is very strongly frequency dependent. An exception here is the compact mass, see (2.30), where the dynamic mass has the same value as the mass of the structure. To make it easier to compare the impedance curves, the magnitudes of some idealised impedances, the associated dynamic masses, the material and the respective geometric dimensions are compiled below. The dimensions are chosen so that all idealised components have a dynamic mass of 10 kg at a frequency of 10 Hz [1].

Compact mass	$m_b = m$
Ideal spring	$m_b = \frac{k_{\mathrm{F}}}{\omega^2}$
Ideal damper	$m_b = \frac{d}{\omega}$
Ideal beam	$m_b = \frac{2\sqrt{2}\cdot\rho\cdot A\cdot\sqrt[4]{\frac{EJ_b}{\rho\cdot A}}}{\sqrt{\omega}}$
Ideal plate	$m_b = \frac{8\cdot\sqrt{\frac{E\cdot h^3}{12(1-\mu^2)}\cdot\rho\cdot h}}{\omega}$

Figure 2.11 shows the continuous curves of the dynamic masses of idealised components as a function of frequency.

From this follows that the resistance of the structures to exciting alternating forces becomes smaller and smaller with increasing frequency. Only in the case of the compact mass (e.g. anvil) this is not the case. The decreasing dynamic masses of the other structures are one reason why it is easier to excite these structures to vibrate with increasing frequency. Relatively heavy machine structures can therefore be excited to vibrate more or less easily, depending on the firing frequency.

Furthermore, it follows from Fig. 2.11 that the dynamic mass can be increased considerably with the aid of a compact mass at the point where the force is introduced, especially if the structures have low dynamic masses (increasing the input impedance).

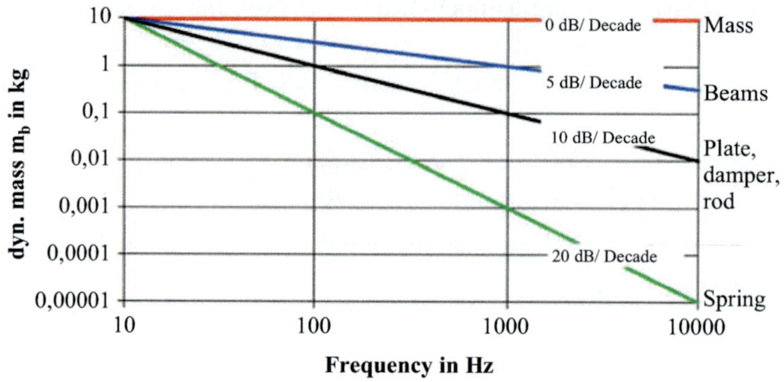

Fig. 2.11 Dynamic mass of idealised components [1]

2.2.4.2 Mechanical Input Impedance of Real Components

The mechanical input impedance of real components differs significantly from the impedance of ideal components. In contrast to the impedance of ideal components, the impedance of real components does not have a continuous curve over the frequency. Impedance curves of real components show dips (resonance points) and peaks (anti-resonance points) caused by the natural oscillation behavior of the components. Figure 2.12 shows the impedance curve of real components schematically. The entire frequency range can be divided into three ranges:

Frequency range I:	Below the 1st natural frequency of the component. In this frequency range, the components behave as a free mass or as a spring, depending on their clamping. Damping measures in this frequency range are ineffective
Frequency range II:	Between the 1st and 5th to 8th natural frequency of the component In this frequency range, depending on the damping of the system, resonance excitations can cause considerable dips in the impedance curve, i.e. very high vibration amplitudes can occur for the same force excitation The damping measures increase the impedances at the resonant frequencies and reduce them at the anti-resonant frequencies, see Fig. 2.12. Therefore, the damping measures lead to significant vibration reductions in this frequency range, especially when the structures or components perform resonant vibrations
Frequency range III:	Above the 5th to 8th natural frequency of the component Depending on the clamping conditions, the impedance curves here can take on the character of a plate or a beam. Due to the higher density of natural frequencies, the probability of resonance excitation in this frequency range is lower than in the II. frequency range. Therefore, the damping measures are not particularly effective in this frequency range for the purpose of vibration reduction

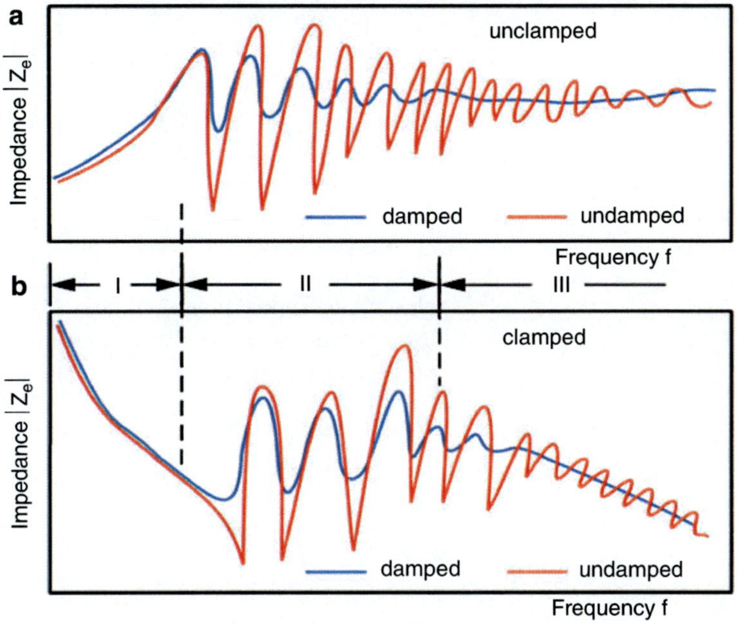

Fig. 2.12 Schematic impedance curve of real components; (**a**) not clamped, (**b**) clamped

The low impedances, as they occur at resonance points and can lead to resonance states there, are often decisive for the noise and vibration development. The impedance dips of real components, which depend not only on the clamping conditions, material properties and component geometry but also on the existing system damping, can only be determined precisely by measurements [3]. Since in many cases, e.g. during the planning phase, it is not possible to determine the impedance by measurement, a simplified model [4] for estimating the mechanical input impedance is discussed below.

λ/4 model for estimating the mechanical input impedance [4]

The magnitude of the mechanical input impedance of real components can be estimated without considering resonance dips according to (2.33) as follows:

$$|Z| \approx \omega \cdot m_{\mathrm{b}} \quad \mathrm{kg/s} \tag{2.34}$$

Here, m_{b} is the co-moved or dynamic mass of the structural location whose input impedance is to be determined. m_{b} corresponds here to the mass within an imaginary sphere at the force application location with radius λ/4, where λ is the governing bending or shear wavelength. In most cases, λ corresponds to the bending wavelength λ_{B}. For plate-like components, such as those found in many sound-radiating machine structures and/or surfaces, the bending wavelength can be determined as follows [1]:

$$c_{\mathrm{B}} = \sqrt{2\pi} \cdot \sqrt[4]{\frac{B'}{m''}} \cdot \sqrt{f} \quad \mathrm{m/s} \tag{2.35}$$

$$\lambda_{\mathrm{B}} = \frac{c_{\mathrm{B}}}{f} = \sqrt{2\pi} \cdot \sqrt[4]{\frac{B'}{m''}} \cdot \frac{1}{\sqrt{f}} \quad \mathrm{m} \tag{2.36}$$

With

c_B	Flexural shaft speed in m/s
B'	Bending stiffness of the plate referred to the length in Nm
m''	Mass of the plate in kg/m^2 related to the area.

For a plate of thickness h is:

$$B' = \frac{E \cdot h^3}{12(1-\mu^2)} \quad \mathrm{Nm} \tag{2.37}$$

$$m'' = \rho \cdot h \quad \mathrm{kg/m^2} \tag{2.38}$$

E	Modulus of elasticity of the plate in N/m^2
ρ	Density of the plate in kg/m^3
h	Slab thickness in m
μ	Poisson or transverse contraction number. For many metals, e.g. steel, is $\mu \approx 0.3$

The mass within the imaginary sphere of radius corresponds, $\lambda_{\mathrm{B}}/4$ for a plane plate, to the mass of a circular disk of radius. $\lambda_{\mathrm{B}}/4$ Using (2.35) (2.36), (2.37) and (2.38), the dynamic mass of the plane plate can be determined as follows:

$$m_{\mathrm{b}} \approx \pi \cdot \left(\frac{\lambda_{\mathrm{B}}}{4}\right)^2 \cdot h \cdot \rho = \frac{\pi^2}{8} \cdot \sqrt{\frac{E \cdot \rho}{12(1-\mu^2)}} \cdot \frac{h^2}{f} \quad \mathrm{kg} \tag{2.39}$$

Using (2.34), we then obtain the magnitude of the mechanical input impedance of a plane plate:

$$|Z| \approx \omega \cdot m_{\mathrm{b}} = \frac{\pi^3}{4} \cdot \sqrt{\frac{E \cdot \rho}{12(1-\mu^2)}} \cdot h^2 \quad \mathrm{kg/s} \tag{2.40}$$

Figure 2.13 shows the measured input impedance and the dynamic mass of a plate, force excitation in the centre of the plate. For comparison, the idealised curves according to (2.39) and (2.40) are also plotted. It follows that the idealised curves are only suitable for estimating the mean values.

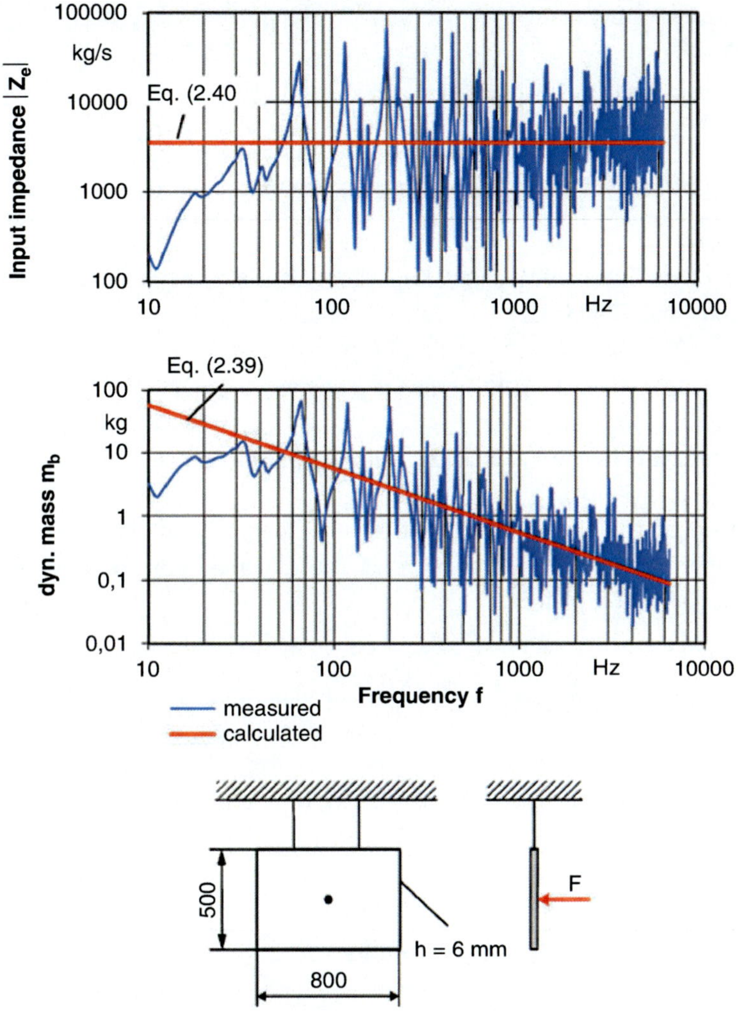

Fig. 2.13 Measured and calculated magnitude of input impedance and dynamic mass of a freely suspended (not clamped!) and centrally excited flat steel plate

As can also be easily seen from Fig. 2.13, it is not possible to determine the dips in the course of the dynamic mass or impedance using the $\lambda/4$ model. This means that the impedances determined with the help of this model can only be used as an approximation for the frequency ranges in which no pronounced natural frequencies occur (frequency range I and III in Fig. 2.12).

The dips in the input impedance or the dynamic mass, which are decisive for the noise development, can best be determined by measurements, see Chap. 5.

Table 2.1 Calculated values for Example 2.1

f	10	100	1000	10000	Hz
c_B	24.3	76.8	242.9	768.2	m/s
λ_B	2.43	0.77	0.24	0.08	m
m_b	54.6	5.5	0.55	0.05	kg
$\lvert Z\rvert$	3429	3429	3429	3429	kg/s

Example 2.1
Estimation of the mechanical input impedance of a $h = 6$ mm thick steel plate at $f =$ 10, 100, 1000 and 10,000 Hz.

$$E_{\text{Stahl}} = 2.1 \cdot 10^{11} \ \text{N/m}^2; \quad \mu = 0.3; \quad \rho_{\text{Stahl}} = 7850 \ \text{kg/m}^3$$

Table 2.1 summarises the calculated values for c_B, λ_B, m_b and according to $|Z|$ (2.35), (2.36), (2.39) and (2.40):

For the simple calculation of the bending wavelength λ_B and the dynamic mass m_b, the following approximate formula can be given as a numerical value equation from (2.36) and (2.39) for a plane steel plate:

$$\lambda_{B,\text{Steel}} \approx 3.14 \cdot \frac{\sqrt{h}}{\sqrt{f}} \quad \text{m} \ (h \text{ in mm}; f \text{ in Hz}) \tag{2.41}$$

$$m_{b,\text{Steel}} \approx 15.16 \cdot \frac{h^2}{f} \quad \text{kg} \ (h \text{ in mm}; f \text{ in Hz}) \tag{2.42}$$

2.2.5 Structure-Borne Sound Power

The indirect sound radiation of machine structures which are excited to mechanical vibrations ("structure-borne sound vibrations") by alternating forces is referred to as secondary or indirect airborne sound (structure-borne sound!). In this case, the air particles are indirectly excited to vibrate by structural vibrations, e.g. of the housing.

Alternating forces occur, for example, in gearboxes due to shocks, rolling and friction processes, imbalances of rotating components, pressure pulsations as well as detachment phenomena during flow processes at deflections and changes in cross-section.

In the simplest case, the intensity (see (2.23) of the radiation has the magnitude:

$$I_S = \overline{\tilde{v}_S^2} \cdot Z \quad \text{W/m}^2 \tag{2.43}$$

Where is the $\overline{\tilde{v}_S^2}$ square of the rms value of the velocity averaged over the radiating surface S and $Z = \rho \cdot c$ is the acoustic characteristic impedance of the air.

Here it was assumed that the sound pressure caused by the structural vibrations and the sound velocity of the air particles in front of the radiating surface are in phase.

Assuming that all parts of the surface oscillate in phase with the mean square of velocity $\overline{\tilde{v}_S^2}$, the sound power radiated fromthis surface can be determined as follows:

$$P_{\mathrm{K,th}} = \overline{\tilde{v}_S^2} \cdot Z \cdot S \quad \mathrm{W} \tag{2.44}$$

Here $P_{\mathrm{K,th}}$ is the theoretically radiated structure-borne sound power, which usually represents a maximum value. The actually radiated structure-borne sound power of the area S is smaller and is taken into account by the factor σ [1]:

$$P_{\mathrm{K}} = \overline{\tilde{v}_s^2} \cdot Z \cdot S \cdot \sigma \quad \mathrm{W} \tag{2.45}$$

$$\overline{\tilde{v}_s^2} = \frac{1}{S} \int_A \tilde{v}_s^2 \mathrm{d}S = \frac{1}{S} \sum_{i=1}^{n} \overline{\tilde{v}_{S_i}^2} \cdot S_i \quad \mathrm{m}^2/\mathrm{s}^2 \tag{2.46}$$

with:

$\tilde{v}_s^2$	Square of the rms value of the velocity on the surface element dS,
$\overline{\tilde{v}_{S_i}^2}$	Square of the rms value of the velocity as the mean for the subarea S_i,
S	Total area of the sound emitting structure,
$\overline{\tilde{v}_s^2}$	mean square of the rms value of the velocity on the radiating surface S (structure-borne sound),
σ	is the so-called radiation factor of the surface S and indicates the proportion of the actually radiated structure-borne sound power.

2.2.6 Degree of Radiation

The structural vibrations which are radiated as airborne sound are referred to as "structure-borne sound radiation". In addition to the vibration velocity v and area S, the structure-borne sound radiation is essentially dependent on the radiation factor "σ". The radiation factor describes the conversion of structural or structure-borne sound vibrations into airborne sound. It indicates the extent to which structure-borne sound is radiated as (secondary) airborne sound at the respective frequencies. Specified in logarithmic measure, the degree of radiation is referred to σ' as the **radiation coefficient**:

$$\sigma = \frac{P_{\mathrm{K}}}{\rho \cdot c \cdot \overline{\tilde{v}_s^2} \cdot S} \tag{2.47}$$

$$\sigma' = 10 \cdot \lg(\sigma) \quad \mathrm{dB} \tag{2.48}$$

Figure 2.14 schematically shows the structure-borne sound radiation of plate-like structures, the usual sound-radiating components of a machine. The airborne sound excited by the structural vibrations, can only be radiated from a cut-off frequency f_{g}, at which the airborne and bending wavelengths are equal ($\lambda_{\mathrm{air}} = \lambda_{\mathrm{B}}$), i.e. $\sigma = 1$.

Using (2.1) and (2.36), the air and bending wavelengths at the cutoff frequency f_{g} can be determined as follows:

$$\lambda_{\mathrm{air},fg} = \frac{c}{f_g} \quad \mathrm{m} \tag{2.49}$$

$$\lambda_{\mathrm{B},f_{\mathrm{g}}} = \sqrt{2\pi} \cdot \sqrt[4]{\frac{B'}{m''}} \cdot \frac{1}{\sqrt{f_{\mathrm{g}}}} \quad \mathrm{m} \tag{2.50}$$

From this it is easy to see that the air and bending wavelengths depend differently on the frequency. Thus, of course, the degree of radiation is also frequency-dependent. By equating equations (2.49) and (2.50), one then obtains the cut-off frequency f_{g}:

$$f_{\mathrm{g}} = \frac{1}{2\pi} c^2 \sqrt{\frac{m''}{B'}} \quad \mathrm{Hz} \tag{2.51}$$

At the frequencies above the cutoff frequency, $\lambda_{\mathrm{air}} < \lambda_{\mathrm{B}}$, the structural vibrations are fully radiated in most cases. For the practical application is:

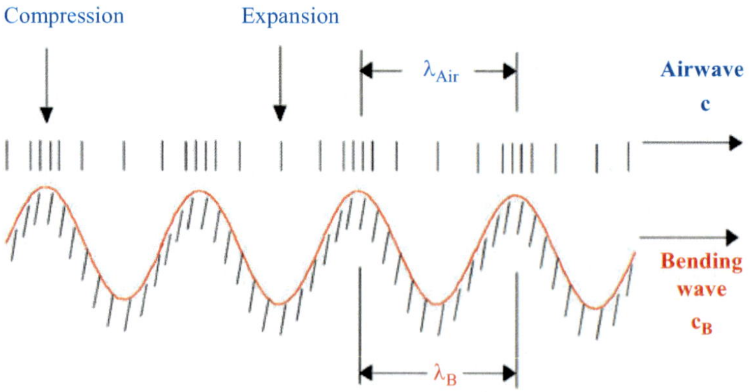

Fig. 2.14 Schematic representation of the structure-borne sound radiation of plate-like structures

$$\left.\begin{array}{lll}\sigma \approx 1 & \text{for} & f \geq f_g \\ \sigma < 1 & \text{for} & f < f_g\end{array}\right\} \quad (2.52)$$

The reason for this behavior is that at the frequencies $f < f_g$, the wavelength of the radiated airborne sound λ_{air} is larger than the bending wavelength λ_B of the vibrating structures. As a result, the larger sound wavelengths in the air prevent the radiation of the smaller bending wavelengths, which are reproduced as airborne sound immediately in front of the vibrating structures. This effect is called an **acoustic short circuit** or **coincidence effect** [5, 6].

The position of the cutoff frequency f_g can be shifted by design measures, e.g. by changing the quotient:

$$\text{Mass occupancy/bending stiffness} = m''/B'$$

Acoustically, it is desirable to have as high a cut-off frequency as possible, since for $f < f_g$, $\sigma < 1$ and less structure-borne sound is radiated.

2.3 Sound Level

Oriented to the human hearing (relative, logarithmic) and due to a very large hearing range of 12 powers of ten, the sound field quantities are always given as levels (logarithm of a ratio) with the unit decibel dB.

The sound level is defined as the 10-fold logarithm of the ratio of a power quantity, e.g. P, I, p^2, v^2 and designated by the letter L. Corresponding indices are used to designate the sound level quantities depending on the physical quantity. Because of the uniformity, the designations of most sound level quantities are standardised.

The reference values are also specified for most level quantities. The numerical values are based on the perception threshold of a healthy person (hearing threshold). In the following, some essential level quantities are given.

Sound power level L_W

$$\left.\begin{array}{lll}L_W & = 10 \cdot \lg \dfrac{P}{P_0} & \text{dB} \\ P_0 & = 10^{-12} & \text{W}\end{array}\right\} \quad (2.53)$$

Sound intensity level L_I

$$\left.\begin{aligned} L_I &= 10 \cdot \lg \frac{I}{I_0} \quad \text{dB} \\ I_0 &= 10^{-12} \quad \text{W/m}^2 \end{aligned}\right\} \tag{2.54}$$

Sound pressure level L_p

$$\left.\begin{aligned} L_p &= 10 \cdot \lg \frac{p^2}{p_0^2} = 20 \cdot \lg \frac{p}{p_0} \quad \text{dB} \\ p_0 &= 2 \cdot 10^{-5} \quad \text{N/m}^2 \end{aligned}\right\} \tag{2.55}$$

Velocity level L_v

$$\left.\begin{aligned} L_v &= 10 \cdot \lg \frac{v^2}{v_0^2} = 20 \cdot \lg \frac{v}{v_0} \quad \text{dB} \\ v_0 &= 5 \cdot 10^{-8} \quad \text{m/s} \end{aligned}\right\} \tag{2.56}$$

Sound characteristic impedance K_0

$$\left.\begin{aligned} K_0 &= -10\lg \frac{\rho \cdot c}{\rho_0 \cdot c_0} \quad \text{dB} \\ \rho_0 \cdot c_0 &= 400 \quad \text{N s/m}^3 \end{aligned}\right\} \tag{2.57}$$

Notice:

In the case of free sound propagation (atmospheric pressure and room temperature), for practical application because of $\rho \cdot c \approx \rho_0 \cdot c_0 = 400 \;\; \text{N s/m}^3$:

$$K_0 \approx 0 \quad \text{dB}$$

Area dimension L_S

$$\left.\begin{aligned} L_S &= 10 \cdot \lg \frac{S}{S_0} \quad \text{dB} \\ S_0 &= 1 \quad \text{m}^2 \end{aligned}\right\} \tag{2.58}$$

The reference values are chosen in such a way that the level values can be compared with each other.

$$\left.\begin{aligned} P_0 &= \frac{p_0^2}{\rho_0 \cdot c_0} \cdot S_0 = 10^{-12} \quad \mathrm{W} \\ I_0 &= \frac{p_0^2}{\rho_0 \cdot c_0} = 10^{-12} \quad \mathrm{W/m^2} \\ v_0 &= \frac{p_0}{\rho_0 \cdot c_0} = 5 \cdot 10^{-8} \quad \mathrm{m/s} \\ \frac{p_0}{v_0} &= \rho_0 \cdot c_0 = Z_0 = 400 \quad \mathrm{Ns/m^3} \end{aligned}\right\} \tag{2.59}$$

2.3.1 Level Addition, Energetic Addition

If several levels are added, the power quantities are added:

$$P_{\text{total}} = \sum_{i=1}^{n} P_i \quad \mathrm{W} \tag{2.60}$$

$$I_{\text{total}} = \sum_{i=1}^{n} I_i \quad \mathrm{W/m^2} \tag{2.61}$$

$$p_{\text{total}} = \sqrt{\sum_{i=1}^{n} p_i^2} \quad \mathrm{N/m^2} \tag{2.62}$$

$$v_{\text{total}} = \sqrt{\sum_{i=1}^{n} v_i^2} \quad \mathrm{m/s} \tag{2.63}$$

The sum level is then obtained:

$$\left.\begin{aligned} L_{W_{\text{total}}} &= 10 \lg \frac{P_{\text{total}}}{P_0} = 10 \lg \sum_{i=1}^{n} 10^{L_{Wi}/10} \\ &\text{bzw.} \\ L_{p_{\text{total}}} &= 10 \lg \frac{p_{\text{total}}^2}{p_0^2} = 10 \lg \sum_{i=1}^{n} 10^{L_{pi}/10} \end{aligned}\right\} \quad \mathrm{dB} \tag{2.64}$$

Equation (2.64) is also called **power-wise** addition.

Example 2.2
We are looking for the sum of the following levels:

(a) 70 dB and 70 dB:

$$L_{\text{total}} = 10 \cdot \lg\left(10^{70/10} + 10^{70/10}\right) = 73.0 \text{ dB}$$

(b) 70 dB and 60 dB:

$$L_{\text{total}} = 10 \cdot \lg\left(10^{70/10} + 10^{60/10}\right) = 70.4 \text{ dB}$$

(c) 90 dB and 80 dB:

$$L_{\text{total}} = 10 \cdot \lg\left(10^{90/10} + 10^{80/10}\right) = 90.4 \text{ dB}$$

It follows that when two equal levels are added, the total level increases by 3 dB. If the individual levels differ by more than 10 dB, the total level corresponds approximately to the larger level and the influence of the smaller level is negligible. It is also curious to note that

$$0\,\text{dB} + 0\,\text{dB} = 3\,\text{dB} \quad \text{is}$$

or

$$L_{\text{total}} = 10 \cdot \lg\left(10^{0/10} + 10^{0/10}\right) = 10 \cdot \lg(1+1) = 3 \text{ dB.}$$

2.3.2 Averaging Level

The energetic or power averaging of *n-levels* is analogous to the level addition:

$$\overline{P} = \frac{1}{n} \sum_{i=1}^{n} P_i \quad \text{W} \tag{2.65}$$

respectively

$$\overline{p} = \sqrt{\frac{1}{n} \sum_{i=1}^{n} p_i^2} \quad \text{N/m}^2 \tag{2.66}$$

The average level is then obtained:

$$\left.\begin{array}{c}\overline{L}=10\cdot\lg\dfrac{\overline{P}}{P_0}=10\cdot\lg\left[\dfrac{1}{n}\sum\limits_{i=1}^{n}10^{L_{Wi}/10}\right]\\ \text{respectively}\\ \overline{L}=10\cdot\lg\dfrac{\overline{p}^2}{p_0^2}=10\cdot\lg\left[\dfrac{1}{n}\sum\limits_{i=1}^{n}10^{L_{pi}/10}\right]\end{array}\right\}\quad\text{dB}\qquad(2.67)$$

By (2.64) it follows from (2.67):

$$\overline{L}=L_{\text{total}}-10\cdot\lg n\quad\text{dB}\qquad(2.68)$$

Example 2.3
The mean value of the levels 70 dB and 60 dB is sought:

$$\overline{L}=10\lg\left[\frac{1}{2}\left(10^{70/10}+10^{60/10}\right)\right]=67.4\ \text{dB}$$

respectively

$$\overline{L}=L_{\text{total}}-10\cdot\lg(2)=70.4-3=67.4\ \text{dB}$$

It follows that the power average is larger than the arithmetic average of (70 + 60)/2 = 65 dB and smaller than the largest level.

2.3.3 Noise Level

The problem of the extraneous level occurs in sound level measurements when the level L_M to be measured is distorted by a simultaneously and continuously acting background or extraneous level L_F, which cannot be eliminated.

To correct the measured total level L_G, which is the sum of L_M and L_F, first measure the total level L_G with the interference level L_F present and then the interference level L_F alone with the machine switched off. By taking the difference, one obtains the correction level ΔL:

$$L_M=L_{G-F}=10\cdot\lg\left(10^{\frac{L_G}{10}}-10^{\frac{L_F}{10}}\right)=L_G+10\cdot\lg\left(1-10^{-\frac{L_G-L_F}{10}}\right)$$

$$=L_G-\Delta L\quad\text{dB}\qquad(2.69)$$

$$\Delta L = -10 \cdot \lg\left(1 - \frac{1}{10^{\frac{L_G - L_F}{10}}}\right) = -10 \cdot \lg\left(1 - 10^{-0.1 \cdot (L_G - L_F)}\right) \quad \text{dB} \tag{2.70}$$

According to ISO 3744 [7], the following designations are given for sound pressure measurements for the above-mentioned quantities:

$$\left.\begin{aligned} L_G &= L'_p \\ L_F &= L''_p \\ \Delta L &= K_1 \end{aligned}\right\} \quad \text{dB} \tag{2.71}$$

L'_p	Sound pressure level at existing noise level
L''_p	Sound pressure level with machine switched off, noise level
K_1	Correction level for extraneous noise

2.3.4 A-weighted Level

The loudness of a sound source perceived by humans is very strongly frequency-dependent (in addition to the dependence on the level). Low frequencies are very strongly attenuated by the ear and the high frequencies are partially amplified. The reason for this lies on the one hand in the transmission system between the outer and inner ear – conversion of pressure fluctuations into mechanical vibrations – and on the other hand in the natural vibration behaviour of the auditory canal. The natural frequency of the external auditory canal lies between 2000 and 4000 Hz and thus forms the most sensitive frequency range in perception [1].

Since it is not possible to measure the loudness directly, for practical applications where the loudness is to be characterised, the sound level values are evaluated with a filter curve (A-curve) specified in DIN 45 633. The A-filter curve corresponds approximately to the frequency response of the human ear in the low level range.

The sound level values measured in this way are given the index "A" and referred to as the A-weighted level. The unit is often also extended with the letter A in brackets, e.g.:

L_{WA}	A-weighted sound power level in dB(A)

or

L_{vA}	A-weighted sound velocity level in dB(A)

In addition to the A filter curve, other filter curves also exist, which are designated as B, C and D curves. Figure 2.15 shows the filter curves as a function of frequency.

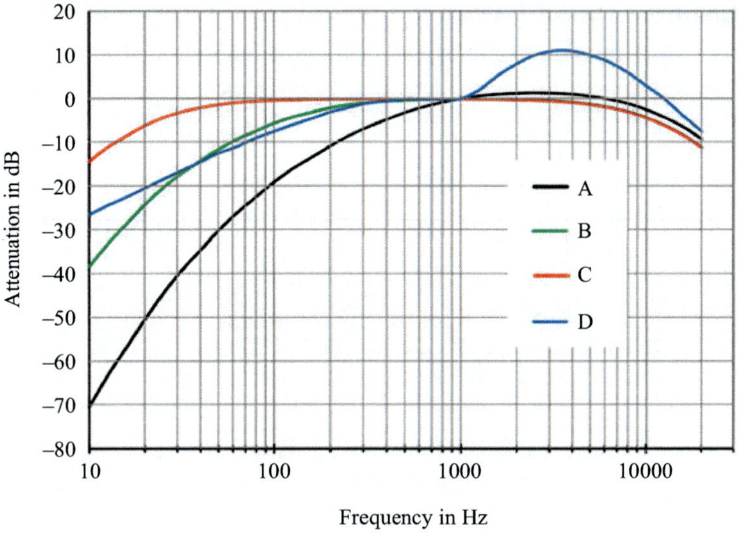

Fig. 2.15 Evaluation curves

The evaluation curves show the frequency response of the ear for narrowband noises in a somewhat simplified way, curve A in the range of less loud noises, curves B and C in the ranges of loud and very loud noises. Curve D applies to aircraft noise and is also no longer relevant to design acoustics. The associated measured quantities are given as A-weighted sound pressure level L_A in dB(A) or as B-, C-, D-weighted sound pressure level L_B, L_C, L_D.

2.3.5 Rating Level

The rating level is used to evaluate the subjective noise annoyance at the place of immission. In addition to the sound pressure level, additional characteristics in the spectral distribution are taken into account, especially when narrow-band components and short-term pulse-like level increases occur.

The rating level L_r (Noise Rating Level) allows to characterise the noise nuisance by specifying a single value quantity usually for a defined period T_r (rating time). The basis for the numerical determination of the rating level is the A-weighted energy-equivalent continuous sound pressure level L_{Aeq} at the place of immission according to (2.72).

$$L_{Aeq} = 10\lg\left[\frac{1}{T}\int_0^T 10^{\frac{L_A(t)}{10}}\mathrm{d}t\right] \quad \mathrm{dB(A)} \tag{2.72}$$

2.3.5.1 Neighbourhood Noise

Commercial noise immissions are assessed in the neighbourhood in the Federal Republic of Germany according to TA-Lärm. According to this, the rating level L_r can be determined as follows [8]:

$$L_{\mathrm{r}} = 10 \cdot \lg \left[\frac{1}{T_{\mathrm{r}}} \cdot \sum_{j=1}^{N} T_i \cdot 10^{0{,}1 \cdot \left(L_{\mathrm{Aeq},j} - C_{\mathrm{met}} + K_{\mathrm{T},j} + K_{\mathrm{I},j} + K_{\mathrm{R},j} \right)} \right] \quad \mathrm{dB(A)} \tag{2.73}$$

$T_{\mathrm{r}} = \sum_{j=1}^{N} T_j = 16$ h	Assessment time for the day: 06:00–22:00 h
$T_{\mathrm{r}} = 1$ h	Assessment time for the night: 22:00–06:00 h The full night hour (e.g. 01:00–02:00 h) with the highest rating level is decisive for the assessment of the night.
T_j	Part time j
N	Number of partial periods
$L_{\mathrm{Aeq},\,j}$	A-weighted energy-equivalent continuous sound pressure level for the part-time T_j
$K_{\mathrm{I},\,j}$	The surcharge for impulsivity is calculated from the difference between the A-weighted 5 s cycle maximum average sound pressure level $L_{\mathrm{AFTeq},\,j}$ and the A-weighted energy equivalent continuous sound pressure level $L_{\mathrm{Aeq},\,j}$ for the respective cycle time T_j [8].
$K_{\mathrm{T},\,j}$	The surcharge for tonality can be determined objectively according to DIN 45681-3/2005. According to TA-Lärm, it is taken into account subjectively, depending on the conspicuousness of the tonal components, by an addition of 3 or 6 dB.
$K_{\mathrm{R},\,j} = 6\,\mathrm{dB}$	Surcharge for times of day with increased sensitivity. 1. on working days: 06:00–07:00, 20:00–02:00 2. on Sundays and public holidays: 06:00–09:00, 13:00–15:00, 20:00–22:00 The rest period surcharge shall be allocated depending on the degree of protection afforded by the place of immission.
C_{met}	Meteorological correction according to ISO 9613-2 [9] C_{met} can be determined with an accuracy of ± 1 dB after an elementary analysis of local weather statistics and can assume values from 0 to 5 dB. In some cases, state-specific regulations for the determination of C_{met} are currently in effect. Since C_{met} leads to a reduction in the assessment level, it is often used $C_{\mathrm{met}} = 0$ so that the result is on the safe side.

2.3.5.2 Work Noise

For occupational noise, the daily noise exposure level (previously referred to as the rating level in the "UVV Lärm" BGV B3) is defined as follows [10]:

$$L_{\mathrm{EX,8\,h}} = L_{\mathrm{Aeq},T_\mathrm{e}} + 10\lg(T_\mathrm{e}/T_0) \quad \mathrm{dB(A)} \tag{2.74}$$

with:

$L_{\mathrm{EX,\,8\,h}}$ Daily noise exposure level in dB(A), referred to 20 μPa. The time-averaged noise exposure level for a nominal eight-hour day.

$$L_{\mathrm{Aeq},T_\mathrm{e}} = 10\lg\left[\frac{1}{T_\mathrm{e}}\sum_{i=1}^{n}\left(T_i \cdot 10^{0.1 \cdot L_{\mathrm{Aeq},T_i}}\right)\right] \quad \mathrm{dB(A)} \tag{2.75}$$

L_{Aeq,T_i} Energy-equivalent continuous sound pressure level in dB(A) acting over the partial time T_i (in h)

T_e Effective exposure time during a working day in h

T_0 Assessment time (=8 h)

In justified circumstances, for activities where noise exposure varies significantly from one working day to another, the weekly noise exposure level may be used instead of the daily noise exposure level to assess the noise levels to which workers are exposed:

$$\overline{L}_{\mathrm{EX,8\,h}} = 10\lg\left[\frac{1}{5}\sum_{i=1}^{n}10^{0.1(L_{\mathrm{EX,8\,h}})_i}\right] \quad \mathrm{dB(A)} \tag{2.76}$$

with:

$\overline{L}_{\mathrm{EX,8\,h}}$	Weekly noise exposure level in dB(A). The time-averaged daily noise exposure level for a nominal week with five eight-hour days is as follows
$L_{\mathrm{EX,\,8\,h,}\,i}$	Daily noise exposure level in dB(A) for the *i*th eight-hour day of the total of *n* working days of the week in question

For the assessment of noise at the workplace, the immission level at the workplace of the person concerned is measured in the area of his ears, alternatively for standing persons at a height of approx. 1.60 m above the standing surface, for seated persons at a height of approx. 0.8 m above the sitting surface. An assessment of the noise level at the workplace with regard to hearing damage is carried out in accordance with the *LärmVibrationsArbSchV* or, taking into account different activities, in accordance with *VDI 2058, Sheet 2 or Sheet 3* [11].

2.4 Temporal and Spectral Representation of Sound Field Quantities

For sound and vibration investigations, knowledge of the total level, e.g. total sound pressure level or total sound power level ($L_{p,\ \mathrm{total}}$, $L_{W,\ \mathrm{total}}$) alone is not sufficient. In addition, detailed information is required, especially about the frequency components contained in the noise. This is obtained by frequency analysis, in which the noise is broken down into its frequency components.

When displaying the amplitude or frequency spectrum, the very large acoustic frequency range of 16–16,000 Hz must be taken into account. In accordance with the human hearing ability, this range is divided logarithmically. In this way, a compression of the frequency range with a uniform resolution is achieved for the benefit of a greater clarity. The result of a first, somewhat coarser division is the so-called octave scale (Fig. 2.16).

The logarithmic unit is the octave step, which represents a doubling of frequency. This results in 10 octave steps on the octave ladder for the entire listening range.

The corner frequencies f_1 and $f_2 = 2 \cdot f_1$ determine the respective octave (Fig. 2.17). It can also be characterised by its centre frequency $f_{\mathrm{m,oct}}$, which geometrically divides the octave step into two equal parts. It is thus:

$$\frac{f_{\mathrm{m_{oct}}}}{f_{1,\mathrm{oct}}} = \frac{f_{2,\mathrm{oct}}}{f_{\mathrm{m_{oct}}}} \tag{2.77}$$

It follows:

$$f_{\mathrm{m,oct}} = \sqrt{f_{1,\mathrm{oct}} \cdot f_{2,\mathrm{oct}}} = \sqrt{2} \cdot f_{1,\mathrm{oct}} = \frac{1}{\sqrt{2}} \cdot f_{2,\mathrm{oct}} \quad \mathrm{Hz} \tag{2.78}$$

The relative bandwidth of octaves is constant and is about 71%:

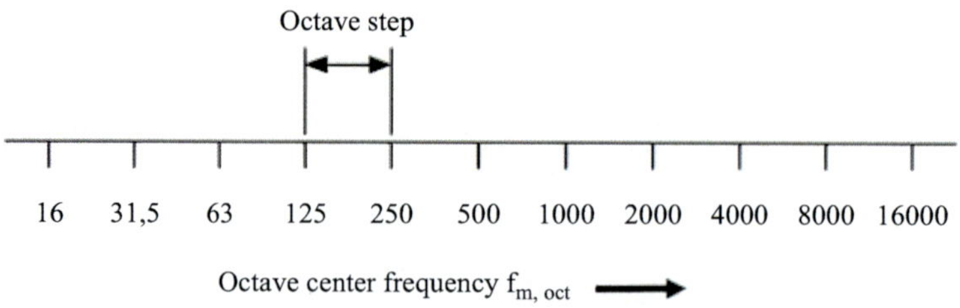

Fig. 2.16 Octave conductor

Fig. 2.17 Frequency ladder with the corner frequencies and the centre frequency

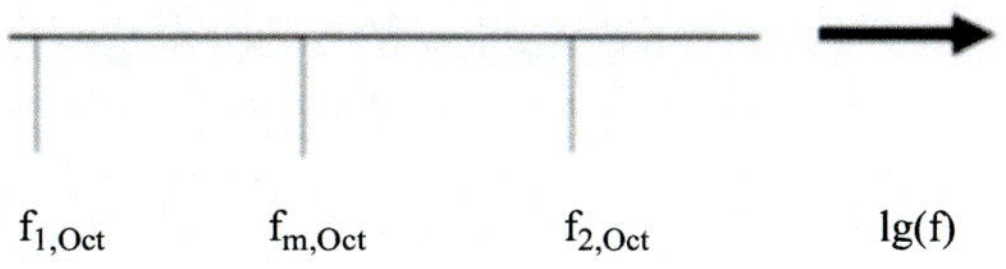

Fig. 2.18 Third-octavesteps steps in an octave

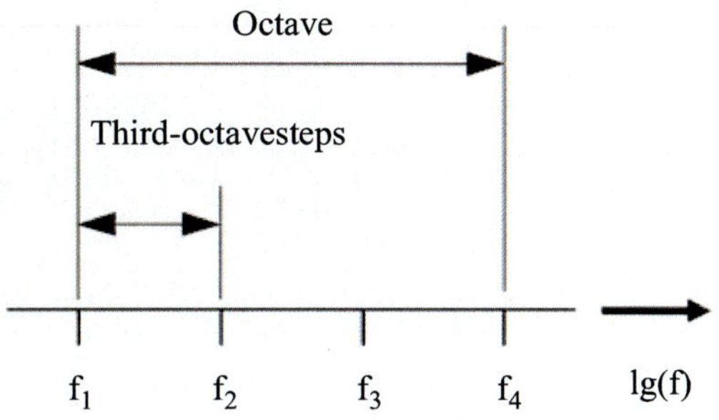

$$\frac{\Delta f_{\text{oct}}}{f_{\text{m,oct}}} = \frac{f_{\text{2,oct}} - f_{\text{1,oct}}}{f_{\text{m,oct}}} = \frac{1}{\sqrt{2}} \approx 0.71 \tag{2.79}$$

It follows that if the centre frequency is $f_m = 100$ Hz, the bandwidth is $\Delta f_{100\,\text{Hz}} \approx 70$ Hz, and if the centre frequency is $f_m = 1000$ Hz, the bandwidth is $\Delta f_{1000\ \text{Hz}} \approx 700$ Hz.

For a finer subdivision, the octave on the logarithmic frequency ladder is divided into three equal parts, so-called third-octavesteps (Fig. 2.18).

$$\frac{f_2}{f_1} = \frac{f_3}{f_2} = \frac{f_4}{f_3} \tag{2.80}$$

By definition:

$$\lg\left(\frac{f_4}{f_1}\right) = \lg(2) \quad \text{and} \quad \lg\left(\frac{f_2}{f_1}\right) = \frac{\lg(2)}{3}$$

It follows then:

$$f_{\text{2,third-oct}} = \sqrt[3]{2} \cdot f_{\text{1,third-oct}} \quad \text{Hz} \tag{2.81}$$

$$f_{\text{m,third-oct}} = \sqrt{f_{\text{1,third-oct}} \cdot f_{\text{2,third-oct}}} = \sqrt{f_{\text{1,third-oct}} \cdot \sqrt[3]{2} \cdot f_{\text{1,third-oct}}} = \sqrt[6]{2} \cdot f_{\text{1,third-oct}} \quad \text{Hz} \tag{2.82}$$

The relative bandwidth of third-octave bands is constant and is:

Table 2.2 Centre frequency f_m, according to EN ISO 266, upper and lower cut-off frequencies, the bandwidth and the attenuation levels of the A-weighting for octaves

$f_{m,oct}$ Hz	$f_{u,oct}$ Hz	$f_{o,oct}$ Hz	ΔF_{oct} Hz	ΔL_{oct} dB
16	11.3	22.6	11.3	−56.4
31.5	22.3	44.5	22.3	−39.5
63	44.5	89.1	44.5	−26.2
125	88.4	176.8	88.4	−16.1
250	176.8	353.6	176.8	−8.6
500	353.6	707.1	353.6	−3.2
1000	707.1	1414.2	707.1	0.0
2000	1414.2	2828.4	1414.2	1.2
4000	2828.4	5656.9	2828.4	1.0
8000	5656.9	11313.7	5656.9	−1.1
16,000	11313.7	22627.4	11313.7	−6.7

$$\frac{\Delta f_{\text{third-oct}}}{f_{\text{m,third-oct}}} = \frac{f_{\text{2,third-oct}} - f_{\text{1,third-oct}}}{f_{\text{m,third-oct}}} = \frac{\sqrt[3]{2} - 1}{\sqrt[6]{2}} \approx 0.23 \tag{2.83}$$

It follows that the bandwidth is approximately 23% of the one-third octave centre frequency and is significantly smaller than the bandwidth of an octave (71%).

Tables 2.2 and 2.4 list the standardised [12] centre frequencies f_m with the associated cutoff frequencies (lower and upper cutoff frequencies f_u and f_o) as well as the bandwidth and attenuation levels of the A-weighting for octaves and third-octave bands. The frequency ranges that are decisive for practice, especially for noise reduction, are marked in color.

If the A-weighted total level is to be determined from an octave or third-octave analysis of a noise, the sum level must be formed from the octave or third-octave levels, taking into account the frequency-dependent attenuation levels of the A-weighting (Tables 2.2 and 2.4). It is

$$L_{A,j} = L_j + \Delta L_j \quad \text{dB(A)} \quad j = f_m \tag{2.84}$$

$$L_A = 10 \cdot \log\left[\sum_{j=1}^{N} 10^{\frac{L_j + \Delta L_j}{10}}\right] \quad \text{dB(A)} \tag{2.85}$$

L_j	Third-octave or octave level at the centre frequency f_j in dB

Table 2.3 Results of the example 2.4

f_m	63	125	250	500	1000	2000	4000	8000	Total	Hz
$L_{p,\text{ oct}}$	68	92	73	86	81	78	75	73	93.5	dB
ΔL_{oct}	−26.2	−16.1	−8.6	−3.2	0.0	1.2	1.0	−1.1	–	dB
$L_{p\text{A, oct}}$	41.8	75.9	64.4	82.8	81.0	79.2	76.0	71.9	87.0	dB(A)

Example 2.4
Given are the measured 1 m octave sound pressure levels $L_{p,\text{ oct}}$ of a machine. The 1 m A-octave and A-total sound pressure levels of the machine are sought: $L_{p\text{A. Oct}}$; $L_{p\text{A}}$. Table 2.3 shows the results.

1 m-A sound pressure level: $L_{p\text{A}} = 87.0\,\text{dB(A)}$.

In the same way that the sound pressure level L_p is frequency-weighted, the sound power level L_W can also be frequency-weighted. The result is the A-weighted sound power level or the A-sound power level $L_{W\text{A}}$. This quantity is particularly suitable for assessing the noise emission of a sound source, e.g. a machine.

Remark:
If the overall A-weighted level is less than the overall linear level, as in the example 2.4, Table 2.3 (87.0 dB(A) < 93.5 dB), this indicates that in the spectrum the frequencies below 1000 Hz are level-determining. Conversely, if the A level is greater than the linear level, it indicates that the frequencies above 1000 Hz are level-determining.

To the extent that the A-sound pressure level is used to assess noise immission, the A-sound power level can be used to assess noise emission.

In addition to frequency analysis with constant relative bandwidth, e.g. octave and third-octave analyses, $\left(\frac{1}{3}\text{Octave}\right)$ narrowband analyses with constant relative bandwidth are also common. $\left(\frac{1}{6}, \frac{1}{12}, \frac{1}{24}\text{ Octave}\right)$

For more precise cause analyses, especially for the localisation of tonal frequency components, frequency analyses with constant bandwidth, for example, are also used. $\Delta f = 1, 2, 4, \ldots \text{Hz}$

If noise of constant power density is present in a certain frequency range, the associated spectra differ by just the ratio of their bandwidths. For octave and third octave spectra, this corresponds to a constant ratio of 70.7/23.2. With the aid of octave, third octave and narrowband filters, periodic functions, for example single tones, can also be analysed (Fig. 2.19).

The amplitude of one of the partial oscillations f_i appears with the same value in all three spectra (Fig. 2.20).

In the following, some physical quantities that are important in practice, e.g. the sound pressure or the alternating force, are given as time and frequency diagrams.

Table 2.4 Centre frequency f_m, according to EN ISO 266, upper and lower cut-off frequencies, the bandwidth and the attenuation levels of the A-weighting for third-octave bands

$f_{m,\text{third-octave}}$ Hz	$f_{u,\text{third-octave}}$ Hz	$f_{o,\text{third-octave}}$ Hz	$\Delta f_{\text{third-octave}}$ Hz	$\Delta L_{\text{third-octave}}$ dB
16	14.3	18.0	3.7	−56.4
20	17.8	22.4	4.6	−50.4
25	22.3	28.1	5.8	−44.8
31.5	28.1	35.4	7.3	−39.5
40	35.6	44.9	9.3	−34.5
50	44.5	56.1	11.6	−30.3
63	56.1	70.7	14.6	−26.2
80	71.3	89.8	18.5	−22.4
100	89.1	112.2	23.2	−19.1
125	111.4	140.3	28.9	−16.1
160	142.5	179.6	37.1	−13.2
200	178.2	224.5	46.3	−10.8
250	222.7	280.6	57.9	−8.6
315	280.6	353.6	72.9	−6.6
400	356.4	449.0	92.6	−4.8
500	445.4	561.2	115.8	−3.2
630	561.3	707.2	145.9	−1.9
800	712.7	898.0	185.3	−0.8
1000	890.9	1122.5	231.6	0.0
1250	1113.6	1403.1	289.5	0.6
1600	1425.4	1795.9	370.5	1.0
2000	1781.8	2244.9	463.1	1.2
2500	2227.2	2806.2	578.9	1.3
3150	2806.3	3535.8	729.4	1.2
4000	3563.6	4489.8	926.3	1.0
5000	4454.5	5612.3	1157.8	0.6
6300	5612.7	7071.5	1458.8	−0.1
8000	7127.2	8979.7	1852.5	−1.1
10,000	8909.0	11224.6	2315.6	−2.5
12,500	11136.2	14030.8	2894.5	−4.3
16,000	14254.4	17959.4	3705.0	−6.7

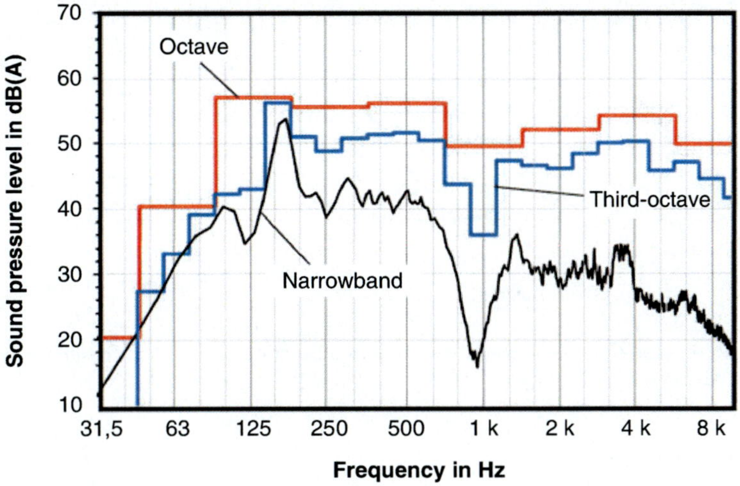

Fig. 2.19 Different frequency analyses of a noise with narrowband components

Fig. 2.20 Schematic representation of various frequency analyses of a tone-containing noise at frequency f_i

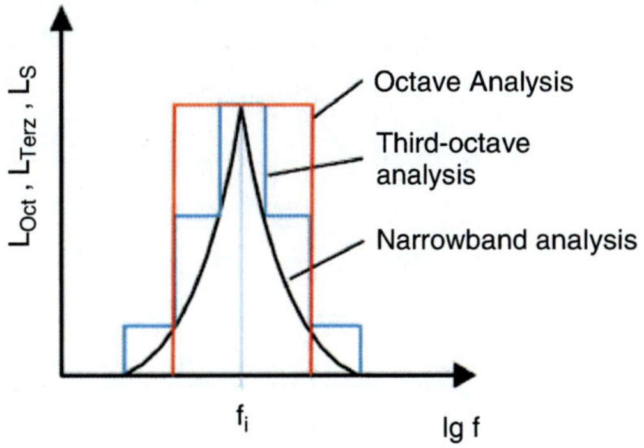

2.4.1 Periodic Time Course

In the simplest case, it is a sinusoidal function or a physical tone with period T_0 (Fig. 2.21). The tone is characterised by the amplitude $\widehat{p}$ and the frequency $f_0 = 1/T_0$. The amplitude determines the loudness and the frequency determines the pitch. The time and frequency characteristics of mechanical vibrations caused by pure unbalance have the same course.

The arbitrary periodic time course (Fig. 2.22), on the other hand, belongs acoustically to the simple or musical sounds, so called because this type of sound is produced primarily by musical instruments.

One can obtain harmonic partial oscillations from the present periodic time diagram by Fourier analysis. They consist of the fundamental f_0 and the harmonics f_1, f_2, etc. with the amplitudes. $\widehat{p}_0, \widehat{p}_1, \widehat{p}_2$. The lowest frequency f_0 determines the pitch of the sound, the interaction of determines the $\widehat{p}_0, \widehat{p}_1, \widehat{p}_2$ strength of the sound.

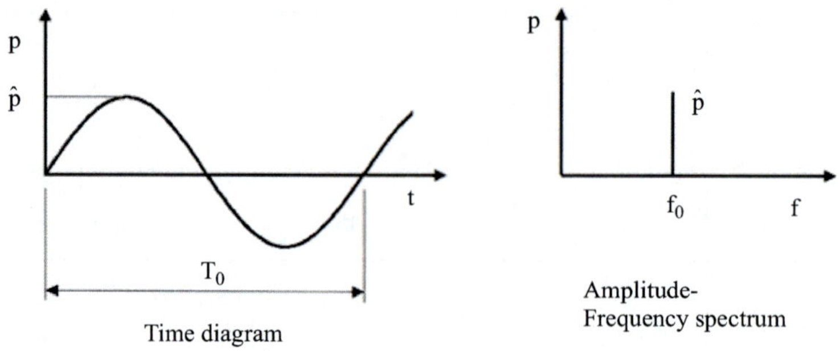

Fig. 2.21 Representation of the sound pressure as a function of time and frequency with a harmonic time characteristic (sinusoidal tone)

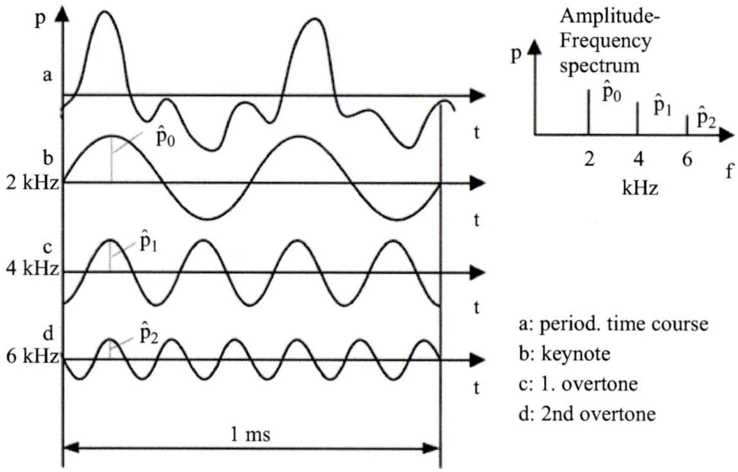

Fig. 2.22 Time history and frequency spectrum of a periodic function [1]

The representation of the result by Fourier analysis in the so-called amplitude spectrum leads to a discrete spectrum or line spectrum, which in the case of a single tone consists of only one line, see Fig. 2.20.

2.4.2 Random, Stochastic Time History, General Noise

This is a random, stochastic time course, e.g. friction processes. Figure 2.23 schematically shows the time diagram of a physical quantity, e.g. the sound pressure $p(t)$, in which no periods can be recognised. The associated frequency spectrum is, in contrast to the line

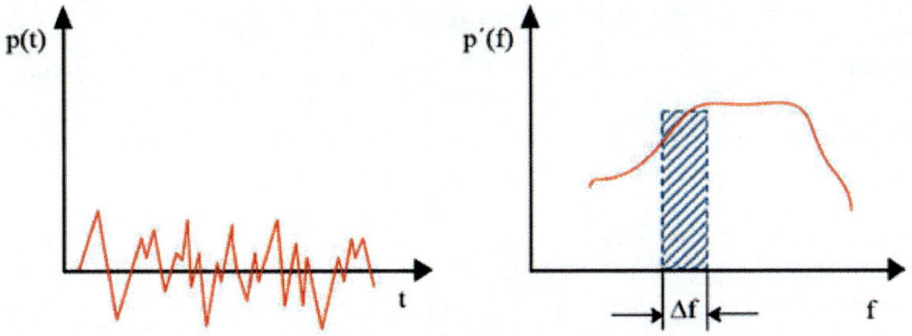

Fig. 2.23 Uncontrolled, stochastic process in the time and frequency diagram

spectrum of periodic processes, a continuous spectrum. The frequency spectrum is broadband distributed. This means that no amplitudes can be specified as with the discrete spectrum, but an amplitude density related to the frequency must be plotted $p'(f)$.

2.5 Level Reduction by Sound Reflection and Sound Absorption

2.5.1 Sound Insulation

The reduction of sound power in the course of sound propagation by reflections from sound-reflecting partition surfaces, e.g. a concrete wall, is referred to as sound attenuation. (Reduction of sound energy by reflection, Fig. 2.24)

The sound insulation is expressed by the sound reduction index R and can be determined as follows (see DIN 4109):

$$R = D + 10 \cdot \lg \frac{S}{A} \quad \text{dB} \tag{2.86}$$

$$D = L_1 - L_2 \quad \text{dB} \tag{2.87}$$

Here are:

D	Sound level difference in dB
L_1	Sound level in the transmitter room in dB
L_2	Sound level in the receiving room in dB
S	Area of the sound-insulating wall in m^2
A	Equivalent absorption area of the receiving room in m^2

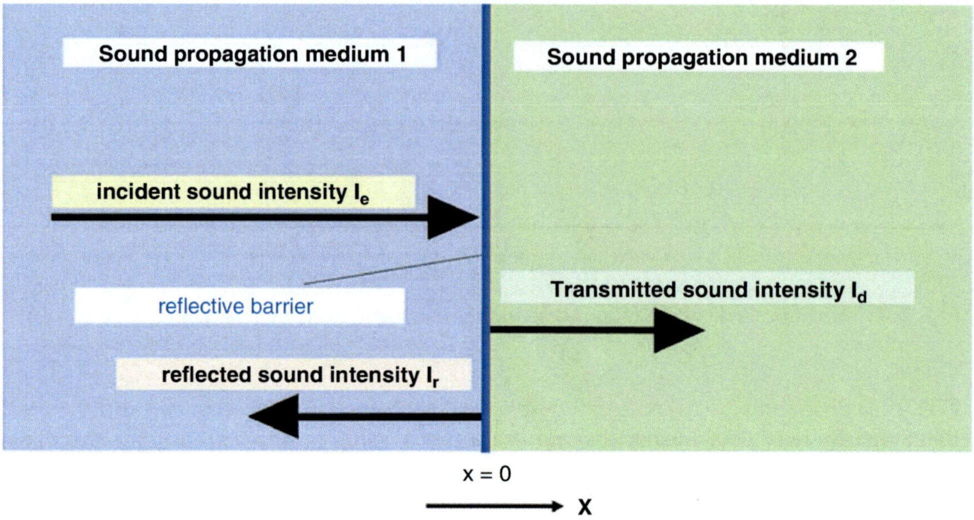

Fig. 2.24 Schematic representation of sound propagation through a sound-reflecting obstacle (without edge transmission and damping)

2.5.2 Sound Attenuation

The absorption of sound energy at sound-absorbing "partitions" (sound-absorbing walls), i.e. the conversion of sound energy into heat, is referred to as sound attenuation (reduction of sound energy by friction Fig. 2.25).

Sound attenuation is expressed by the sound **absorption coefficient** α and can be determined as follows:

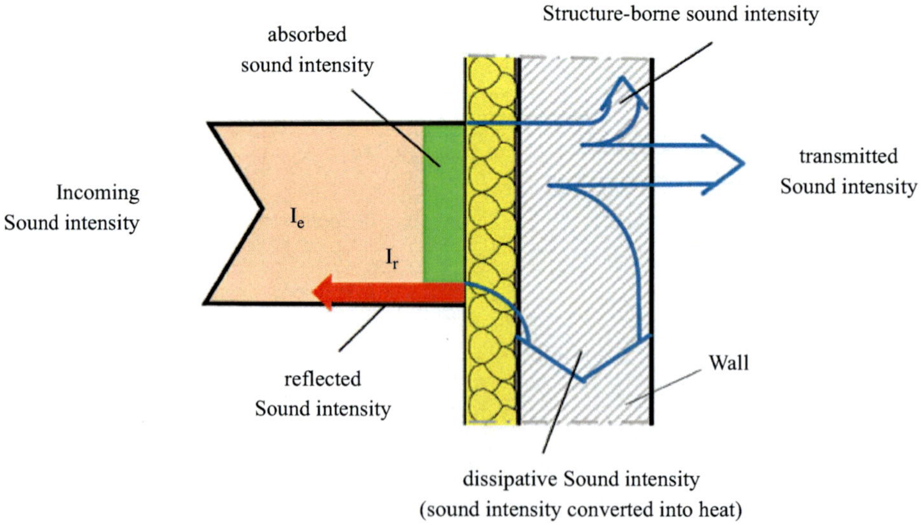

Fig. 2.25 Schematic representation of sound propagation through a sound-absorbing obstacle

$$\alpha = \frac{I_e - I_r}{I_e} \tag{2.88}$$

With

I_e	incident sound intensity in W/m^2
I_r	reflected sound intensity in W/m^2
$I_e - I_r$	absorbed sound intensity in W/m^2

Only a part of the incident energy is reflected back into the transmitting room. The non-reflected part is called the absorbed part (sound attenuation).

2.5.3 Equivalent Absorption Area *A*

The absorption capacity of a closed room is composed of the absorption of the bounding walls, ceiling and floor and all other sound-absorbing elements present in the room. These are surface sound absorbers which, in addition to the natural ones, may also contain additionally arranged pore or resonance absorbers. Their absorption capacity depends on the absorption coefficient α_i and the area S_i of the individual surface elements. Thereby, for each of these surface elements S_i the absorption is

$$\overline{A}_i = \alpha_i \cdot S_i \quad \mathrm{m}^2 \tag{2.89}$$

The absorption has $\overline{A}_i$ the unit of an area and can also be taken as an equivalent substitute area of S_i for which $\alpha = 1$, i.e. the equivalent absorption area A indicates in m^2 how large a surface with the absorption coefficient $\alpha = 1$ would have to be in order to absorb the same sound energy per unit time as the actually existing surface in m^2 with the real absorption coefficient α does. The equivalent sound absorption area is determined from the reverberation time of the room in question, which can be determined relatively easily by measurement.

2.5.3.1 Reverberation Time *T*

The reverberation time is defined as the time in s that elapses in the room after the sound source is switched off until the sound pressure has decayed to the thousandth part or the sound pressure level by 60 dB. The reverberation time T can be determined, based on a simple energy consideration in a diffuse sound field, as follows [1]:

$$T = \frac{V}{\overline{A}_{\text{total}}} \frac{24 \cdot \ln(10)}{c} = \frac{V}{\overline{A}_{\text{total}}} \frac{55.26}{c} \quad \mathrm{s} \tag{2.90}$$

With

$\overline{A}_{\text{total}}$	Average total absorption of a room in m^2
V	Room volume in m^3
c	Speed of sound in m/s

Especially for air ($c = 340$ m/s), the reverberation time can be calculated according to the following numerical equation, the so-called "Sabine reverberation time T_S" [13]:

$$T = T_S = 0.163 \cdot \frac{V}{\overline{A}_{\text{total}}} \quad \text{s} \tag{2.91}$$

Herein, T is to be put in s, V *is* to be put in m^3, $\overline{A}_{\text{total}}$ is to be put in m^2.

The reverberation time and consequently also the equivalent total absorption area of the room are frequency-dependent. Usually, the reverberation time is determined in one-third octave or octave bands by measurements. For this purpose, e.g. octave or third octave bands are filtered out of the white noise of a noise generator and radiated into the room by a loudspeaker. The sound pressure level is also filtered and recorded after the noise generator is switched off.

For measurement reasons, half the reverberation time $T/2$ is often determined, since the drop in the excited sound pressure level usually does not cover the full 60 dB. Usually, the drop of the excited sound pressure level does not cover the full 60 dB, but already reaches the interference sound pressure level beforehand.

As indicated in Fig. 2.26, the time it takes for the sound pressure level to decay from −5 dB to −35 dB is equal to half the reverberation time $T/2$. Doubling this gives the reverberation time T we are looking for.

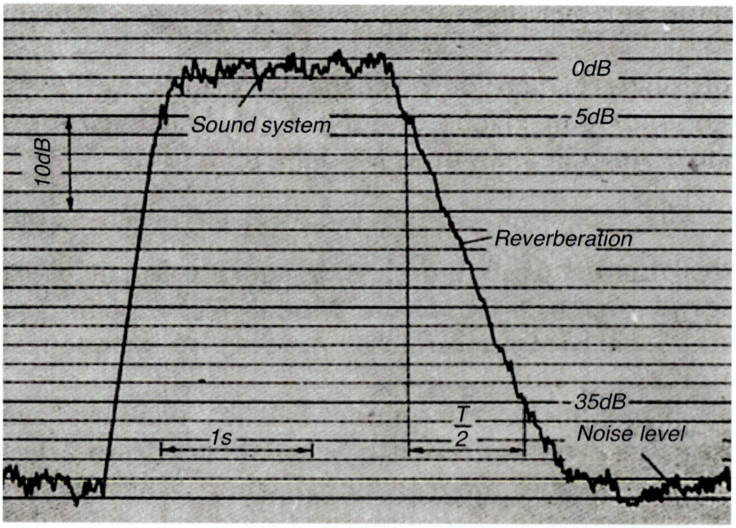

Fig. 2.26 Measured reverberation time of a room at the octave centre frequency $f_m = 1000$ Hz

In room acoustics, the reverberation time determines the audibility of a room. If the reverberation times are determined for octave or third-octave bands, the total absorption of the room can be represented as a $\overline{A}_{\text{total}}$ function of frequency for octave or third-octave bands. One can still replace the latter by T_S in all acoustic relations of the room in which the magnitude is included.

It should be noted that the equivalent absorption area represents all surfaces and also the airborne sound attenuation of the measurement room in addition to the attenuation. Due to the low airborne attenuation properties, this only needs to be taken into account for very large rooms or standard measurements in reverberation rooms [1].

2.5.3.2 Reverberation Radius r_H

The sound field generated by a point source, a 0th-order omnidirectional radiator, in a real or semi-diffuse room changes from direct sound to the diffuse field at a distance r_H, see Fig. 2.6. This distance is called the reverberation radius. Figure 2.27 illustrates schematically the sound fields in enclosed spaces. The sound field in the room inside the sphere ($r < r_H$) has the character of a free field, i.e. the intensity decreases with the square of the distance:

$$I(r) = \frac{P_Q}{4\pi \cdot r^2} \quad \text{respectively} \quad I(r_H) = \frac{P_Q}{4\pi \cdot r_H^2} \quad \text{W/m}^2 \tag{2.92}$$

The sound field at $r > r_H$ has diffuse character, i.e. the intensity is constant (see Fig. 2.6) and can be determined as follows [1]:

$$I_H = \frac{4 \cdot P_Q}{\overline{A}_{\text{total}}} \quad \text{W/m}^2 \tag{2.93}$$

The reverberation radius $r = r_H$ can be determined by equating the intensities ($I(r_H) = I_H$) at the interface:

$$r_H \approx \sqrt{\frac{\overline{A}_{\text{total}}}{16 \cdot \pi}} = 0.141 \cdot \sqrt{\overline{A}_{\text{total}}} \quad \text{m} \tag{2.94}$$

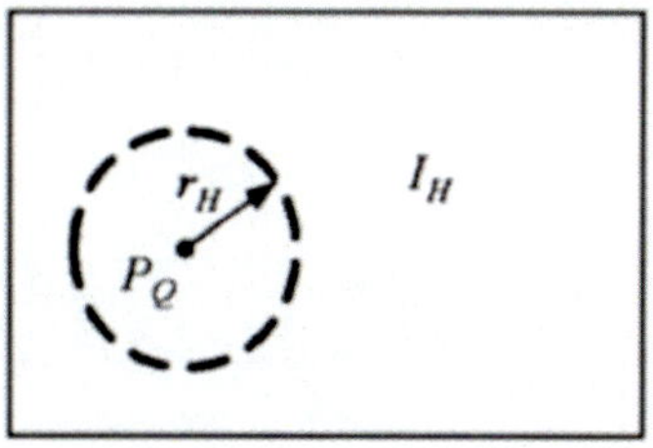

Fig. 2.27 Schematic representation of the reverberation radius in an enclosed space

It can be seen that the reverberation radius r_H depends only on the total absorption area $\overline{A}_{total}$. With (2.91) the reverberation radius can be given as a function of the reverberation time:

$$r_H \approx 0.057\sqrt{\frac{V}{T_S}} \quad \text{m} \tag{2.95}$$

In the numerical value equation (2.95), r_H in m is obtained when V is substituted in m^3 and T_S in s.

The reverberation radius is a parameter that is used, among other things, in room acoustics. Room acoustic measures, e.g. increasing the absorption area by means of an acoustic ceiling, lead to a reduction of the reverberation time and an increase of the reverberation radius. This means that by increasing the room absorption, the room reflections can be reduced and the average room or reverberation level can be lowered. However, the room acoustic measures are ineffective for the immission locations (e.g. workplaces) which are located within the sphere with the radius r_H.

2.6 A-sound Power Level L_{WA}

The sound emission or noise development of a machine is best characterised by the A-sound power level as a total value and the associated frequency spectrum (L_{WA,f_m} third-octave or octave spectrum). The task of primary noise reduction or low-noise design is to reduce the sound emission of a machine. The effectiveness of measures can also be clearly demonstrated by determining the A-sound power levels with and without measures.

The metrological determination of the sound power level is internationally standardised. Some standards are listed in Table 2.5.

In the Federal Republic of Germany, DIN 45635, Part 1, continues to apply, together with the numerous subsequent sheets for special applications. In the following, based on DIN 45635, Part 1 (ISO 3744), the essential relationships in the determination of the A-sound power level (L_{WA} and L_{WA,f_m}) are explained.

Oriented at (2.25), in which the sound power of a source P_Q is given as a function of sound pressure p, area S, and impedance ($\rho \cdot c$), the sound power level L_W of the source can be calculated as follows:

$$L_W = 10 \cdot \lg\left(\frac{P_Q}{P_0}\right) = 10 \cdot \lg\frac{p^2}{p_0^2} + 10\lg\frac{S}{S_0} - 10 \cdot \lg\frac{\rho \cdot c}{\rho_0 \cdot c_0} \quad \text{dB} \tag{2.96}$$

$$L_W = L_p + L_S + K_0 \quad \text{dB} \tag{2.97}$$

with

Table 2.5 Some measurement standards for the determination of sound power [7, 14–21]

Measurement method	Measuring environment	Standard
Emissions – Sound pressure level at the workplace	Free field over reflecting surface indoors or outdoors	ISO 11202
	No special measuring environment	ISO 11203 ISO 11204
Original sound source, enveloping surface method	Hard-walled measuring chamber	ISO 3743-1
	Low-reflection measuring chamber, accuracy class 1	ISO 3745
	Outdoors or in large rooms, accuracy class 2	ISO 3744
	No special measuring environment. Accuracy class 3	ISO 3746
Intensity measurement	No special measuring environment	ISO 9614-1 ISO 9614-2

L_p	Sound pressure level according to (2.55)
L_S	Area measure according to (2.58)
K_0	Sound characteristic impedance according to (2.57). For practical application, $K_0 \approx 0$ (can be neglected).

Using (2.97), one can determine the sound power level of a machine under free-field conditions.

If the sound power is determined in enclosed spaces in the presence of extraneous noise and room reflections, then the determination equation for the sound power level L_W according to ISO 3744 follows from (2.97):

$$L_W = \overline{L_{pf}} + 10\lg\left(\frac{S}{S_0}\right) \quad \text{dB} \tag{2.98}$$

$$\overline{L_{pf}} = \overline{L'_p} - K_1 - K_2 \quad \text{dB} \tag{2.99}$$

$$\overline{L'_p} = 10\lg\left(\frac{1}{N}\sum_{i=1}^{N} 10^{0.1L'_{pi}}\right) \quad \text{dB} \tag{2.100}$$

$$K_1 = -10\lg\left(1 - 10^{-0.1\Delta L}\right) \quad \text{dB} \tag{2.101}$$

$$\Delta L = \overline{L'_p} - \overline{L''_p} \quad \text{dB} \tag{2.102}$$

$$\overline{L''_p} = 10\lg\left(\frac{1}{N}\sum_{i=1}^{N} 10^{0.1L''_{pi}}\right) \quad \text{dB} \tag{2.103}$$

$$K_2 = 10\lg\left(1 + \frac{4 \cdot S}{\overline{A}_{\text{total}}}\right) \quad \text{dB} \tag{2.104}$$

with:

$\overline{L_{\text{pf}}}$	Measurement area sound pressure level in dB, sound pressure level averaged over measurement area *S*, corrected for extraneous noise and ambient noise
$\overline{L'_p}$	Sound pressure level averaged over the measurement area *S* during operation of the source under investigation in the presence of extraneous noise
K_1	Correction for extraneous noise
$\overline{L''_p}$	Extraneous noise level averaged over the measurement area *S*
L'_{pi}, L''_{pi}	Sound pressure level or extraneous noise level measured at the *i-th* microphone position
K_2	Ambient correction for room reflections. The frequency dependence of the ambient correction K_2 is determined by the frequency dependence of the equivalent absorption area or the reverberation time of the measurement room
S	Measuring area (enveloping area) in m^2; $S_0 = 1\ \text{m}^2$
$\overline{A}_{\text{total}}$	Equivalent sound absorption area of the room in m^2 according to (2.91)
N	Number of measuring points

Except for the measurement area *S*, all specified quantities are *A-weighted* or specified in frequency bands (third-octave or octave bands).

If the measured values, $\overline{L_{\text{pf}}}$, $\overline{L'_p}$, K_1 and K_2 are present in the frequency bands $f_{\text{m}j}$, the A-weighted sound power level can also be determined at the frequency band $f_{\text{m}j}$ as follows:

$$L_{\text{W}f_{\text{m}j}} = \overline{L'_{p,f_{\text{m}j}}} - K_{1,f_{\text{m}j}} - K_{2,f_{\text{m}j}} + L_S \quad \text{dB} \tag{2.105}$$

$$L_{\text{WA},f_{\text{m}j}} = L_{\text{W}f_{\text{m}j}} + \Delta L_{f_{\text{m}j}} \quad \text{dB} \tag{2.106}$$

or

$$L_{WA,f_{mj}} = \overline{L'_{pA,f_{mj}}} - K_{1,f_{mj}} - K_{2,f_{mj}} + L_S \quad \text{dB(A)} \tag{2.107}$$

With

$\overline{L'_{pA,f_{mj}}}$	Averaged A-sound pressure level at the centre frequency and f_{mj} in the presence of extraneous noise
$\Delta L_{f_{mj}}$	Correction level of the A-weighting at the centre frequency f_{mj}. Tables 2.2 and 2.4 show the correction levels for octave and third octave centre frequencies.

The A-total sound power level L_{WA} can be determined as follows:

$$L_{WA} = 10\lg\left[\sum_{j=1}^{M} 10^{\frac{L_{WA,f_{mj}}}{10}}\right] \quad \text{dB(A)} \tag{2.108}$$

M	Number of frequency bands

Figure 2.28 shows the measurement or enveloping surface with the arrangement of measurement points for the determination of the sound power level for a cuboid sound source.

The area measure L_S can be determined as follows:

$$L_S = 10 \cdot \lg\frac{S}{S_0} \quad \text{dB} \quad S_0 = 1 \quad \text{m}^2 \tag{2.109}$$

$$S = 2 \cdot a \cdot c + 2 \cdot b \cdot c + a \cdot b \quad \text{m}^2 \tag{2.110}$$

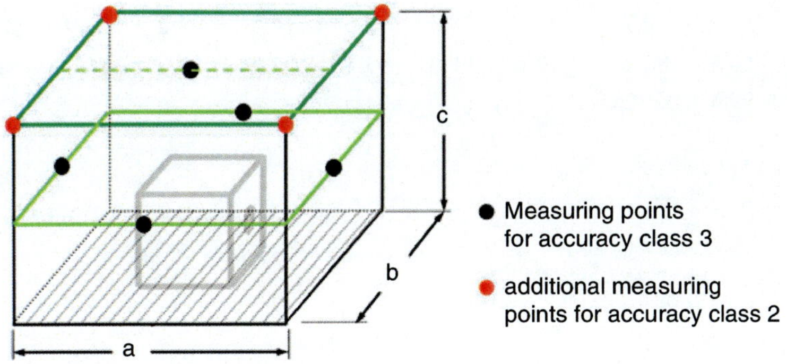

Fig. 2.28 Measuring or enveloping surface S with arrangement of measuring points for the determination of the sound power level for a cuboid sound source

Example 2.5

The A-octave and A-total sound power levels of the machine given in the example should be determined according to ISO 3744 in the presence of noise and room reflections.

Dimensions of the measuring chamber:	$L_R \times B_R \times H_R = 16 \times 12 \times 6$ m
Machine dimensions:	$L_M \times B_M \times H_M = 2.3 \times 1.8 \times 1.6$ m
Measuring distance:	$r = 1$ m $K_0 = 0$ dB

Tables 2.6 and 2.7 show the results of the example.

Table 2.6 Enveloping surface; surface area; room volume

L_M	2.3	m	a	4.3	m	L_R	16	m
B_M	1.8	m	b	3.8	m	B_R	12	m
H_M	1.6	m	c	2.6	m	H_R	6	m
measuring distance r =	1.0	m						
S =	**58.5**	**m²**	**L_s=**	**17.7**	**dB**	**V =**	**1152**	**m³**

Table 2.7 Measurement data and calculation results

f_m in Hz	63	125	250	500	1000	2000	4000	8000	Total	
$\overline{L_{pA',oct}}$	41.8	75.9	64.4	82.8	81.0	79.2	76.0	71.9	87.0	dB(A)
$\overline{L_{pA'',oct}}$	40.1	70.4	55.0	76.6	78.0	74.3	68.9	62.5	82.0	dB(A)
ΔL	1.7	5.5	9.4	6.2	3.0	4.9	7.1	9.4	–	dB
K_1	4.9	1.4	0.5	1.2	3.0	1.7	0.9	0.5	–	dB
T	1.2	1.1	1.2	0.9	1.0	0.8	0.7	0.6	–	S
$\overline{A}_{total}$	156	171	156	209	188	235	268	313	–	m²
K_2	4.0	3.7	4.0	3.3	3.5	3.0	2.7	2.4	–	dB
$\overline{L_{pA,oct}}$	32.9	70.7	59.9	78.3	74.5	74.5	72.3	68.9	82.1	dB(A)
$L_{WA.oct}$	**50.6**	**88.4**	**77.6**	**96.0**	**92.1**	**92.2**	**90.0**	**86.6**	**99.8**	**dB(A)**
L_{WA}	**99.8**	**dB(A)**								

References

1. Sinambari, Gh.R., Sentpali, S.: Ingenieurakustik, 6. Aufl. Springer Vieweg Verlag, Wiesbaden (2020)
2. Skudrzyk, E.: Die Grundlagen der Akustik. Springer-Verlag, Wien (1954)
3. Thorn, U.: Vorausbestimmung der Schall- und Körperschallreduzierung durch Einfügen von mechanischen Impedanzen bei realen Strukturen. Dipl. Arbeit FH Bingen (1995)
4. Heckl, M.: Eine einfache Methode zur Abschätzung der mechanischen Impedanz, Tagungsband zur DAGA 80, pp. 827–830. VDE-Verlag (1980)
5. Cremer, L., Heckl, M.: Körperschall, 2. Aufl. Springer Verlag (1996), Möser, M., Kropp, W: 3. Aufl. 2009
6. Cremer, L.: Die wissenschaftlichen Grundlagen der Raumakustik, 2. Aufl., 1. Teil. Hirzel-Verlag, Stuttgart (1978)
7. ISO 3744, 2010: Acoustics – Determination of sound power levels and sound energy levels of noise sources using sound pressure – Engineering methods for an essentially free field over a reflecting plane
8. Technische Anleitung zum Schutz gegen Lärm – TA Lärm: Sechste Allgemeine Verwaltungsvorschrift zum Bundes-Immissionsschutzgesetz vom 26.08.1998
9. ISO 9613-2,1996: Acoustics – Attenuation of sound during propagation outdoors – Part 2: General method of calculation
10. IFA-LSA-01-400,2019: Lärmschutz-Arbeitsblatt, Beurteilung der Lärmexposition nach der Lärm- und Vibrations-Arbeitsschutzverordnung
11. VDI 2058 Blatt 2: Beurteilung von Lärm hinsichtlich Gehörgefährdung, 1988; Blatt 3: Beurteilung von Lärm am Arbeitsplatz unter Berücksichtigung unterschiedlicher Tätigkeiten, 1999, (Entwurf, 04/2013)
12. IEC 61260-1, 2014: Electroacoustics – Octave-band and fractional-octave-band filters – Part 1: Specifications
13. Schmidt, H.: Schalltechnisches Taschenbuch, VDI Verlag, 1996
14. ISO 11202,2010: Acoustics – Noise emitted by machinery and equipment – Determination of emission sound pressure levels at a work station and at other specified positions applying approximate environmental corrections
15. ISO 11204,2010: Acoustics – Noise emitted by machinery and equipment – Determination of emission sound pressure levels at a work station and at other specified positions applying accurate environmental corrections
16. ISO 3743-1,2010: Acoustics – Determination of sound power levels and sound energy levels of noise sources using sound pressure – Engineering methods for small, movable sources in reverberant fields – Part 1: Comparison method for a hard-walled test room
17. ISO 3745, 2012+Amd 1,2017: Acoustics – Determination of sound power levels and sound energy levels of noise sources using sound pressure – Precision methods for anechoic rooms and hemi-anechoic rooms
18. ISO 3744,2010: Acoustics – Determination of sound power levels and sound energy levels of noise sources using sound pressure – Engineering methods for an essentially free field over a reflecting plane
19. ISO 3746, 2010: Acoustics – Determination of sound power levels and sound energy levels of noise sources using sound pressure – Survey method using an enveloping measurement surface over a reflecting plane
20. ISO 9614-1,1993: Acoustics – Determination of sound power levels of noise sources using sound intensity – Part 1: Measurement at discrete points
21. ISO 9614-2,1996: Acoustics – Determination of sound power levels of noise sources using sound intensity – Part 2: Measurement by scanning

Sound Generation

For the development of primary, constructive noise reduction measures, knowledge of the sound generation mechanisms is decisive. In the following, the mechanisms of sound generation by flow processes and mechanical vibrations, which are usually responsible for noise generation in machines and systems, are given.

The principles of sound generation can be summarised as follows:

- Sound occurs wherever energy transfer or energy conversion takes place for a limited period of time.
- The converted energy is a measure of the amplitude of the sound produced – "**sound intensity**".
- The time in which the energy is converted is a measure of the frequency content of the sound produced – "**pitch**".

Sound generation can be divided into two types:

(a) **Primary airborne sound**

Primary or **direct sound generation** is primarily caused by flow processes, whereby the air particles are directly excited to vibrate. For this, there must be a direct connection, e.g. openings, between the flow area and the environment. Examples of this are: fan, free jet noise, wind instruments.

G. R. Sinambari, *Design Acoustics*,
https://doi.org/10.1007/978-3-658-40183-2_3

(b) **Secondary airborne or structure-borne noise**

Secondary or **indirect sound generation** occurs through structure-borne sound radiation from structures that are excited to vibrate by alternating forces. The alternating forces can be generated both mechanically and fluidically. Examples are: gearbox, transformer, housing radiation of fans and pumps, string instruments.

In the case of machine noise, both types of sound generation usually occur together, so that the sum of the two components (airborne and structure-borne sound power, P_L and P_K, or their levels L_{WL} and L_{WK}) is always responsible for the total noise generation (total sound power "P_{total}").

Figure 3.1 schematically shows the sound generation mechanisms [1].

With

$$L_{W,\text{total}} = 10 \cdot \lg\left[\sum 10^{L_{WL,i}/10} + \sum 10^{L_{WK,j}/10}\right] \quad \text{dB} \tag{3.1}$$

$$L_{WL} = 10 \cdot \lg \sum 10^{L_{WL,i}/10}; \quad L_{WL,i} = 10 \cdot \lg \frac{P_{L,i}}{P_0} \quad \text{dB} \tag{3.2}$$

$$L_{WK} = 10 \cdot \lg \sum 10^{L_{WK,j}/10}; \quad L_{WK,j} = 10 \cdot \lg \frac{P_{K,j}}{P_0} \quad \text{dB} \tag{3.3}$$

Here are:

P_L, $P_{L,\,i}$	Total and partial airborne sound power in W
P_K, $P_{K,\,j}$	Total and partial structure-borne sound power in W
$L_{W,\,\text{total}}$	Total sound power level in dB
L_{WL} $L_{WL,\,i}$	Total and partial airborne sound power level in dB
L_{WK} $L_{WK,\,i}$	Total and partial structure-borne sound power level in dB
i	Index for airborne sound sources, openings
j	Index for structure-borne sound sources, structural surfaces

Although the receiver – ear and/or microphone – always perceives the total noise development as airborne sound, the distinction between direct and indirect sound generation is of critical importance, above all because of the noise reduction measures to be worked out. For this purpose, among other things, the total and partial sound power levels are to be determined independently of each other and from this the airborne and structure-borne noise generation is to be determined [2, 3]. In Chap. 4, the methodical procedure for the determination of the parameters given in (3.1), (3.2) and (3.3) is explained.

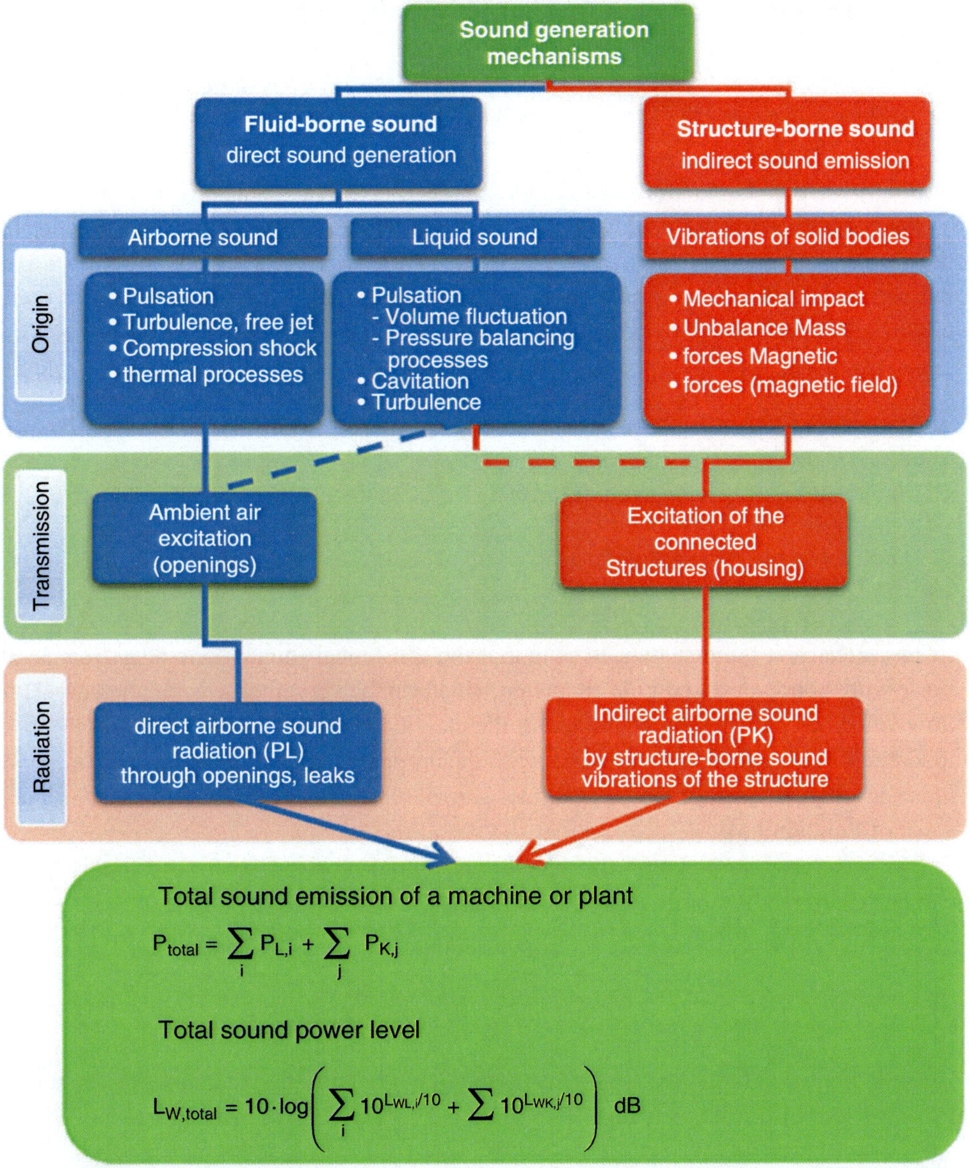

Fig. 3.1 Schematic representation of the sound generation mechanisms

3.1 Sound Generation Through Flow Processes

Air-regenerated noise is caused by temporal fluctuations in a gas or liquid flow, e.g. sudden pressure changes (pressure pulsation), flowing around obstacles, changes in cross-section, deflections and outflow from openings. The surrounding air is excited to vibrations, the resulting pressure fluctuations can then be radiated or perceived as direct airborne sound.

Air-regenerated noise usually has a broadband frequency spectrum. As a rule, flow noises have a broadband frequency spectrum, which is more or less overlaid by discrete frequency components (e.g. pulsation or rotational sounds of flow machines). Based on theoretical considerations, the fuelling mechanisms of such noises can be divided into three main source types:

1. Monopoly sources (I)	temporal change of the volume supply
2. Dipole sources (II)	Influence of alternating forces on the environment of the flow (space limitation!)
3. Quadrupole sources (III)	temporal fluctuations of the shear stresses, especially at the interface between the flowing medium and the stationary environment

Historically, it is interesting to note that the theoretical foundations of air-regenerated noise with monopole and dipole character, which can be very intense even at relatively low flow velocities, are already contained in the classical work of Lord Rayleigh (1877). The theoretical foundations of flow noises with quadrupole character, which only lead to large noise development at high flow velocities, were published in the relatively recent past in the work of Leighthill [4].

The general wave equation for undisturbed sound propagation for the sound pressure can be represented according to (2.2) as follows:

$$\frac{1}{c^2} \cdot \frac{\partial^2 p}{\partial t^2} = \Delta p \quad \mathrm{N/m^4} \tag{3.4}$$

Considering the external perturbations on the flow domain, the general wave equation for flow processes [5] follows from (3.4):

$$\frac{1}{c^2} \cdot \frac{\partial^2 p}{\partial t^2} = \Delta p + \frac{\partial \dot{m}'}{\partial t} - \frac{\partial F_i'}{\partial x_i} + \frac{\partial^2 (\rho \cdot v_i \cdot v_j)}{\partial x_i \partial x_j} \quad \mathrm{N/m^4} \tag{3.5}$$

$\dot{m}'$ on $\mathrm{kg/(s\,m^3)}$	Mass flow per volume: **Source type (I)**
F' on $\mathrm{N/m^3}$	External alternating force per volume **Source type (II)**
$\rho \cdot v_i \cdot v_j$ on $\mathrm{Nm/m^3}$	Kinetic energy of the velocity fluctuations related to the volume **Source type (III)**

Fig. 3.2 Schematic representation of a flow area with direct sound radiation according to (3.5)

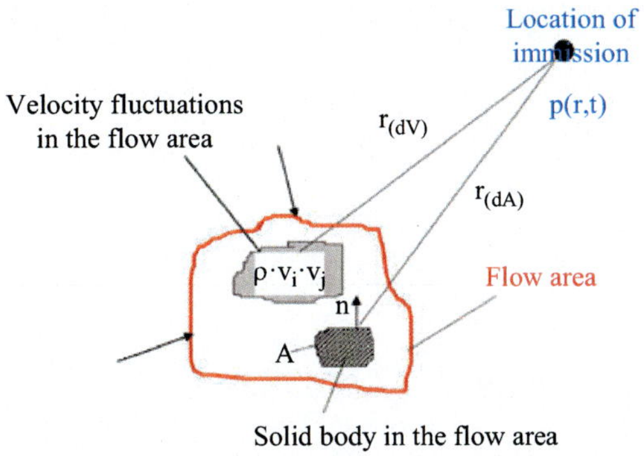

With $\dot{m}' = \text{constant}$, and $F' = \text{constant}$, the general wave equation of sound propagation in a medium at rest $\rho \cdot v_i \cdot v_j = \text{constant}$ follows according to (3.4).

Figure 3.2 schematically shows the flow area, the external disturbance variables according to (3.5), and the sound pressure caused by the flow area at the place of immission under free-field conditions (direct sound radiation) [6].

If the flow area is located in a closed housing, as is the case, for example, with turbomachinery, the sound is radiated by structure-borne sound radiation from the machine housing and/or connected pipelines (indirect sound radiation).

An exact analytical solution of (3.5) is not possible. By representing (3.5) in integral form [5, 6] an approximate formula can be given depending on the disturbance variables and certain conditions.

3.1.1 Sound Generation Due to Velocity Fluctuations

Assuming that only the velocity fluctuations of the flow processes (free turbulence) are responsible for the sound generation, $\rho \cdot v_i \cdot v_j \neq \text{constant}$ it follows from (3.5):

$$\frac{1}{c^2} \cdot \frac{\partial^2 p}{\partial t^2} - \Delta p = \frac{\partial^2 (\rho \cdot v_i \cdot v_j)}{\partial x_i \partial x_j} \quad \text{N/m}^4 \tag{3.6}$$

c	Speed of sound outside the flow area in m/s
Δp	Laplace operator for the sound pressure p

$$\Delta p = \frac{\partial^2 p}{\partial x^2} + \frac{\partial^2 p}{\partial y^2} + \frac{\partial^2 p}{\partial z^2} \quad \text{N/m}^4 \tag{3.7}$$

With the help of (3.6) one obtains in good approximation the sound power generated by direct sound radiation, caused by flow processes [6]:

$$P_{\mathrm{L}} = K_n \cdot \rho \cdot u^3 \cdot S \cdot \mathrm{Ma}^n \quad \mathrm{W} \tag{3.8}$$

$$\mathrm{Ma} = \frac{u}{c} \tag{3.9}$$

Here are:

u	Flow velocity in m/s
c	Speed of sound in the fluid in m/s
ρ	Density of the fluid in kg/m^3
S	Flow cross section in m^2
K_n	Proportionality factor
n	Exponent
Ma	Mach number

Equation (3.8) states that the airborne sound power *PL* generated by flow events is proportional to the flow power ($\rho \cdot u^3 \cdot S$) and the Mach number Ma^n. The proportionality factor K_n is approximately a constant for each type of source and is best determined by measurement. Examples are at constant volume or mass flow and constant velocity:

- Free jet noise,
- Pipeline noise,
- Fans, control valves,
- Flow around obstacles,
- Wind noise on vehicles at constant vehicle speed,
- Propeller and jet aircraft.

For flow machines, such as fans, with constant volume flow $\dot{V}$, it is useful to specify (3.8) as follows:

$$P_{\mathrm{L}} = 2 \cdot K_n \cdot \dot{V} \cdot \Delta p \cdot \mathrm{Ma}^n c \quad \mathrm{W} \tag{3.10}$$

$$P_{\mathrm{S}} = \dot{V} \cdot \Delta p \quad \mathrm{W} \tag{3.11}$$

With

$\dot{V} = u \cdot S$	Average volume flow in m^3/s
$\Delta p = \frac{\rho}{2} \cdot u^2$	Static pressure rise in N/m^2 Note: Not to be confused with the Laplace operator, Equation (3.7)
P_S	Flow power of a turbomachine in W or kW

The sound power level is then obtained:

$$L_{WL} = 10 \cdot \log \frac{P_L}{P_0} \quad \text{dB} \qquad P_0 = 10^{-12} \quad \text{W} \tag{3.12}$$

P_L	Airborne sound power generated by flow processes in W
L_{WL}	Airborne sound power level in dB

Depending on the flow process, the **exponent *n*** can assume a value between approx. 1 and 5:

***n* ≈ 1 "monopoly sources":**
Temporal volume change, e.g. propeller, compressor, pulsating discharge from the exhaust ($P_L \sim u^4$)
***n* ≈ 3 "dipole sources":**
Effect of alternating forces, e.g. pipe flows,
Cross-section change, deflections ($P_L \sim u^6$)
***n* ≈ 5 "quadrupole sources":**
free turbulence, e.g. jet engines, reducing and blow-off valves ($P_L \sim u^8$)
Figure 3.3 summarises the main characteristics of different types of sources.

If one substitutes the Mach number, (3.9) in (3.8), the effect of the various influencing parameters can be better seen:

$$P_L = K_n \cdot \rho \cdot u^3 \cdot S \cdot \frac{u^n}{c^n} \quad \text{W} \tag{3.13}$$

If one takes into account that, with regard to sound generation by flow processes, the density and sound velocity can be regarded as approximately constant for certain media and operating parameters (pressure and temperature), it follows from (3.13) that sound generation is decisively influenced by the flow velocity u. The flow velocity u is the most important parameter for sound generation.

Depending on the source type ($n = 1$ to 5), the airborne sound power can increase with the 4th to 8th power of the flow velocity. The flow-induced sound power can be significantly reduced if it is possible to reduce the highest occurring velocity. A halving of the flow velocity leads to a reduction of the sound power level by 12–24 dB, depending on the source type.

Other important parameters influencing the generation of sound by flow processes are:

K_n	Depending on the source type, the proportionality factor is mainly dependent on the flow guidance and the surface condition of the flow guides. This variable can essentially be determined experimentally.
S	An increase in the flow cross-section at the points with the highest velocity (constrictions) leads to a reduction in sound power. This is done primarily by reducing the flow velocity.

Source type	Monopoly sources	Dipole Sources	Quadrupole Sources
Sound generation	Temporal volume change, displacement effect	interaction	turbulent free flow
Fictitious model of the air particles	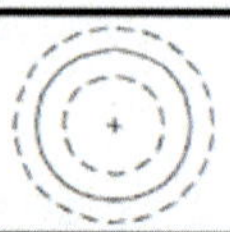	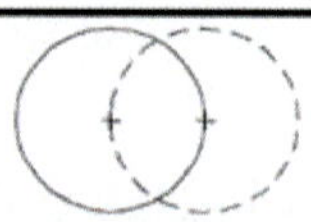	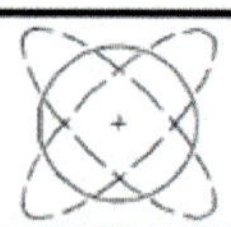
Features	low-frequency noise	→	High frequency noise
Flow velocity u	u < 5 - 10 m/s	u < 100 m/s	u > 100 m/s
Exponent n	n ≈ 1 - 2	n ≈ 3 - 4	n ≈ 5 - 6
Examples	Propellers, fans, compressors, exhaust pipes, tread grooves Car tyres	Pipe flows e.g.: Cross-section changes, measuring nozzle edges, deflections	Jet engines, reducing and blow-off valves, free jets

Fig. 3.3 Overview of source types for air-regenerated noise

3.1.2 Sound Generation Due to Fluctuations in Flow Rate and Alternating Forces

In turbomachinery, especially in pumps such as gear pumps, vane pumps or axial piston pumps, the noise generation is mainly determined by two external influencing factors (disturbance variables):

(a)	Fluctuations in the flow rate $\left(\frac{\partial \dot{m}'}{\partial t} > 0\right)$
(b)	Alternating force caused by impulse-like pressure equalisation processes $\left(\frac{\partial F_i'}{\partial x_i} > 0\right)$

If one takes into account that the noise development due to speed fluctuations (turbulence) can be neglected in comparison to the above-mentioned influencing factors, then it follows from (3.5):

$$\frac{1}{c^2} \cdot \frac{\partial^2 p}{\partial t^2} - \Delta p \approx \frac{\partial \dot{m}'}{\partial t} - \frac{\partial F_i'}{\partial x_i} \quad \mathrm{N/m^4} \tag{3.14}$$

The effect of the individual disturbance variables on the overall noise development is determined separately below.

3.1.2.1 Noise Generation Due to Fluctuations in the Flow Rate

Assuming that density ρ, cross-section *S* and speed *n* are constant, the time variations of the mass flow are periodic and proportional to the volume flow variations $\dot{V}(t)$. The volume flow fluctuations can be determined from the product of the sound velocity *v*(*t*) in the pressure pipe with the pipe cross-section *S:*

$$\dot{V}(t) = S \cdot v(t) \quad \mathrm{m^3/s} \tag{3.15}$$

The ratio of the volume flow fluctuations to the average volume flow is referred to as the relative change in volume flow over time with μ.

$$\mu = \frac{\dot{V}(t)}{\dot{V}} \tag{3.16}$$

In the case of plane shafts (*p* and *v* are in phase), which are usually present in the discharge lines of a pump, the alternating pressure in the discharge line can be calculated using (3.15), (3.16) and (2.18) as follows [8]:

$$p_{\Delta V} = \rho \cdot c \cdot v(t) = \rho \cdot c \cdot \frac{\dot{V}(t)}{S} = \rho \cdot c \cdot \frac{\dot{V}}{S} \cdot \mu \quad \mathrm{N/m^2} \tag{3.17}$$

The relative temporal change μ of the volume flow is essentially dependent on the pump type.

For vane and axial piston pumps up to $\mu \approx (25) \cdot 10^{-3}$, for gear pumps $\mu \approx 2 \cdot 10^{-2}$.

The sound power generated in the pressure line due to the time variations of the flow rate can be determined for a plane shaft as follows s. (2.24):

$$P_{\Delta V} = I \cdot S = p_{\Delta V} \cdot v \cdot S = \frac{p_{\Delta V}^2}{\rho \cdot c} \cdot S \quad \mathrm{W} \tag{3.18}$$

By (3.17), it follows from (3.18):

$$P_{\Delta V} = \frac{p_{\Delta V}^2}{\rho \cdot c} \cdot S = \rho \cdot c \cdot \frac{\dot{V}^2}{S} \cdot \mu^2 \quad \mathrm{W} \tag{3.19}$$

The internal sound power level, caused by the temporal fluctuations of the flow rate, then results:

$$L_{\mathrm{Wi},\Delta V}=10\cdot \lg\frac{P_{\Delta V}}{P_0}=10\cdot \log\left(\frac{\rho\cdot c\cdot \frac{\dot{V}^2}{S}\cdot \mu^2}{10^{-12}\,\mathrm{W}}\right)\quad \mathrm{dB} \tag{3.20}$$

3.1.2.2 Sound Generation Through Impulse-like Pressure Equalisation Processes

Due to sudden pressure equalisation processes, for example, when the delivery volume V_0 of the suction side is abruptly connected to the pressure side, the fluid experiences the compression due to the external forces ΔV.

Analogous to Hooke's Law [9]:

$$\varepsilon=\frac{\Delta l}{l}=\frac{\sigma}{E} \tag{3.21}$$

ε	Stretch
σ	Voltage in N/m^2
E	Modulus of elasticity in N/m^2

an equation is obtained for the volume compression ΔV:

$$\frac{\Delta V}{V_0}=\frac{\Delta p}{E_F} \tag{3.22}$$

V_0	Delivery volume of the suction side before pressure compensation in m^3
E_F	Volume elasticity of the fluid in N/m^2 $E_{\mathrm{Water}} \approx 2\cdot 10^9\,\mathrm{N/m^2}=20000$ bar $E_{\mathrm{Öl}} \approx 1\cdot 10^9\,\mathrm{N/m^2}=10000$ bar $E_{\mathrm{Air}} \approx 1.4\cdot 10^5\,\mathrm{N/m^2}=1.4$ bar

The reciprocal of the elasticity E is called the compression modulus $K = 1/E$.

With

$$c_F=\sqrt{\frac{E_F}{\rho_F}}\,\mathrm{m/s}\Rightarrow E_F=\rho_F\cdot c_F^2\quad \mathrm{N/m^2} \tag{3.23}$$

c_F, ρ_F	The speed of sound and density in the fluid

from (3.22) the volume compression (volume change) follows due to the sudden pressure difference or pressure equalisation Δp:

$$\Delta V = V_0 \cdot \frac{\Delta p}{c_F^2 \cdot \rho_F} \quad \text{m}^3 \tag{3.24}$$

Δp	Static pressure difference in volume V_0 before and after pressure equalisation in N/m²

Due to the pressure compensation or the compression of the volume by ΔV, some fluid flows back from the pressure side into the suction kidney. The resulting velocity in the fluid is then:

$$v(t) = \frac{1}{\Delta t} \cdot \frac{\Delta V}{S} \quad \text{m/s} \tag{3.25}$$

Δt	Duration of compression or pulse duration during pressure equalisation in s

With

n	Speed in revolutions per minute 1/min
z	Number of delivery volumes per revolution (e.g. for a gear pump this corresponds to the number of teeth)

is the period of the single compression T in s:

$$T = \frac{60}{z \cdot n} \quad \text{s} \tag{3.26}$$

Figure 3.4 shows the time and frequency characteristics of intermittent pulses, as they occur, for example, in a first approximation during the pressure equalisation processes in a gear pump or vane pump, using the example of the sound velocity.

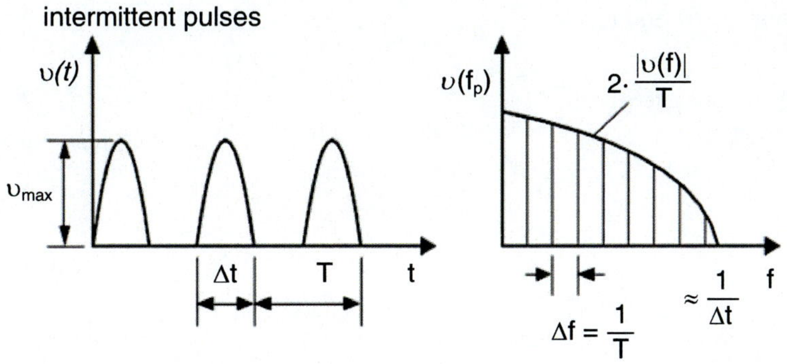

Fig. 3.4 Time and frequency characteristics of intermittent pulses

The frequency spectrum of the single pulse determines the envelope of the line spectrum, see Fig. 3.4. The partial amplitude of the individual lines in the spectrum has the magnitude [7, 25]:

$$|v(f_p)| = \frac{2}{T} \cdot \left| \int_0^T v(t) \cdot e^{-j2\pi \frac{t}{T}} \cdot dt \right| = 2 \cdot \frac{|v(f)|}{T} \quad \text{m/s} \tag{3.27}$$

Assuming that the pressure equalisation processes, as they occur in liquid pumps, can be represented in a first approximation by rectangular pulses, the internal sound power level of the pump is calculated theoretically below based on its operating data.

The Fourier spectrum of a rectangular pulse $v(f)$, Fig. 3.5, can be determined as follows [7]:

$$|v(f)| = \left| \int_{-\frac{\Delta t}{2}}^{+\frac{\Delta t}{2}} v(t) \cdot e^{-j2\pi \cdot f \cdot t} dt \right| = J_S \cdot \left| \frac{\sin(\pi \cdot f \cdot \Delta t)}{\pi \cdot f \cdot \Delta t} \right| \quad \text{m} \tag{3.28}$$

J_S	The area under the curve $v(t)$ in the time domain Δt, see Fig. 3.5
$\lvert v(f) \rvert$	Magnitude of the Fourier spectrum of the vibration velocity

Using (3.25), J_S can be determined as follows:

$$J_S = \int_{-\frac{\Delta t}{2}}^{\frac{\Delta t}{2}} v(t)\, dt = v(t) \cdot \Delta t \approx \frac{\Delta V}{S} \quad \text{m} \tag{3.29}$$

Figure 3.6 shows the normalised frequency spectrum of the single rectangular pulse according to (3.28).

For intermittent square-wave pulses, the partial amplitude of the sound velocity of the individual lines in the spectrum (pulsation frequency and its harmonics) is obtained from (3.27) using (3.28) and (3.29):

Fig. 3.5 Time characteristic of a rectangular pulse

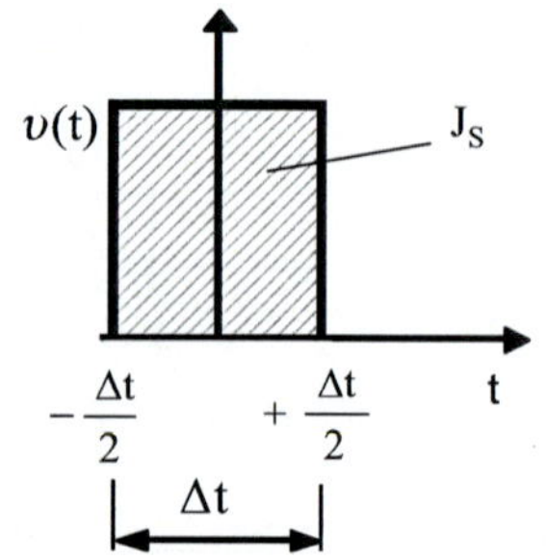

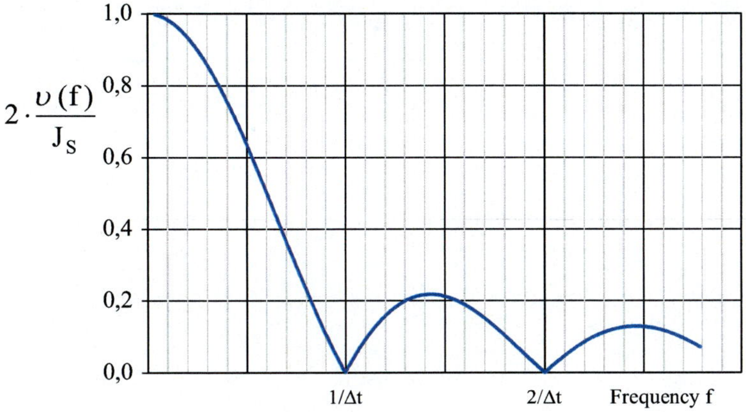

Fig. 3.6 Frequency spectrum of a rectangular pulse

$$|v(f_p)| = 2 \cdot \frac{|v(f)|}{T} = J_S \cdot \frac{2 \cdot K_p}{T} \approx \frac{2 \cdot \Delta V}{T \cdot S} \cdot K_p \quad \text{m/s} \tag{3.30}$$

with

$$K_p = \left| \frac{\sin(\pi \cdot f_p \cdot \Delta t)}{\pi \cdot f_p \cdot \Delta t} \right| \tag{3.31}$$

$$f_p = \frac{p+1}{T} = \frac{n \cdot z}{60}(p+1) \text{ Hz} \quad p = 0, 1, 2, 3, \ldots \tag{3.32}$$

For $p = 0$, (3.32) gives the so-called pulsation frequency (f_0). For $p > 0$ one obtains the harmonics of the pulsation frequency f_p.

Figure 3.7 shows the normalised frequency spectrum of a rectangular pulse (envelope) and of intermittent rectangular pulses at the same $\Delta t = 0.4$ ms.

From this it is easy to see that the amplitudes become smaller with increasing frequency and are zero at frequencies ($k/\Delta t$ $k = 1, 2, 3\ldots$).

For plane waves, the sound pressure at the pulsation frequency f_p due to the pulse-like pressure equalisation processes of a fluid pump can be given from (3.30) as follows:

$$p_{\Delta p}(f_p) \approx \rho_F \cdot c_F \cdot v(f_p) \approx (\rho_F \cdot c_F) \cdot \frac{2 \cdot \Delta V}{T \cdot S} \cdot K_p \quad \text{N/m}^2 \tag{3.33}$$

According to (3.24), ΔV can be determined as follows:

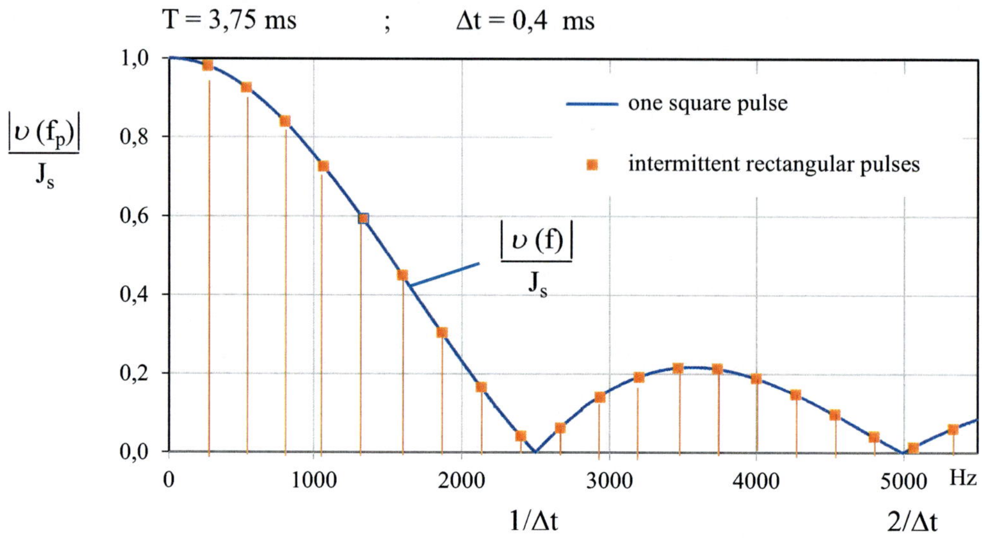

Fig. 3.7 Frequency spectrum of a rectangular pulse (envelope) and of intermittent rectangular pulses

$$\Delta V = V_0 \cdot \frac{\Delta p}{c_F^2 \cdot \rho_F} = \dot{V} \cdot T \cdot \frac{\Delta p}{c_F^2 \cdot \rho_F} \tag{3.34}$$

By (3.34), it follows from (3.33):

$$p_{\Delta p}(f_p) = \frac{2 \cdot \dot{V} \cdot \Delta p}{S \cdot c_F} \cdot K_p \tag{3.35}$$

The total internal sound pressure p_{total} is obtained from the sum of the fundamental frequency of the pulsation f_0 ($p = 0$ at (3.32)) and its harmonics:

$$p_{\Delta p,\text{total}}^2 = \sum_{p=0}^{N} p_{\Delta p}(f_p)^2 = 4 \cdot \left(\dot{V} \cdot \frac{\Delta p}{c_F \cdot S}\right)^2 \cdot \sum_{p=0}^{N} K_p^2 \quad \text{N}^2/\text{m}^4 \tag{3.36}$$

The internal sound power $P_{i,\,\Delta p}$ generated in the pump due to the pressure equalisation processes can be determined for a plane shaft as follows:

$$P_{i,\Delta p} = \frac{p_{\Delta p,\text{total}}^2}{\rho_F \cdot c_F} \cdot S = \left[4 \cdot \left(\dot{V} \cdot \frac{\Delta p}{c_F \cdot S}\right)^2 \cdot \sum_{p=0}^{N} K_p^2\right] \cdot \frac{S}{\rho_F \cdot c_F} \quad \text{dB} \tag{3.37}$$

The internal sound power level is then obtained:

$$L_{Wi,\Delta p} = 10 \cdot \lg \frac{P_{i,\Delta p}}{P_0} = 10 \cdot \lg \left\{ \frac{\left[4 \cdot \left(\dot{V} \cdot \frac{\Delta p}{c_F \cdot S} \right)^2 \cdot \sum_{p=0}^{N} K_p^2 \right] \cdot \frac{S}{\rho_F \cdot c_F}}{10^{-12}\,\mathrm{W}} \right\} \quad \mathrm{dB} \tag{3.38}$$

The total internal sound power level $L_{Wi,\ \mathrm{total}}$ of the pump caused by the time variations of the flow rate (3.20) and the pressure equalisation processes (3.38) can be determined as follows:

$$L_{Wi,\mathrm{total}} = 10 \cdot \lg \left[10^{\left(L_{Wi,\Delta V} + L_{Wi,\Delta p}\right)/10} \right] \quad \mathrm{dB} \tag{3.39}$$

The resulting internal sound power or sound power level propagates in the connected pressure line as fluid sound and can excite more distant components to vibrate. In addition, they can also indirectly excite the connected structures, e.g. pipelines or pump housings, to structure-borne sound vibrations.

At higher static pressures of the pump, approx. $\Delta p > 20$ bar, the total level of the internal sound power level is generally determined decisively by the pressure compensation processes. The fluctuations of the flow rate are essentially noticeable at lower static pressures of the pump at approx. $\Delta p < 20$ bar.

With the aid of (3.39), the total internal sound pressure level $L_{pi,\ \mathrm{total}}$, measured variable in the connected pressure line, can also be determined [7]:

$$L_{pi,\mathrm{total}} = L_{Wi,\mathrm{total}} - L_S + 10 \cdot \lg \frac{\rho_F \cdot c_F}{\rho_0 \cdot c_0} + \Delta L_z \quad \mathrm{dB} \tag{3.40}$$

$\Delta L_z \approx 0\ldots6$ dB	Level surcharge for reflections in the pressure line
$L_S = 10 \cdot \lg \frac{S}{S_0}$ dB	Square measure
S	Cross-section of the pressure line in m^2 ($S_0 = 1$ m^2)
$\rho_F \cdot c_F$	Impedance of the pumped medium in N s/m^3
$\rho_0 \cdot c_0 = 400$ N s/m^3	Reference impedance

The internal sound pressure level at the pulsation frequency and the harmonic of the pulsation frequency can be determined using (3.31), (3.32) and (3.37) as follows:

$$L_{pi}(f_p) = L_{Wi}(f_p) - L_S + 10 \cdot \lg \frac{\rho_F \cdot c_F}{\rho_0 \cdot c_0} + \Delta L_z \quad \mathrm{dB} \tag{3.41}$$

With

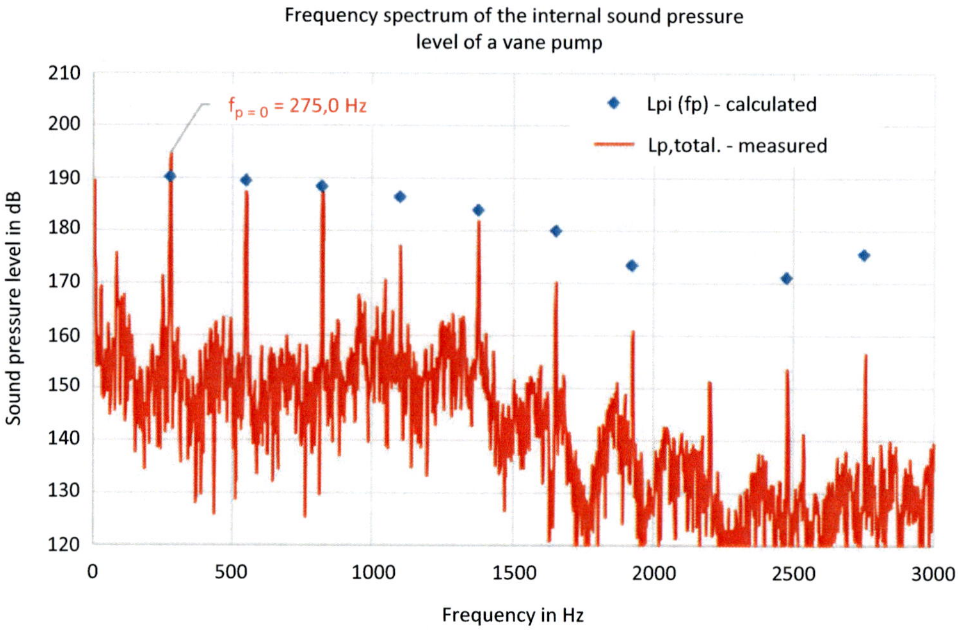

Fig. 3.8 Measured total sound pressure level at the pump outlet of a vane pump and calculated sound pressure level at the pulsation frequency of the pump f_0 and its harmonics f_p

$$L_{Wi}(f_p) = 10 \cdot \lg \frac{P_i(f_p)}{P_0} = 10 \cdot \lg \left\{ \frac{\left[4 \cdot \left(\dot{V} \cdot \frac{\Delta p}{c_F \cdot S} \right)^2 \cdot \left| \frac{\sin(\pi \cdot f_p \cdot \Delta t)}{\pi \cdot f_p \cdot \Delta t} \right|^2 \right] \cdot \frac{S}{\rho_F \cdot c_F}}{10^{-12}\,\mathrm{W}} \right\} \ \mathrm{dB} \quad (3.42)$$

Figure 3.8 shows the measured total internal sound pressure level at the pump outlet of a vane pump and the calculated sound pressure level at the pulsation frequency and its harmonics for the following operating data.

$$n = 1650\ 1/\min; \quad \rho_{Öl} = 900\ \mathrm{kg/m^3}; \quad E_{Öl} = 10^9\ \mathrm{N/m^2}$$
$$d_i = 12\ \mathrm{mm}; \quad z = 10\ \text{(number of leaves)}; \quad \Delta t = 0.45\ \mathrm{ms}$$
$$\Delta p \approx 60\ \mathrm{bar}; \quad \dot{V} \approx 22\ 1/\min; \quad \Delta L_z = 6\ \mathrm{dB}$$

$$f_0 = 1650 \cdot 10/60 = 275.0\ \mathrm{Hz}\ \text{(pulsation frequency!)}; \quad T = 1/275 = 0.0036\ \mathrm{s}$$

It is pointed out that the relatively good agreement for the pulsation frequency and its harmonics, which are also level-determining, is mainly due to the assumptions for the time course of the pulses and their action time Δt, which were partly determined by measurements. Other effects, e.g. flow noise, friction, volume flow fluctuations etc. are obviously responsible for the low levels.

3.1.3 Aeropulsive Noise

They are based on the generation of alternating pressure in the air by displacement and are consequently well approximated by 0th order radiators (monopole).

On the one hand, there are pulsating flow processes in which a limited volume of air is rhythmically expelled or sucked in again. The resulting pressure equalisation processes with the immediate environment alternately build up overpressure and underpressure. These pressure fluctuations propagate through the air at the speed of sound, and the associated noises have a rattling effect and can be very loud.

Examples of this are ejection noises at the exhaust of internal combustion engines, at the discharge valve of compressors and compressed air devices. They can also be displacement processes in which air is expelled intermittently, such as from the tread grooves of rolling car tires or from the common gap between two periodically intermeshing machine parts (e.g. gear pairings).

The main part of the radiated sound energy is in the range of the pulsation frequency. The fundamental frequency of pulsation f_0 can be determined as follows:

$$f_0 = \frac{n \cdot z}{60} \quad \text{Hz} \tag{3.43}$$

n	Speed in revolutions per minute 1/min
z	Number of pulsations per revolution

In addition to the fundamental frequency f_0, the harmonics f_k according to (3.44) also appear in the spectrum, depending on the time functions of the individual compressions:

$$f_k = f_0 \cdot (1 + k) \ \text{Hz} \quad k = 1, 2, 3, \ldots \tag{3.44}$$

The distribution of the pulsation energy to the above mentioned frequencies is decisively influenced by the vibration behaviour of the medium in the connected pipeline.

On the other hand, it is a matter of air displacement by moving, especially by rotating body elements. When rotating, the latter push a positive pressure field in front of them and pull a negative pressure field behind them (hydrodynamic near field without local sound radiation). For an observer at rest in relation to the moving element, however, this means a pressure change within one revolution. Consequently, in the case of rotating wheels composed of spokes, blades, and vanes, a periodic alternating pressure with the governing fundamental frequency f_0 arises in the surrounding air. It can be calculated according to (3.43) with:

z	Number of leaf elements on the perimeter

The associated sound is called a rotating sound. Such a rotational sound can be perceived on rotating propellers, fans, turbine rotors, etc.

Tracking sounds are related to rotational sounds. However, they have a dipole character and arise from the interaction between a wake zone and a moving near field. Behind an obstacle in a flow, a wake field is established, also called slipstream or deadwater area, with an average velocity lower than that of the undisturbed flow. If a body of approximately the same size moves transversely in the immediate vicinity, its hydrodynamic near field is disturbed as it passes through the slipstream zone. This disturbance propagates in the surrounding, still air at the speed of sound.

If a wheel with regularly arranged blade elements rotates through the slipstream of a flowing obstacle or if a blade element rotates through the slipstream of a group of regularly arranged and flowing obstacles, the so-called siren sound is produced. It has a line spectrum with discrete frequencies which are multiples of the fundamental frequency, cf. (3.44).

Such a siren sound can occur, for example, on turbomachinery from the interaction of the fixed guide vanes and the rotating impeller blades.

3.1.4 Noise Generation Due to Vortex Formation

When flowing around obstacles, noises with dipole character occur due to detachment phenomena. Depending on the Reynolds number, a laminar or turbulent flow occurs behind the obstacle "vortex zone".

The limit at which the flow changes from laminar to turbulent occurs at a so-called critical Reynolds number "$\mathrm{Re}_{\mathrm{krit}}$", which can assume different values depending on the flow form. As the Reynolds number increases, the flow becomes more turbulent. From about $\mathrm{Re} = 10^6$, one speaks of fully turbulent flow, i.e. in the vortex zone there are almost exclusively random velocity fluctuations.

This excites both structure-borne sound in the obstacle and airborne sound, which is indicated by pressure fluctuations in the vortex zone. The airborne sound then propagates in the surrounding air at the speed of sound and is perceived as noise in the audible range. Examples of this are wind noise on vehicles and buildings, noise from turbomachinery and in pipelines, which occurs when flowing around fixtures, obstacles and changes in cross-section [5, 12–14].

For small Reynolds numbers, $\mathrm{Re} = \mathrm{Re}_{\mathrm{krit}} \approx 400$, a laminar flow with periodic vortex shedding at the obstacle is established behind the obstacle, the so-called Kármán vortex street [8, 10].

The associated sounds, vortex shedding, take on an increasingly narrow-band character as the Reynolds number becomes smaller. The fundamental frequency is equal to the frequency of vortex shedding and is:

$$f_{\text{vortex frequency}} = \text{St} \cdot \frac{U_0}{d} \quad \text{Hz} \tag{3.45}$$

U_0	Incoming flow velocity in m/s
d	Width of the obstacle in m
St	Strouhal number

The Strouhal number generally represents a constant, in the case of obstacle flow around we have St ≈ 0.2, e.g. singing of wires in the wind [27].

As the incident flow velocity increases, the frequency of the vortex tones increases linearly, and the intensity of the sound power radiated by vortex formation increases with the 5th to 6th power of U_0. This means that with a doubling of the incident flow velocity, the sound power level increases by approx. 18 dB.

The notion of vortex frequency can also be applied to closed rotating machine parts. In (3.45), U_0 is then to be replaced by the circumferential speed ($r \cdot \omega$) at a distance *r from* the axis of rotation and d by the width b of the rotating part. One thus obtains for the rotor between $0 < r < R$ (R outer radius) a frequency band of the stroke excitation $0 < f_{\text{vortex frequency}} < 0.2 \cdot \frac{R \cdot \omega}{b}$, with preference for the higher frequencies. The associated noise is called rotational noise in contrast to the narrowband rotational sound.

Finally, vortices and thus noises ranging from slap-tones to broadband noise can also be generated in pipelines and general flow channels, such as in shell-and-tube heat exchangers, when flow passes around internals in the pipelines [11] or when the flow separates at sudden changes in cross-section and abrupt deflections [12].

Finally, vortex shedding occurs at sharp edges protruding into the flow, at cutting edges and at non-flush measuring spigots. These are then referred to as cutting edge or edge tones with the same properties as those of the slice tones.

3.1.5 Noise Generation by Free Jets

When a jet of air emerges from an opening into the adjacent, stationary air space, a boundary zone of violent, turbulent mixing between two areas of air with very different velocities naturally arises. In the vast majority of cases, the so-called turbulent free jet is formed, as shown in Fig. 3.9 [13, 14]

For $\text{Re} < \text{Re}_{\text{krit}}$, such as when exiting a narrow gap, a laminar free jet is formed (see Fig. 3.9b). It shows the flow behavior of a fully formed vortex street with periodic vortex shedding, similar to Kármán's vortex street in the laminar flow around an obstacle. The associated sounds, the *gap sounds*, are narrowband. The fundamental frequency is equal to the frequency of the vortex shedding and is:

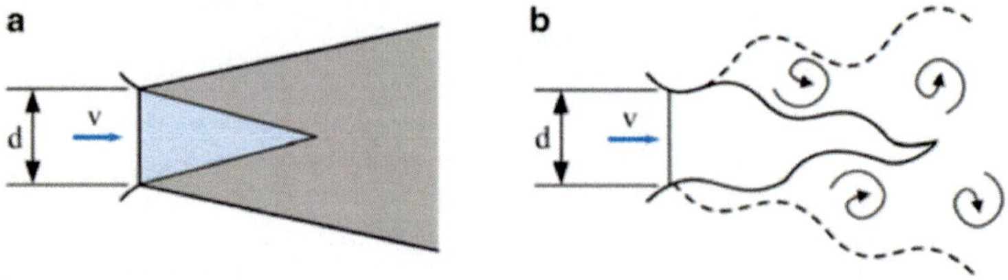

Fig. 3.9 (**a**) Turbulent free jet, (**b**) Laminar free jet

$$f_{gap} = \mathrm{St} \cdot \frac{U_0}{d_{gap}} \ \mathrm{Hz} \tag{3.46}$$

Herein, U_0 is the undisturbed outflow velocity, d_{gap} is the gap width, and St is the Strouhal number with the value of St $\approx$ 0.04. As the Reynolds number ($\mathrm{Re} > \mathrm{Re}_{krit}$) increases, higher and higher velocity fluctuations are superimposed on the vortices, which have smaller effective diameters near the nozzle and larger effective diameters downstream. They lead to a rapid decay especially of the smaller vortices, the larger vortices exist somewhat longer. In the mixing zone in particular, this leads to random velocity fluctuations and consequently to pressure fluctuations of the same magnitude. The latter propagate in the surrounding air at the speed of sound and are perceived in the auditory range as broadband noise. Its high-frequency components arise from the decay of the smaller vortices, i.e. in the area of the mixing zone close to the nozzle, while the low-frequency components arise from the decay of the larger vortices, i.e. in the area further away from the nozzle.

Due to the deformations of the turbulence balls in the mixing zone, the radiation characteristic of free jet noise can best be approximated by the 2nd-order spherical radiator (quadrupole). The jet noise, which has a pronounced directional characteristic (Fig. 3.10) [7], has its radiation maximum at a Strouhal number, (3.47) St $\approx$ 0.2 up to 0.3. From it, the frequency of the maximum can be calculated directly, which is generally high.

The Strouhal number of free jet noise is a dimensionless index and is defined as follows:

$$\mathrm{St} = \frac{f \cdot d}{u} \tag{3.47}$$

d	Nozzle diameter in m
u	Outlet velocity in m/s
f	Frequency in Hz

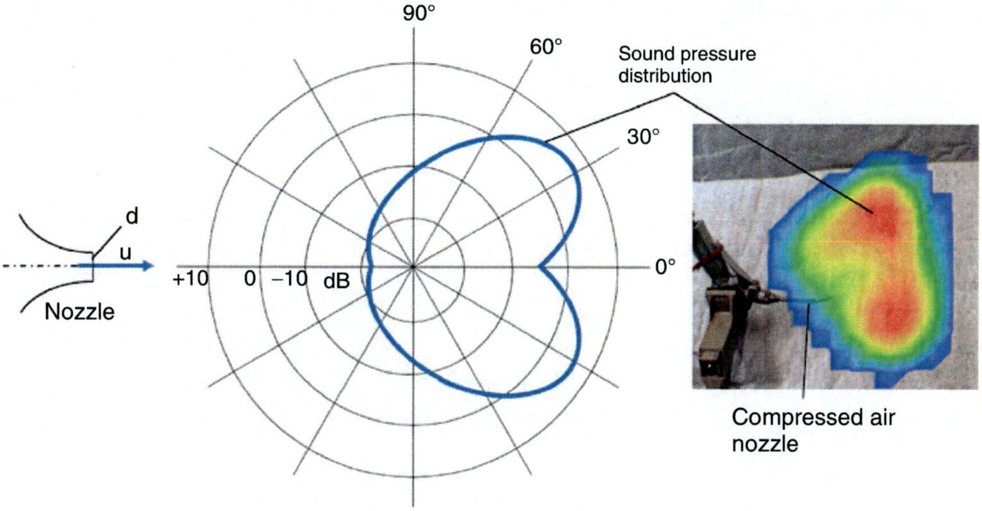

Fig. 3.10 Directional characteristics of turbulent free jet noise

Figure 3.11 shows the experimentally determined relative octave spectrum of the free-jet noise [7, 8].

The relative octave spectrum is defined as follows:

$$\Delta L_{W,\text{oct}} = 10 \cdot \lg \frac{P_{\text{octave}}}{P_{\text{total}}} = L_{W,\text{oct}} - L_{W,\text{total}} \quad \text{dB} \tag{3.48}$$

The relative octave spectrum of turbulent free jet noise can also be calculated according to the spectrum experimentally determined in Fig. 3.11 by the following empirical formula:

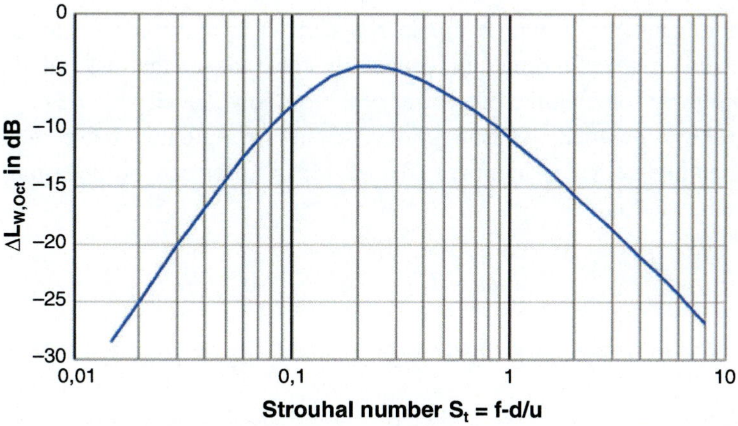

Fig. 3.11 Relative octave spectrum of turbulent free jet noise

$0.01 \le \mathrm{St} \le 0.25$

$$\Delta L_{W,\mathrm{Oct}} = -9.6 \cdot [\lg(\mathrm{St}_{\mathrm{Oct}})]^3 - 45.8 \cdot [\lg(\mathrm{St}_{\mathrm{Oct}})]^2 - 45.8 \cdot [\lg(\mathrm{St}_{\mathrm{Oct}})] - 17.4 \quad \mathrm{dB} \tag{3.49}$$

$0.25 \ge \mathrm{St} \ge 10$

$$\Delta L_{W,\mathrm{Oct}} = 3.1 \cdot [\lg(\mathrm{St}_{\mathrm{Oct}})]^3 - 6.0 \cdot [\lg(\mathrm{St}_{\mathrm{Oct}})]^2 - 14.8 \cdot [\lg(\mathrm{St}_{\mathrm{Oct}})] - 10.6 \quad \mathrm{dB} \tag{3.50}$$

Note
Considering that free-jet noise is very broadband, (3.49) and (3.50) can also be used to estimate the relative one-third octave spectrum of free-jet noise to a good approximation as follows:

$$\left.\begin{aligned} &\mathbf{St \le 0.25} \\ &\Delta L_{W,\text{third-oct}} = -9.6 \cdot [\lg(\mathrm{St}_{\text{third-oct}})]^3 - 45.8 \cdot [\lg(\mathrm{St}_{\text{third-oct}})]^2 \\ &\qquad - 45.6 \cdot [\lg(\mathrm{St}_{\text{third-oct}})] - 17.4 - 10 \cdot \lg(3) \\ &\mathbf{St \ge 0.25} \\ &\Delta L_{W,\text{third-oct}} = 3.1 \cdot [\lg(\mathrm{St}_{\text{third-oct}})]^3 - 6.0 \cdot [\lg(\mathrm{St}_{\text{third-oct}})]^2 \\ &\qquad - 14.8 \cdot [\lg(\mathrm{St}_{\text{third-oct}})] - 10.6 - 10 \cdot \lg(3) \end{aligned}\right\} \mathrm{dB} \tag{3.51}$$

The free jet noise is best represented by quadrupole sources at higher Mach numbers ($0.6 < \mathrm{Ma} < 1.6$). According to (3.8), the sound power is proportional to the 8th power of the exit velocity ($n = 5$, see Fig. 3.3). This also means that, for example, doubling the exit velocity increases the sound power level by 24 dB.

At lower Mach numbers, vortices can occur in the area of the nozzle, which interact with the nozzle to cause noise with a dipole character. The noise centre moves closer to the nozzle, there is more of a dipole influence. This also means that the sound power is then only proportional to about the 6th power of the outflow velocity ($n = 3$).

Using (3.8), the total acoustic power of the free-radiated noise, with quadrupole character, at higher Mach numbers ($0.6 < \mathrm{Ma} < 1.0$) can be determined to a good approximation as follows [6]:

$$P_{\mathrm{total}} = K_{\mathrm{Q}} \cdot \rho \cdot S \cdot u^3 \cdot \mathrm{Ma}^n \quad \mathrm{W} \tag{3.52}$$

$$K_{\mathrm{Q}} \approx 5 \cdot 10^{-5}; \quad \mathrm{Ma} = \frac{u}{c} \qquad \text{Mach number}$$

$$n = 5 \quad 0.6 > \mathrm{Ma} > 1.0 \qquad \text{Quadrupole}$$

With

$$\rho = \frac{p_{\mathrm{sta}}}{R \cdot T} \quad \mathrm{kg/m^3} \tag{3.53}$$

Density of the fluid outside the free jet[1]

$$c = \sqrt{\kappa \cdot R \cdot T} \quad \mathrm{m/s} \tag{3.54}$$

Speed of sound outside the free jet[1] follows from (3.52):

$$P_{\mathrm{total}} = K_{\mathrm{Q}} \cdot \frac{p_{\mathrm{sta}}}{R \cdot T} \cdot S \cdot \frac{u^8}{(\kappa \cdot R \cdot T)^{5/2}} = K_{\mathrm{Q}} \cdot \frac{\mathrm{p_{sta}}}{(R \cdot T)^{3.5} \cdot k^{2.5}} \cdot S \cdot u^8 \quad \mathrm{W} \tag{3.55}$$

From this, the total sound power level of turbulent free jet noise:

$$L_{W,\mathrm{total}} = 10 \cdot \lg \frac{P_{\mathrm{total}}}{P_0} \quad \mathrm{dB} \qquad P_0 = 10^{-12} \quad \mathrm{W}$$

$$L_{W,\mathrm{total}} \approx -48 + 80 \cdot \lg \frac{u}{u_0} + 10 \cdot \lg \frac{S}{S_0} + 10 \cdot \lg \frac{p_{\mathrm{sta}}}{p_{\mathrm{sta},0}} - 35 \cdot \lg \frac{R \cdot T}{R_0 \cdot T_0}$$

$$-25 \cdot \lg \frac{k}{k_0} \quad \mathrm{dB} \tag{3.56}$$

With

$K_{\mathrm{Q}} \approx 5 \cdot 10^{-5}$	Proportionality factor of the quadrupole sources
u	Flow velocity in m⁄s
$u_0 = 1$ m/s	Reference speed
$S = \pi \cdot d_{\mathrm{i}}^2/4$	Outlet cross section in m^2 (d= Outlet diameter in m)
$S_0 = 1$ m^2	Reference area
p_{sta}	Static pressure of the fluid
$p_{\mathrm{sta},\,0} = 10^5$ N/m^2	Static reference pressure
R	Specific gas constant in J/kg K
$R_0 = 287$ J/kg K	Reference gas constant
T	Absolute temperature of the fluid in K
$T_0 = 273$ K	Reference temperature
$\kappa = c_p/c_v$	Adiabatic exponent
$\kappa_0 = 1.4$	Reference adiabatic exponent

[1] Here, it was assumed that the mean temperatures (T) inside and outside the free jet did not differ much.

Note

The reference values u_0, $p_{sta,0}$, R_0, T_0, κ_0, can in principle be freely chosen. Other reference values only change the constant level (−48 dB) in (3.56)!

The specified proportionality factor K_Q represents a rough average value and is dependent on many factors, e.g. speed, opening shape, roughness of the nozzle, Mach number etc.. Therefore, it is recommended that this quantity be determined experimentally for specific boundary conditions and tasks. Equation (3.56) is only conditionally suitable for low Mach numbers Ma < 0.6, since in addition to quadrupole sources, dipole and monopole sources can also occur, especially in the area of the nozzle.

In the case of supercritical free jet expansion and an undesigned Laval nozzle or in the case of simple nozzle exits, a deflagration-like expansion with almost regular formation of compression and dilution waves occurs. The latter can lead to intermittent banging noises (sonic boom), examples of free jet noises include the nozzle noise of jet engines and the noise in the expansion zone of reducing and blow-off valves, [13]. Both noises can be very loud and annoying.

Example 3.1

The A-total and A-octave sound power levels of the outflow noise from a well-rounded nozzle into the open air are sought.

$$
\begin{aligned}
&R = 287\ \text{J/kg K};\quad t = 40\ {}^\circ\text{C} \Rightarrow \text{T} = 313\ \text{K};\quad \kappa = 1.4;\\
&p_{sta} = 1.05\ \text{bar} \Rightarrow p_{sta} = 1.05 \cdot 10^5\ \text{N/m}^2\\
&u = 250\ \text{m/s};\quad d_i = 0,1\ \text{m} \Rightarrow S = 0.0078\ \text{m}^2\\
&c = \sqrt{\kappa \cdot R \cdot T} = 343.1\ \text{m/s};\quad \text{Ma} = 0.73
\end{aligned}
$$

According to (3.56) the total sound power level is obtained:

$$L_{W,\text{total}} = 120.9\ \text{dB}$$

With (3.48), (3.49) and (3.50) and attenuation level of the A-weighting according to Table 2.2, the A-octave sound power level, see Table 3.1, can be calculated:

The A-total sound power level is then obtained:

$$L_{WA} = 10\lg \sum 10^{0.1\,L_{WA,\text{Oct}}} = 119.1\ \text{dB(A)}$$

Table 3.1 Results of Example 3.1

$f_{m,oct}$	63	125	250	500	1000	2000	4000	8000	Hz
$L_{W,total}$					120.9				dB
St_{oct}	0.03	0.05	0.10	0.20	0.40	0.80	1.60	3.20	1/m
$\Delta L_{W,oct}$	−22.3	−14.5	−8.0	−4.7	−5.8	−9.2	−13.8	−19.2	dB
$\Delta L_{,oct}$	−26.2	−16.1	−8.6	−3.2	0.0	1.2	1.0	−1.1	dB
$L_{WA,oct}$	72.4	90.3	104.2	113.0	115.1	112.9	108.1	100.6	dB (A)

3.1.6 Noise Generation-due to Intake Noise

Intake noise, which occurs in many areas (e.g. air-conditioning systems, medical technology, vacuum cleaners, etc.), is caused by speed fluctuations. The sound power can be determined analogously to (3.10) as follows [7]:

$$P_A = K_n \cdot \dot{V} \cdot \Delta p_V \cdot \mathrm{Ma}^n \quad \mathrm{W} \tag{3.57}$$

With

P_A	Sound power of the intake noise in W
$\dot{V} = u_m \cdot S$	Volume flow in m³/s
$u_m = \frac{\dot{V}}{S}$	Mean flow velocity in m/s
$S = \frac{\pi \cdot d^2}{4}$	Nozzle or pipe cross-section in m²
$\Delta p_V = \zeta \cdot \frac{\rho}{2} \cdot u_m^2$	Pressure loss of the nozzle in N/m²
ζ	Pressure loss coefficient of the nozzle
$\mathrm{Ma} = u_m/c$	Mach number
n	Exponent of Mach number
K_n	Proportionality factor at exponent n

With this information, it follows from (3.57):

$$P_A = K_n \cdot u_m \cdot S \cdot \zeta \cdot \frac{\rho}{2} u_m^2 \cdot \mathrm{Ma}^n = K_A \cdot \rho \cdot S \cdot u_m^3 \cdot \mathrm{Ma}^n \quad \mathrm{W} \tag{3.58}$$

$$K_A = \frac{K_n \cdot \zeta}{2} \quad \text{Proportionality factor of the intake noise} \tag{3.59}$$

The total sound power level is then obtained:

$$L_{W,\text{total}} = 10 \cdot \lg \frac{P_A}{P_0} = 10 \cdot \lg \frac{K_A \cdot \rho \cdot S \cdot u_m^3 \cdot \text{Ma}^n}{10^{-12}\,\text{W}} \quad \text{dB} \tag{3.60}$$

Through experimental studies, the proportionality factor K_A and the exponent of Mach number n were empirically determined [15]:

$$K_A \approx 1.35 \cdot 10^{-3} \quad \text{and} \quad n \approx 2.83$$

Using the experimentally determined quantities n and K_A, an estimation formula is obtained from (3.60) with (3.53) and (3.54) for determining the total intake noise levels for compressible media [16]:

$$\begin{aligned} L_W = \; & 21.0 + 58.3 \cdot \lg \frac{u_m}{u_0} + 10 \cdot \lg \frac{S}{S_0} + 10 \cdot \lg \frac{p}{p_0} \\ & - 24.2 \cdot \lg \frac{R \cdot T}{R_0 \cdot T_0} - 14.2 \cdot \lg \frac{\kappa}{\kappa_0} \quad \text{dB} \end{aligned} \tag{3.61}$$

At this:

$u_0 = 1$ m/s	Reference speed
$S_0 = 1$ m^2	Reference area
$p_0 = 10^5$ Pa	Reference pressure
$R_0 = 287$ N m/kg K	Reference gas constant
$T_0 = 273$ K	Reference temperature
$\kappa_0 = 1.4$	Reference adiabatic exponent

The experimental investigations have shown that the proportionality factor K_A according to (3.59) is a constant to a good approximation, although the pressure loss coefficient of the nozzle ζ and the proportionality factor K_n, see (3.58), are very different depending on the nozzle shape and velocity. Moreover, the results show that the investigated nozzle shapes have only a minor influence on the sound generation [15].

Compared to the pure pipe flow, the intake noises are approx. 14 dB higher than the flow noises in the pipe. The reason for the higher levels is most likely that the flow velocity in the inlet area of the nozzle is significantly higher than the average flow velocity in the pipeline due to constriction during inflow.

Figure 3.12 shows an example of the velocity distributions (CFD analyses) at one of the nozzles investigated [16].

From this it can be seen that the flow velocity at the nozzle inlet is significantly higher than the mean flow velocity in the pipe.

As can be easily seen from (3.61), the sound generation is decisively determined by the flow velocity. Therefore, the nozzle inlet is certainly a possible source of sound generation.

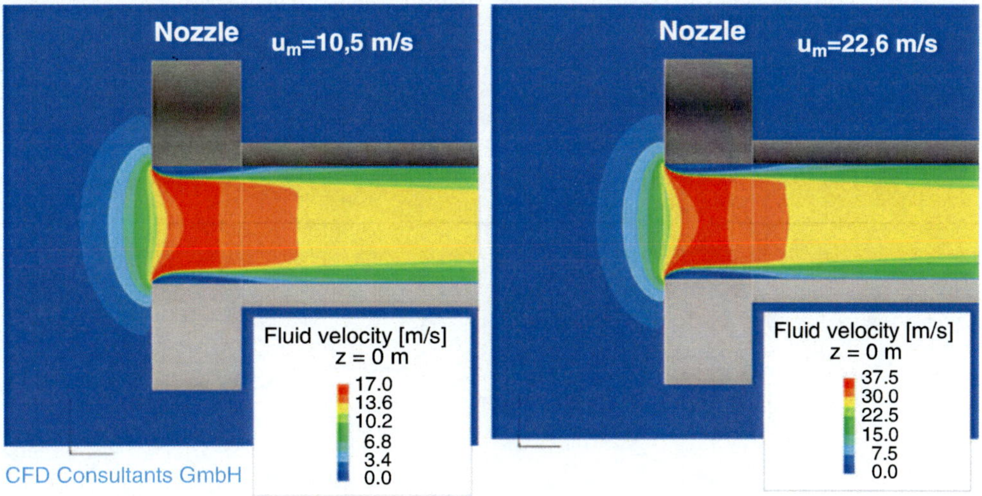

Fig. 3.12 Velocity distribution during suction with the aid of a nozzle at average flow velocities: and $u_m = 10.5$ m/s $u_m = 22.6$ m/s

The relative octave spectrum of the intake noise can be determined using (3.62) and (3.63). These equations were determined empirically based on experimental investigations and a regression analysis as a function of the Strouhal number (St_{Oct}), (3.47) [15, 16]:

$0.1 \leq S_t \leq 2.0$

$$\Delta L_{W,\text{Oct}} = 6.8 \cdot [\lg(\text{St}_{\text{Oct}})]^3 - 21.8 \cdot [\lg(\text{St}_{\text{Oct}})]^2 - 13.6 \cdot [\lg(\text{St}_{\text{Oct}})] - 7.7 \quad \text{dB} \tag{3.62}$$

$2 \leq S_t \leq 40$

$$\Delta L_{W,\text{Oct}} = -4.2 \cdot [\lg(\text{St}_{\text{Oct}})]^3 - 1.9 \cdot [\lg(\text{St}_{\text{Oct}})]^2 + 11.0 \cdot [\lg(\text{St}_{\text{Oct}})] - 14.6 \quad \text{dB} \tag{3.63}$$

The A-octave and A-total sound power levels can be determined as follows:

$$L_{WA,\text{Oct}} = L_W + \Delta L_{W,\text{Oct}} + \Delta L_{A,\text{Oct}} \quad \text{dB(A)} \tag{3.64}$$

$\Delta L_{A,\text{ Oct}}$	A-weighting at the octave centre frequencies, see Table 2.2

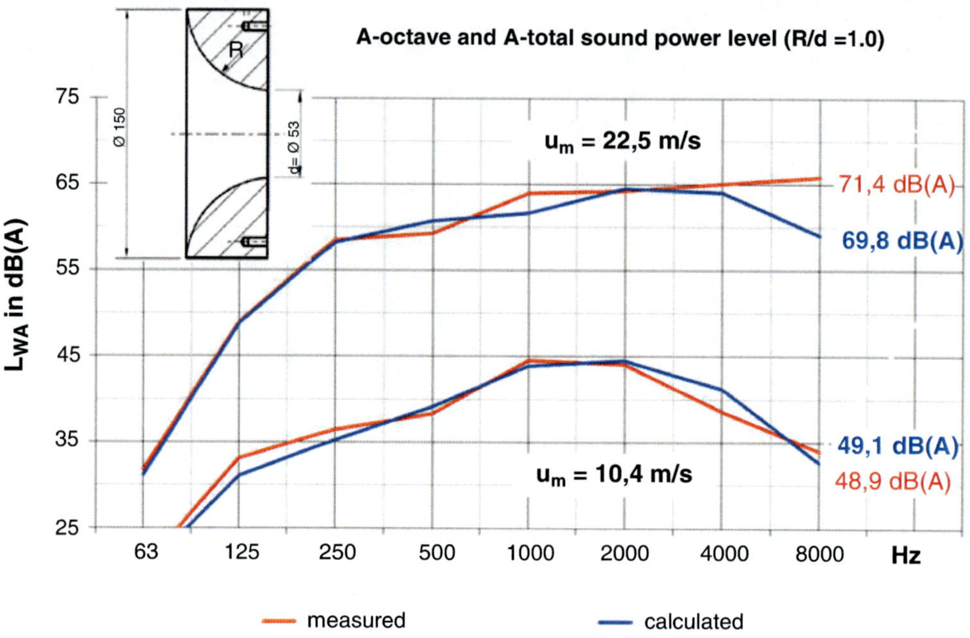

Fig. 3.13 Measured and calculated A-sound power levels of a nozzle ($R/d = 1.0$) at different speeds

$$L_{WA} = 10 \cdot \lg \sum_{63\,\text{Hz}}^{8000\,\text{Hz}} 10^{0.1 \cdot L_{WA,\text{Oct}}} \quad \text{dB(A)} \tag{3.65}$$

Figures 3.13 and 3.14 compare the measured and calculated A-total and A-octave sound power levels for two investigated nozzles at different mean flow velocities as examples. For further results, data and test setups see in [15, 16].

It follows from this that the estimation formulas given here can be used to determine the total and octave sound power levels of the intake noises quite well. Of course, this statement is only valid without restrictions for the investigated nozzle shapes and speed ranges. The verification of the estimation formulas by the company IBS[2] on some examples from practice has also provided a good agreement.

[2]IBS Ingenieurbüro für Schall- und Schwingungstechnik GmbH, D-67227 Frankenthal/Pfalz, www.ibs-akustik.de

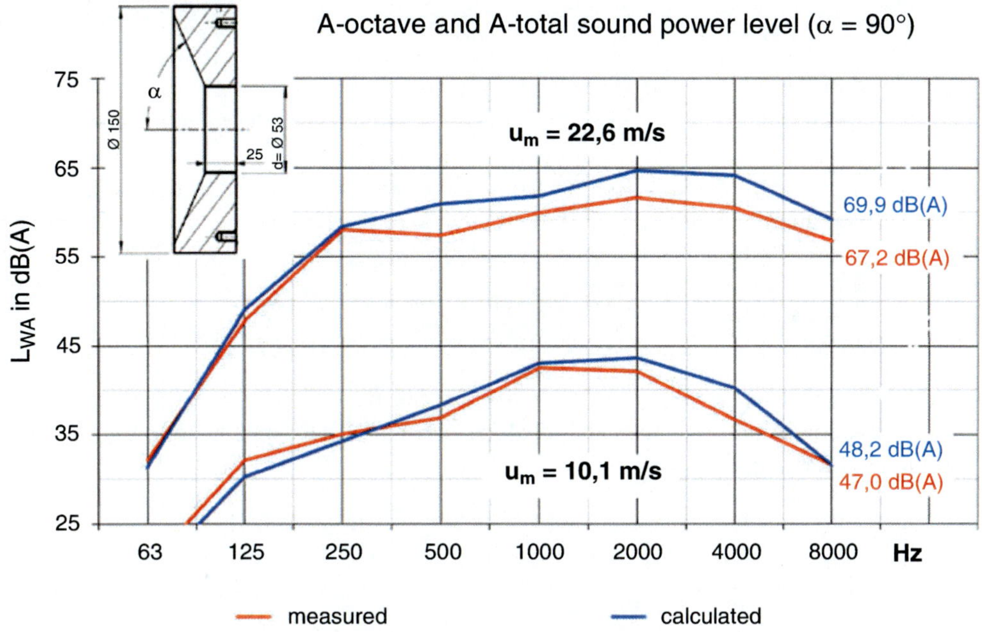

Fig. 3.14 Measured and calculated A-sound power levels of a nozzle ($\alpha = 90°$) at different speeds

3.1.7 Noise Generation in Turbulent Boundary Layers

The cause of all fluid noises discussed so far are noticeable disturbances in the flow. Therefore, practically no noise is perceptible even in undisturbed laminar flows. However, this is no longer true for a turbulent base flow in a straight pipeline. Noise occurs in it even without such disturbances. The place of noise generation is the turbulent boundary layer at the wall. Therefore, this type of fluid noise is always associated with the generation and radiation of structure-borne noise. The cause of this fluid and structure-borne sound are strong pressure fluctuations in the boundary layer, which are caused by the fact that small turbulence balls of the order of magnitude of the boundary layer constantly disintegrate and re-emerge in the boundary layer, whereby impulse shocks are continuously exerted.

The sound power increases with increasing flow velocity and turbulence (degree of turbulence). Noticeable structure-borne sound radiation can also occur if the boundary is formed from lighter walls and there are also discontinuity points in the flow. In the following, the effect of some essential influencing parameters in sound generation is shown using the example of pipe flow.

(a) **Pipe flow with compressible media**

The sound power of gas-flowing straight pipelines can be determined according to (3.13) as follows [7, 12]:

$$P_D = K_D \cdot \rho \cdot u_m^3 \cdot S \cdot Ma^3 \quad W \tag{3.66}$$

$K_D \approx (2\ldots12) \cdot 10^{-5}$	Proportionality factor of the dipole sources

Using (3.53) and (3.54), it follows from (3.66), analogously to the free-radiated noise, that the total internal sound power P_D:

$$P_D = K_D \cdot \frac{p_{sta}}{(R \cdot T)^{5/2} \cdot \kappa^{3/2}} \cdot u^6 \cdot S \quad W \tag{3.67}$$

Considering that K_D is also a function of the mean flow velocity u_m, the internal sound power level in straight pipelines and ideal gases can be determined as follows, taking into account experimental investigations:

$$L_{W_i} = 10\lg\frac{P_D}{P_0}\,\text{dB}; \quad P_0 = 10^{-12}\,\text{W}$$

$$\left.\begin{array}{l} L_{W_i} = K + 60\lg\frac{u_m}{u_0} + 10\lg\frac{S}{S_0} + 10\lg\frac{p_{sta}}{p_{sta_0}} - 25\lg\left(\frac{R \cdot T}{R_0 \cdot T_0}\right) - 15\lg\left(\frac{\kappa}{\kappa_0}\right)\text{dB} \\ K \approx 8 - 0.16 \cdot u_m\,\text{dB} \qquad \text{Correction level} \end{array}\right\} \tag{3.68}$$

The reference values according to VDI 3733 [12] are:

$u_0 = 1$ m/s	Reference speed
$S_0 = 1$ m^2	Reference area
$p_0 = 101325$ Pa	Reference pressure = atmospheric pressure
$R_0 = 287$ N m/kg K	Reference gas constant
$T_0 = 273$ K	Reference temperature
$\kappa_0 = 1.4$	Reference adiabatic exponent

(b) **Pipe flow with incompressible media**

The internal sound power level of straight pipes through which liquid flows can also be determined according to (3.66);

$$L_{WD,i} = 10 \cdot \log\frac{P_D}{P_0} \quad \text{dB}; \qquad P_0 = 10^{-12} \quad \text{W}$$

With $n = 3$ and the speed of sound given by (3.23), it follows from (3.66):

$$P_D = K_D \cdot \rho_F \cdot S \cdot u_m^3 \cdot \frac{u_m^3}{c_F^3} = K_D \cdot \rho \cdot S \cdot u_m^6 \cdot \left(\frac{\rho_F}{E_F}\right)^{\frac{3}{2}} \quad W \tag{3.69}$$

$$L_{\mathrm{WD,i}} = 120 + 10 \cdot \lg(K_{\mathrm{D}}) + 60 \cdot \lg \frac{u_{\mathrm{m}}}{u_0} + 10 \cdot \lg \frac{S}{S_0} + 25 \cdot \lg \frac{\rho_{\mathrm{F}}}{\rho_0} - 15 \cdot \lg \frac{E_{\mathrm{F}}}{E_0} \quad \mathrm{dB} \tag{3.70}$$

With the freely selected reference values:

u_m	Mean flow velocity in m/s
$u_0 = 1$ m/s	Reference speed
$S = \pi \cdot d_i^2/4$	Pipe cross-section in m^2 (d_i= Pipe inside diameter in m)
$S_0 = 1$ m^2	Reference area
ρ_{F}	Density of the liquid in kg/m^3
$\rho_0 = 1$ kg/m^3	Reference density
E_F	Modulus of elasticity of the liquid in N/m^2
$E_0 = 1$ N/m^2	Reference modulus of elasticity
$K_{\mathrm{D}} \approx (2\ldots12) \cdot 10^{-5}$	Proportionality factor

(c) **Frequency spectrum of pipeline noise**

The noise generated by the boundary layer is basically broadband. The frequency spectrum of flow noise in straight pipelines has its maximum at low frequencies and then falls continuously with increasing frequency. This applies in good approximation to both compressible and incompressible media.

The relative octave spectrum (Fig. 3.15) of turbulent boundary layer noise in straight pipelines can be determined as follows [12]:

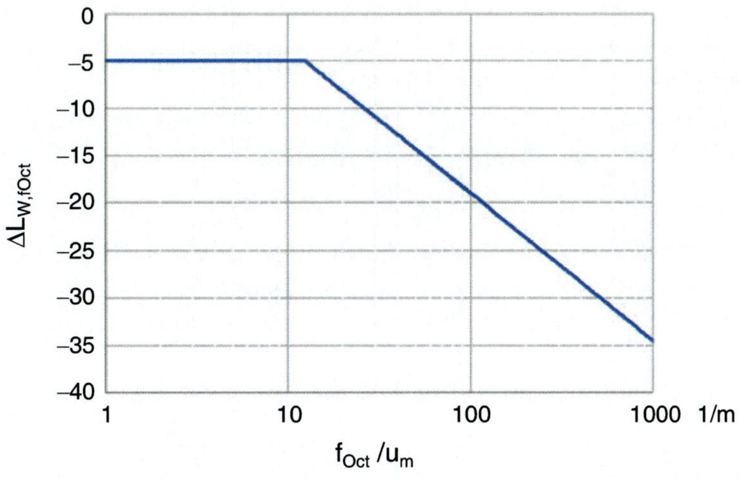

Fig. 3.15 Relative octave spectrum of turbulent boundary layer noise in straight pipelines

$$\left.\begin{array}{l}\Delta L_{W,\mathrm{oct}}=12-15.5\cdot \lg\dfrac{f_{\mathrm{oct}}}{u_m}\,\mathrm{dB}\quad \text{for}\quad \dfrac{f_{\mathrm{oct}}}{u_m}>12.5\\ \dfrac{f_{\mathrm{oct}}}{u_m}\leq 12.5\Rightarrow \Delta L_{W,\mathrm{oct}}=-5\ \mathrm{dB}\end{array}\right\} \tag{3.71}$$

Note
Since the turbulent boundary layer noises are very broadband, one can also estimate the third-octave spectrum of the boundary layer noises in a good approximation, similar to the free jet noises, see (3.49), (3.50) and (3.51). For this purpose, $f_{\mathrm{m,Oct}}$ is replaced by $f_{\mathrm{m,third\text{-}oct}}$ for the determination of $\Delta L_{W,\ \mathrm{third\text{-}oct}}$ in (3.71) and subtracted $10\cdot\ \lg(3)$ from the result.

The internal sound power level can also be determined by measuring the internal sound pressure level in the pipe [6, 12]:

$$L_{p_{\mathrm{i}}}=20\lg\frac{\widetilde{p}_{\mathrm{i}}}{p_0}\quad \mathrm{dB} \tag{3.72}$$

The internal sound power level can then be given as follows:

$$L_{W_{\mathrm{i}}}=L_{p_{\mathrm{i}}}+10\lg\frac{S}{S_0}+K_0-K_{\mathrm{D}}\quad \mathrm{dB} \tag{3.73}$$

or in frequency bands:

$$L_{W\mathrm{i},f_{\mathrm{m}}}=L_{p_{\mathrm{i},f_{\mathrm{m}}}}+10\lg\frac{S}{S_0}+K_{0,\mathrm{i}}-K_{\mathrm{D,fm}}\quad \mathrm{dB} \tag{3.74}$$

Equations (3.73) and (3.74) are similar in structure to (2.98) and (2.99) (sound power determination in enclosed spaces). In the case of sound-hard boundary, the cross-sectional area S corresponds to the absorption area inside the pipe.

$$K_{0,\mathrm{i}}=-10\cdot\lg\frac{Z_{\mathrm{i}}}{Z_0}=-10\cdot\lg\frac{\rho_{\mathrm{F}}\cdot c_{\mathrm{F}}}{\rho_0\cdot c_0}\quad \mathrm{dB} \tag{3.75}$$

$K_{0,\ \mathrm{i}}$ is a correcting element for the case that the impedance of the medium Z_{i} in the pipe is $Z_0=400\ \ \mathrm{N\,s/m^3}$ not equal to the reference impedance.

The correction level K_{D} corresponds to the ambient correction K_2 for room reflections, see Sect. 2.6, and takes into account the level increases of the sound pressure in the pipe due to reflections. Sound propagation in ducts occurs as plane waves when:

$$\lambda_\mathrm{i} \geq 1.72 \cdot d_\mathrm{i} \quad \text{Tubes with circular cross section} \tag{3.76}$$

$$\lambda_\mathrm{i} \geq 2 \cdot a_\mathrm{i} \quad \text{Rectangular tube} \tag{3.77}$$

With

$$\lambda_\mathrm{i} = \frac{c_\mathrm{F}}{f} \quad \mathrm{m} \tag{3.78}$$

d_i	Pipe inside diameter in m
a_i	Largest edge length of a rectangular duct in m
c_F	Speed of sound in the fluid in m/s

In the case of plane sound propagation in straight and long $(l >> \lambda_i)$ pipelines, the reflection influence is negligible, i. e $K_\mathrm{D} \approx 0$.

Experimental investigations have shown that the sound field in long straight pipes can be regarded as approximately diffuse if the wavelength is smaller than 25% of the pipe diameter [6]:

$$\frac{\lambda_\mathrm{i}}{d_\mathrm{i}} = \frac{c_\mathrm{F}}{d_\mathrm{i} \cdot f} \leq 0.25 \tag{3.79}$$

If pipe bends or short pipelines with internals such as orifices, valves, etc. are present, the influence of the reflection cannot be neglected even in the case of plane shafts. This also applies in the presence of standing waves.

For the pure diffuse field, K_D, similar to the sound propagation in closed rooms, corresponds to the ambient correction $K_2 = 10\lg\left(4S/\overline{A}_\mathrm{total}\right)$. For is $S = \overline{A}_\mathrm{total}$ dB. $K_\mathrm{D} = 10 \cdot \lg 4 = 6$

Using (3.79), K_D can be determined as follows [7]:

$$K_{\mathrm{D},f_\mathrm{m}} = 8 \cdot \left(1 - \frac{c_\mathrm{F}}{d_\mathrm{i} \cdot f_\mathrm{m}}\right) \quad \mathrm{dB} \qquad \left(0 \leq K_\mathrm{D,fm} \leq 6\,\mathrm{dB}\right) \tag{3.80}$$

Figure 3.16 shows (3.80) graphically.

Using (3.71), it is also possible to estimate the sound pressure level for the frequency band at the centre frequency f_m from the calculated total internal sound power levels, see (3.68) or (3.70):

$$L_{p\mathrm{i},f_\mathrm{m}} = L_{W\mathrm{i,total}} + \Delta L_{W,f_\mathrm{m}} - 10\lg\frac{S}{S_0} - K_{0_\mathrm{i}} + K_{\mathrm{D},f_\mathrm{m}} \quad \mathrm{dB} \tag{3.81}$$

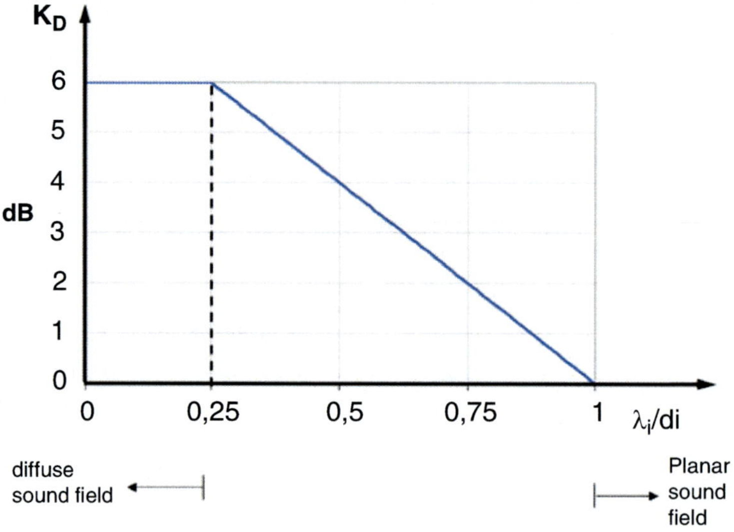

Fig. 3.16 Level magnitude K_D as a function of the ratio λ/d_i

Here are:

$L_{Wi,\text{ total}}$	The total internal sound power level in dB
$\Delta L_{W,f_m}$	The relative frequency spectrum at the centre frequency f_m in dB

Using the attenuation level of the A-weighting according to Tables 2.2 and 2.4, the A-weighted internal sound pressure level at the centre frequency f_m can also be determined from (3.81):

$$L_{piA,f_m} = L_{pi,f_m} + \Delta L_{A,f_m} \quad \text{dB(A)} \tag{3.82}$$

3.1.7.1 Sound Radiation from Pipelines

For the indirect sound radiation of a pipeline the external sound power P_a or the associated sound power level L_{Wa} is primarily responsible for the internal sound power P_{Qi}, the insulation of the pipe wall R and the pipe length l. The noise generated by many turbomachines is largely determined by the sound radiation of the connected pipelines and flow ducts.

The external sound power level L_{Wa} of a straight pipe of length l, taking into account the attenuation along the length of the pipe (Fig. 3.17), can be determined as follows [6, 7, 12]:

$$L_{Wa,l,f} = L_{Wi,m,f} - R_f + 10\lg\frac{l}{d_i} + 6 + K_{a,f} \quad \text{dB} \tag{3.83}$$

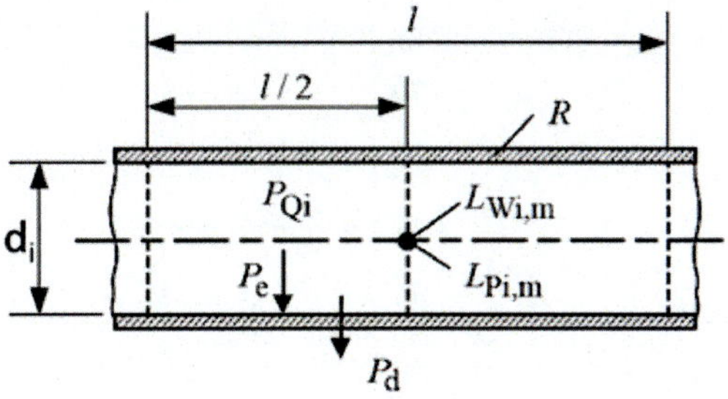

Fig. 3.17 Schematic diagram for the determination of the external sound power level at a pipe section of length l

Sound insulation value of the pipe wall

$$R_f = 10 \cdot \lg \frac{P_e(f)}{P_d(f)} \quad \text{dB} \tag{3.84}$$

Correction element for consideration the sound power reduction within the pipe length l at the frequency f

$$K_{\alpha,f} = 10 \lg \frac{\sinh B}{B} \quad \text{dB} \tag{3.85}$$

Auxiliary variable

$$B = \frac{l}{d_i} \left(2 \cdot 10^{-\frac{R}{10}} + \frac{\alpha \cdot d_i}{8{,}69} \right) \tag{3.86}$$

α	Damping constant in $\frac{dB}{m}$
$L_{Wi,\ m,\ f}$	Internal sound power level in dB

Figure 3.18 shows K_α as a function of the sound insulation of the pipe for different l/d_i. The calculation was computed for an average pipe attenuation, $\alpha \cdot d_i = 0.05$ dB. From this it can be seen that for larger sound insulation ($R > 25$ dB), the level magnitude K_α assumes very small values and it can be neglected for practical applications ($K_\alpha \approx 0$). For smaller sound insulation, $R < 25$ dB, and long pipelines, $l/d_i > 10$, the level magnitude K_α is not negligible.

The external sound power level calculated according to (3.83) represents only that part of the structure-borne sound radiation of the pipeline which is being excited by the internal sound power level. The internal sound power level always consists of the sum of the background noise caused by the pure pipe flow and the portion caused by external sources, such as fans, pumps, etc..

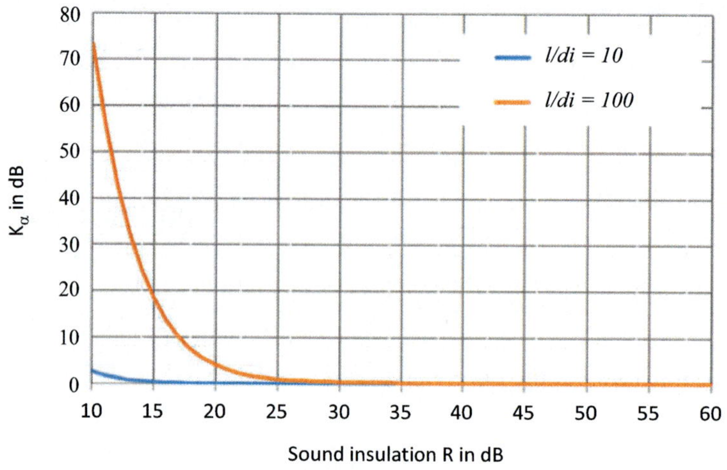

Fig. 3.18 K_α as a function of the sound insulation as a ratio l/d_{i}

Due to the coupling of the pipeline to external sources, there is in principle also the possibility that structure-borne sound is directly transmitted from the external source to the pipeline and radiated from the pipeline as an external sound power level. This part of the external sound power level of the pipeline is not considered by (3.83). If, for example, the external sound power level of a pipeline section determined by measurements is significantly higher than calculated according to (3.83), this is an indication that, among other things, the pipeline was additionally excited to structure-borne sound vibrations by external sources.

The sound insulation value of the pipe at frequency f can be determined as follows [7, 12, 17]:

$$\left.\begin{aligned} R_f &= 10\lg\frac{c_{\mathrm{De}}\cdot\rho_{\mathrm{w}}\cdot h}{(c_{\mathrm{F}}\cdot\rho_{\mathrm{F}})\cdot d_{\mathrm{i}}} + K_f + K = R_{\mathrm{m}} + K_f + K \quad \mathrm{dB} \\ K_f &= -20\lg\frac{f}{f_{\mathrm{R}}} \quad \mathrm{dB} \qquad (f \le f_{\mathrm{R}}) \\ K_f &= 30\lg\frac{f}{f_{\mathrm{R}}} \quad \mathrm{dB} \qquad (f \ge f_{\mathrm{R}}) \end{aligned}\right\} \tag{3.87}$$

Here are:

Average sound reduction index in dB

$$R_{\mathrm{m}} = 10\lg\frac{c_{\mathrm{De}}\cdot\rho_{\mathrm{w}}\cdot h}{(c_{\mathrm{F}}\cdot\rho_{\mathrm{F}})\cdot d_{\mathrm{i}}} \tag{3.88}$$

expansion wave velocity in m⁄s

$$c_{De} = \sqrt{\frac{E_W}{\rho_W \cdot (1 - \mu^2)}} \tag{3.89}$$

Ring expansion frequency of the pipeline

$$f_R = \frac{c_{De}}{\pi \cdot d_i} \text{ Hz} \tag{3.90}$$

Here are:

ρ_W Density of the pipe wall in kg/m^3
E_W Modulus of elasticity of the pipe wall in N/m^2
μ Poisson's ratio ($\mu_{Stahl} \approx 0.3$)
h Wall thickness in m
ρ_F Density of the medium in the pipeline in kg/m^3
c_F Speed of sound of the medium in the pipeline in m/s
d_i Pipe inside diameter in m

K	Correction level in dB

The quantity K is not a constant and is most likely a function of the pipe dimensions, especially the pipe diameter [6, 12, 17]. In the VDI guideline 3733 [12], an average quantity of $K = 10$ dB is recommended for practical applications.

3.1.8 Example: Noise Generation of a straight Pipeline

In the following, using the example of the noise development of a straight pipe section, Fig. 3.19, which is excited by the pipe flow and an external source, the handling of the theoretical principles given here will be explained. The external source is a "centrifugal fan with backward curved blades", according to VDI 3731, sheet 2 [19]. In addition, the example of the "centrifugal fan with backward curved blades" will be used to show how certain characteristics can be determined experimentally [1].

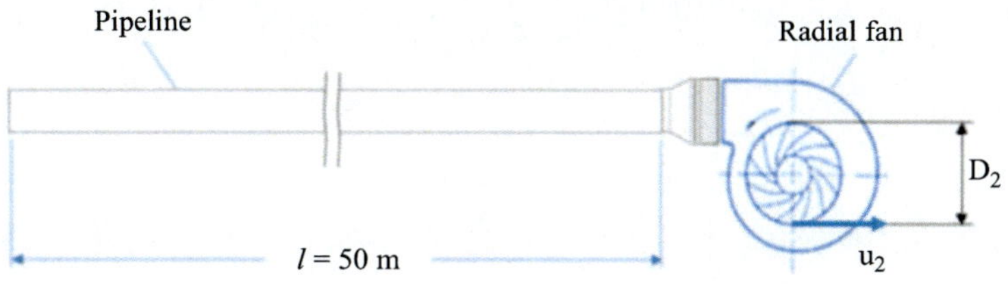

Fig. 3.19 Centrifugal fan with backward curved blades

3.1.8.1 Total Internal Sound Power Level of a Centrifugal Fan

The sound power of a fan can be determined according to (3.10) using the following approach:

$$P_{\mathrm{L}} = K_n \cdot \dot{V} \cdot \Delta p_V \cdot \mathrm{Ma}^n \quad \mathrm{W}; \qquad \mathrm{Ma} = \frac{u_2}{c_{\mathrm{F}}} \tag{3.91}$$

Here are:

$\dot{V}$	Volume flow in m³/s
$u_2 = \frac{D_2}{2} \cdot \omega$	Peripheral speed of the impeller in m⁄s
D_2	Impeller diameter in m
$\omega = 2 \cdot \pi \cdot \frac{n}{60}$	Circular frequency in 1⁄s
n	Impeller speed in 1⁄min
c_F	Speed of sound of the medium in m⁄s
Δp_V	Total pressure drop of the fan in N/m²

$$\Delta p_V = \Delta p_{\mathrm{th}} - \Delta p_{\mathrm{t}} = \frac{\Delta p_{\mathrm{t}}}{\eta} - \Delta p_{\mathrm{t}} = \Delta p_{\mathrm{t}} \cdot \left(\frac{1}{\eta} - 1\right) \tag{3.92}$$

Δp_{th}	Theoretical pressure increase of the fan in N/m²
Δp_{t}	Total pressure increase of the fan in N/m²
η	Fan efficiency
K_n	Proportionality factor

By (3.92) it follows from (3.91):

$$P_{\mathrm{L}} = K_n \cdot \dot{V} \cdot \Delta p_{\mathrm{t}} \cdot \left(\frac{1}{\eta} - 1\right) \cdot \left(\frac{u_2}{c_{\mathrm{F}}}\right)^n \quad \mathrm{W} \tag{3.93}$$

The sound power level is then calculated:

$$L_W = 10 \cdot \lg \frac{P_{\mathrm{L}}}{P_0} = L_{W,\mathrm{spec}} + 10 \cdot \lg\left[\frac{\dot{V}}{\dot{V}_0} \cdot \frac{\Delta p_{\mathrm{t}}}{\Delta p_0} \cdot \left(\frac{1}{\eta} - 1\right)\right] + 10 \cdot n \cdot \lg\left(\frac{u_2}{c_{\mathrm{F}}}\right) \quad \mathrm{dB} \tag{3.94}$$

with

$$L_{W,\mathrm{spec}} = 10 \cdot \log(K_n) \ \mathrm{dB}; \quad \dot{V}_0 = 1 \ \mathrm{m^3/s}; \quad \Delta p_0 = 1 \ \mathrm{N/m^2}; \quad P_0 = 10^{-12} \ \mathrm{W}$$

L_W	Duct sound power level of the fan
$L_{W,\mathrm{spec}}$	Specific sound power level, depending on fan type

For the determination of the unknown quantities here, the specific sound power level $L_{W,\mathrm{spec}}$ and the exponent n, (3.94) is rearranged:

$$L_W - 10 \cdot \lg\left[\frac{\dot{V}}{V_0} \cdot \frac{\Delta p_t}{\Delta p_0} \cdot \left(\frac{1}{\eta} - 1\right)\right] = L_{W,\mathrm{spec}} + 10 \cdot n \cdot \lg\left(\frac{u_2}{c_F}\right) \quad \mathrm{dB} \tag{3.95}$$

The left-hand side of (3.95) can be determined by measuring the total sound power level L_W and the operating data. If one enters the normalised levels determined in this way as a function of the circumferential Mach number u_2/c_F, one can determine the unknown quantities $L_{W,\mathrm{spec}}$ and n empirically by performing a regression analysis if sufficient measured values are available. Figure 3.20 shows an example of the measured total sound power levels for centrifugal fans with backward curved blades according to [19].

The measured values were determined for different fans and power ranges. The function determined by regression calculation, the mean straight line, is:

$$L_W - 10 \cdot lg\left[\frac{\dot{V}}{V_0} \cdot \frac{\Delta p_t}{\Delta p_0} \cdot \left(\frac{1}{\eta} - 1\right)\right] = 85.2 + 15.3 \cdot lg\left(\frac{u_2}{c_F}\right) \quad \mathrm{dB} \tag{3.96}$$

It follows:

$$L_{W,\mathrm{spec}} = \mathbf{85.2}\,\mathrm{dB} \quad \text{and} \quad \mathbf{n = 1.53}$$

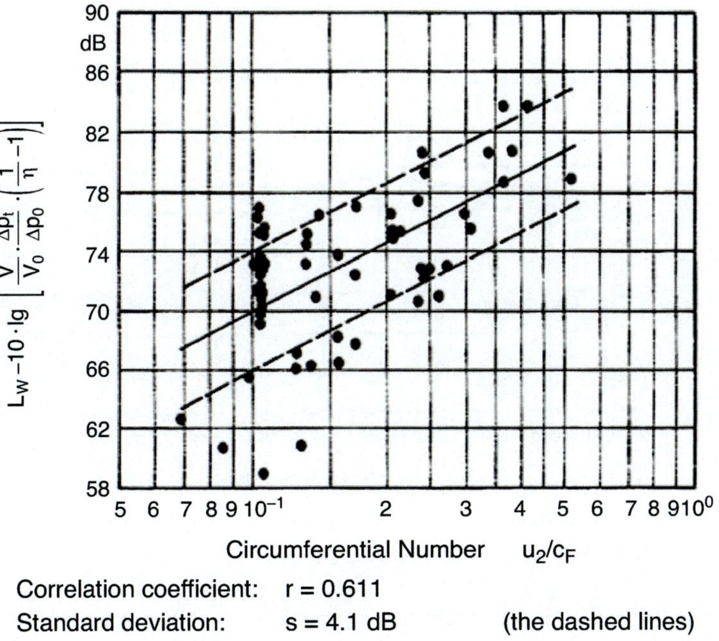

Fig. 3.20 Measured linear sound power level L_W of centrifugal fans with backward curved blades as a function of the circumferential Mach number [19] u_2/c_F

The octave spectrum of the sound power level of centrifugal fans with backward curved blades determined according to (3.95) can be determined approximately for high-pressure fans as follows [19]:

$$\Delta L_{W,\mathrm{oct}}(\mathrm{St}) = -\left[6 + 12 \cdot (\lg(\mathrm{St}) - 0.13)^2\right] \quad \mathrm{dB} \tag{3.97}$$

$$\mathrm{St} = \frac{f_{\mathrm{oct}} \cdot D_2}{u_2} \quad \text{Strouhal number} \tag{3.98}$$

f_{Oct}	Octave centre frequency in Hz

Figure 3.21 shows the relative octave spectrum of centrifugal fans with backward curved blades for high pressure fans, according to VDI 3731, B. 2.

For high-pressure fans, the so-called "high-speed coefficient σ" should be within a certain limit: $0.07 < \sigma < 0.20$.

The fast mode number σ is defined as follows:

$$\sigma = \phi^{1/2} \cdot \psi^{-3/4} \quad \text{fast mode number} \tag{3.99}$$

$$\phi = \frac{\dot{V}}{\left(\pi \cdot D_2^2/4\right) \cdot u_2} \quad \text{flow number} \tag{3.100}$$

$$\psi = \frac{\Delta p_t}{\rho_F \cdot u_2^2/2} \quad \text{pressure number} \tag{3.101}$$

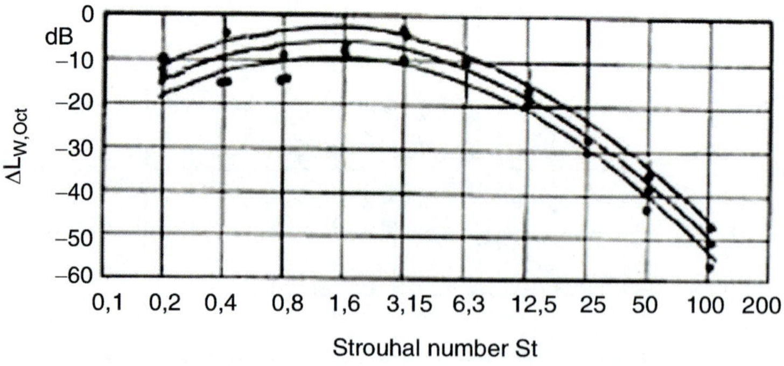

Standard deviation: s = 3,3 dB

Blade frequency power level excess: $\Delta L_{W,d}$ = 0 bis 8 dB

Fig. 3.21 Standardised octave spectrum of centrifugal fans with backward curved blades for high-pressure fans [19]

The level supplement "$\Delta L_{W,\,\mathrm{d}}$" applies to the octaves in which the blade frequency (rotary sound) is located!

Determination of the internal sound power level of the fan

Volume flow $\dot{V}$	5750 m³/h = 1.597 m³/s
Total pressure increase Δp_t	80 mbar = 8000 N/m²
Exit temperature t	25 ° C; T = 298 K
Inlet pressure p	978 mbar = 97800 N/m²
Efficiency η	0.89
Impeller diameter D_2	0.850 m
Number of blades z	9
Speed n	3000 1/min
Gas constant R	287 N m/kg K
Adibatic exponent κ	1.4

The density at the pressure side and the sound velocity can be determined as follows, see (3.53) and (3.54):

$$\rho_{\mathrm{F}} = \frac{\mathrm{p_{sta}}}{R \cdot T} = \frac{97800 + 8000}{287 \cdot 298} = 1.237 \ \mathrm{kg/m^3}$$
$$c_{\mathrm{F}} = \sqrt{\kappa \cdot R \cdot T} = \sqrt{1.4 \cdot 287 \cdot 298} = 346 \ \mathrm{m/s}$$

The rotary sound of the fan is, see (3.43):

$$f_0 = \frac{n \cdot z}{60} = \frac{3000 \cdot 9}{60} = 450 \ \mathrm{Hz}$$

The rotary sound is in the range of the octave with the centre frequency $f_{\mathrm{m}} = 500$ Hz, see Table 2.2.

The circumferential speed and the circumferential Mach number of the impeller are:

$$u_2 = \frac{D_2}{2} \cdot \omega = \frac{D_2}{2} \cdot 2\pi \cdot \frac{n}{60} = \frac{D_2 \cdot \pi \cdot n}{60} = 133.5 \ \mathrm{m/s}$$
$$\mathrm{Ma} = 133.5/346 = 0.386$$

The fast mode number σ of the fan can be determined using (3.99) (3.100) and (3.101):

$$\phi = 0.021$$
$$\psi = 0.726$$
$$\sigma = 0.185 \quad \Rightarrow \text{High pressure fan } (0.07 < \sigma < 0.20)$$

The total sound power level of the fan is then obtained according to (3.95):

$$\begin{aligned} L_{W,\text{total}} &= 10 \cdot \lg\left[\frac{\dot{V}}{V_0} \cdot \frac{\Delta p_t}{\Delta p_0} \cdot \left(\frac{1}{\eta} - 1\right)\right] + 85.2 + 15.3 \cdot \lg\left(\frac{u_2}{c}\right) \\ &= 10 \cdot \lg\left[\frac{1.597}{1} \cdot \frac{8000}{1} \cdot \left(\frac{1}{0.89} - 1\right)\right] + 85.2 + 15.3 \cdot \lg(0.386) = 110.9 \text{ dB} \end{aligned}$$

The total and octave internal A-weighted sound power levels of the backward curved blade centrifugal fan can then be determined as follows:

$$L_{WA,\text{Oct}} = L_{W,\text{total}} + \Delta L_{W,\text{Oct}} + \Delta L_{\text{Oct}} \, \text{dB(A)}.$$

Table 3.2 shows the calculation results for the octave spectrum. A level surcharge of $L_{W,d} = 8$ dB (Fig. 3.21) was taken into account for the octave centre frequency $f_m = 500$ Hz

The inner A-octave sound power levels determined in this way apply to the excitation of the pipeline as an external source. These values can also be determined by measurements and should be provided by the manufacturers.

Table 3.2 Internal sound power level of the fan

Frequency in Hz	63	125	250	500	1000	2000	4000	8000	Total	Dim.
St. to Eq. (3.98)	0.4	0.8	1.6	3.2	6.4	12.7	25.5	50.9	–	–
$\Delta L_{W,oct}$ according to Eq. (3.97)	−9.3	−6.6	−6.1	−7.7	−11.4	−17.4	−25.5	−35.8	–	dB
$L_{W,oct}$, without level supplement	101.5	104.2	104.8	103.2	99.4	93.5	85.3	75.0	109.6	dB
$L_{W,oct}$ with level supplement	101.5	104.2	104.8	111.2	99.4	93.5	85.3	75.0	113.1	dB
$\Delta L_{,oct}$ according to Table 2.2	−26.2	−16.1	−8.6	−3.2	0.0	1.2	1.0	−1.1	–	dB
$L_{WAoct,Fan}$	**75.3**	**88.1**	**96.2**	**108.0**	**99.4**	**94.7**	**86.3**	**73.9**	**109.0**	dB(A)

3.1.8.2 Total Internal Sound Power Level due to Pipe Flow

The pipeline is a steel pipe with the following data:

$$
\begin{aligned}
d_i &= 0.28\ \text{m}\varnothing; \quad S = 0.062\ \text{m}^2 \\
l &= 50\ \text{m} \\
h &= 3\ \text{mm} \quad \text{(wall thickness)} \\
E_W &= 2.1 \cdot 10^{11}\,\text{N/m}^2 \\
\rho_W &= 7850\ \text{kg/m}^3 \\
c_{De} &= 5421.9\ \text{m/s} \\
f_R &= \frac{c_{De}}{\pi \cdot d_i} = 6183.8\ \text{Hz}
\end{aligned}
$$

The volume flow, the static pressure, the temperature, the sound velocity and the medium characteristics correspond to the values on the pressure side of the fan:

$$
\begin{aligned}
\dot{V} &= S \cdot u_m = 1.597\ \text{m}^3/\text{s} \\
u_m &= \frac{\dot{V}}{S} = \frac{1.597 \cdot 4}{\pi \cdot 0.28^2} = 25.9\ \text{m/s} \\
p_{sta} &= 105800\ \text{N/m}^2 \\
\rho_F &= 1.237\ \text{kg/m}^3 \\
c_F &= 346\ \text{m/s} \\
R &= 287\ \text{Nm/kgK} \\
\kappa &= 1.4
\end{aligned}
$$

The total internal sound power level L_{W_i} of the pure air-regenerated noise can be determined according to (3.68):

$$
\begin{aligned}
L_{W_i} &= K + 60\lg\frac{u_m}{u_0} + 10\lg\frac{S}{S_0} + 10\lg\frac{p_{sta}}{p_{sta_0}} - 25\lg\left(\frac{R \cdot T}{R_0 \cdot T_0}\right) - 15\lg\left(\frac{\kappa}{\kappa_0}\right) \\
&= 76.0\ \text{dB} \\
K &= 3.8\ \text{dB}
\end{aligned}
$$

Tables 3.3, 3.4 and 3.5 show the results of the calculation, noise development of a straight pipeline:

It follows that the pure flow noise can be neglected in comparison to the noise generated by the external source in the present example.

Table 3.3 Internal sound power levels of pure air-regenerated noise

Frequency	63	125	250	500	1000	2000	4000	8000	Total	
f_m/u	2.4	4.8	9.6	19.3	38.6	77.1	154.2	308.4	–	1/m
ΔL_W	−5.0	−5.0	−5.0	−7.9	−12.6	−17.2	−21.9	−26.6	–	dB
$\Delta_{L,oct}$	−26.2	−16.1	−8.6	−3.2	0.0	1.2	1.0	−1.1	–	dB
$L_{wiA,flow}$	**44.8**	**54.9**	**62.4**	**64.9**	**63.4**	**60.0**	**55.1**	**48.3**	**69.4**	**dB(A)**

Table 3.4 Total internal sound power levels, as sum of external source (radial fan) and air-regenerated noise

$L_{WA,Oct,Fan}$	75.3	88.1	96.2	108.0	99.4	94.7	86.3	73.9	109.0	dB(A)
$L_{wiA,flow}$	44.8	54.9	62.4	64.9	63.4	60.0	55.1	48.3	69.4	dB(A)
$L_{WiA,total}$	**75.3**	**88.1**	**96.2**	**108.0**	**99.4**	**94.7**	**86.3**	**73.9**	**109.0**	**dB(A)**

Table 3.5 External sound power levels of the pipeline

f_m in Hz	63	125	250	500	1000	2000	4000	8000	Total	Hz
$L_{WiA,\ oct} = L_{WiA,total}$	75.3	88.1	96.2	108.0	99.4	94.7	86.3	73.9	109.0	dB(A)
K_f	39.8	33.9	27.8	21.8	15.8	9.8	3.8	3.4	–	dB
R_{oct}	80.1	74.1	68.1	62.1	56.1	50.1	44.0	43.7	–	dB
$L_{WaA,\ oct}$	**23.8**	**42.5**	**56.6**	**74.4**	**71.9**	**73.1**	**70.8**	**58.8**	**78.9**	**dB(A)**

Considering that a reduction of the internal sound power level in the pipeline below the flow noise is physically not possible, the flow noise indicates the potential of noise reduction in the pipeline.

With the help of the total internal sound power level, the external sound power level of the pipeline is obtained according to (3.83). Table 3.5 shows the calculation results for the octave spectrum. Here it was assumed that $K = 10$ dB and dB $K_{\alpha f} \approx 0$, see Fig. 3.18, and $R_m = 30.3$, according to (3.88).

Due to the large sound-radiating surface, the pipelines are often the main noise sources of fluidic systems.

Furthermore, it was assumed that no structure-borne sound vibrations are transmitted from the external source to the pipeline.

One must also add the external sound power levels of the external source, in this example the fan, to the external A-octave and A-total sound power levels of the pipeline determined in the way described above.

3.2 Sound Generation by Mechanical Vibrations, Structure-Borne Sound

The indirect sound radiation of machine structures which are excited to mechanical vibrations ("structure-borne sound vibrations") by **alternating forces** is referred to as secondary or indirect airborne sound (structure-borne sound!). In this case, the air particles are indirectly excited to vibrate by structural vibrations, e.g. machine housing.

Alternating forces occur, for example, in gearboxes due to shocks, rolling and friction processes, imbalances of rotating components, pressure pulsations, detachment phenomena during flow processes at deflections and changes in cross-section.

The radiated sound power of a vibrating structure (e.g. machine housing), hereinafter correctly referred to as "structure-borne sound radiation", can be represented according to (2.45) as follows [18]:

$$P_K(f) = \sum_{j=1}^{M} P_{K,j}(f) = \sum_{j=1}^{M} \rho \cdot c \cdot v_j^2(f) \cdot \sigma_j(f) \cdot S_j \quad \text{W} \tag{3.102}$$

P_K	Radiated total structure-borne sound power
$P_{K,j}$	Partial structure-borne sound power of a component or of the *j-th* partial area in W
$\rho \cdot c$	Impedance of the surrounding air in N/s/m³
v_j^2	Mean square of velocity on the *j-th* subplane m²/s²
M	Total number of sub-areas
σ_j	Radiance of the *j-th* subarea
S_j	*j-th* partial area of the structure in m²
(f)	Means that the specified quantities are frequency-dependent

In order to determine the decisive influencing parameters similar to the flow processes, (3.102) is transformed. Here, the emergence of the oscillation velocity of the structure "v" is of particular importance.

The machine structures are excited to vibrate by alternating mechanical forces. As a rule, the partial surfaces can be excited to vibrations by several alternating forces, which are also introduced at different points of the machine. Figure 3.22 shows schematically the vibration excitations of the partial surface "j" of a model housing [2, 20–22].

For the vibration excitations of the j-th subarea S_j by the k-th alternating force F_k, the initiation point characterised by the mechanical input impedance given by $Z_{e,k}$ (2.26) is of primary importance. The vibration velocity $v_{e,k}$ caused at the initiation point k is then defined as follows:

$$v_{e,k} = \frac{F_k}{Z_{e,k}} \quad \text{m/s} \tag{3.103}$$

$v_{e,k}$	Vibration velocity (velocity) at the discharge point k
$Z_{e,k}$	Mechanical input impedance of the *k-th* introduction point in kg/s
F_k	Alternating force at the *k-th* discharge point in N

The vibrations introduced into the structure are in turn responsible for the vibration excitation of the entire structure. The vibration velocity of the *j-th* partial surface can now be determined as follows:

$$v_{k,j} = v_{e,k} \cdot h_{ü_{k,j}} = F_k \cdot \frac{1}{Z_{e,k}} \cdot h_{ü_{k,j}} \quad \text{m/s} \tag{3.104}$$

$v_{k,j}$	Mean velocity of the *j*-th partial surface caused by the *k*-th alternating force
$h_{ü_{k,j}}$	Transfer function between the *k*th discharge point and the *j*th subarea or component

By (3.103) and (3.104), it follows for the mean square of velocity on the *jth* subarea as the sum over all the alternating forces:

$$v_j^2(f) = \sum_{k=1}^{K} F_k^2(f) \cdot \frac{1}{Z_{e,k}^2(f)} \cdot h_{ük,j}^2(f) \quad (\text{m/s})^2 \tag{3.105}$$

With (3.105), the total structure-borne sound power of a machine can be given as a function of various influencing parameters:

$$P_K(f) = \underbrace{(\rho \cdot c) \cdot \sum_{j=1}^{N} S_j \cdot \sigma_j(f)}_{\text{Radiation}} \cdot \underbrace{\sum_{k=1}^{K} F_k^2(f) \cdot \frac{1}{Z_{e,k}^2(f)} \cdot h_{ük,j}^2(f)}_{\text{Emergence and transmission}} \quad \text{W} \tag{3.106}$$

It follows from this that the total structure-borne sound power is made up of the sum of all acting alternating forces (summation over *k*) and the sum of all components excited to vibrate (summation over j). The total structure-borne sound power is responsible for the radiation of the secondary airborne sound, the sound pressure at distance *r from* the machine, see Fig. 3.22. The total structure-borne sound power level of a machine is then obtained:

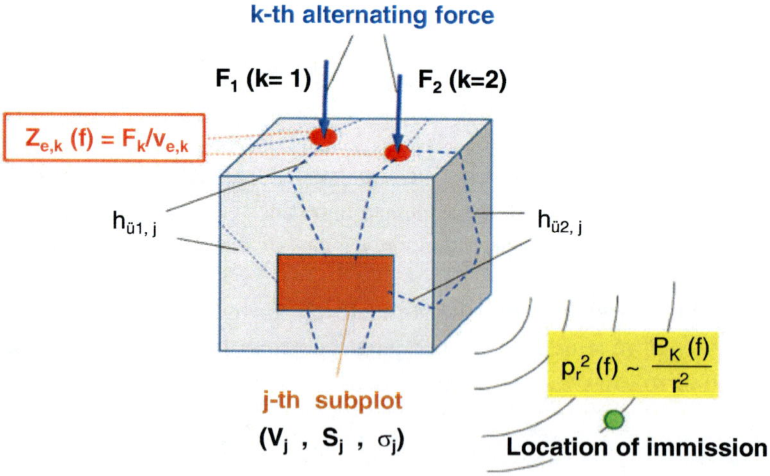

Fig. 3.22 Schematic representation of sound generation by alternating forces on a model enclosure

$$L_{WK}(f) = 10 \cdot \lg \sum_{j=1}^{N} \frac{P_{K,j}(f)}{P_0} = 10 \cdot \lg \sum 10^{\frac{L_{WK,j}}{10}} \quad \text{dB} \tag{3.107}$$

With

$$L_{WK,j}(f) = 10 \cdot \lg \frac{P_{K,j}(f)}{P_0} = -K_0 + L_{vj}(f) + L_{Sj} + \sigma'_j$$

$$\approx L_{vj}(f) + L_{Sj} + \sigma' \quad \text{dB} \tag{3.108}$$

If one takes into account that the structure-borne sound radiation of the machines usually takes place outdoors and the impedance of the air is approximately $Z_{air} = (\rho \cdot c)_{air}$ equal to the reference impedance $\rho_0 \cdot c_0 = 400\ \text{N s/m}^3$, then one can neglect the influence of the air temperature and the air pressure on the structure-borne sound radiation, i. e.:

$$K_0 = -10 \cdot \lg \frac{\rho \cdot c}{\rho_0 \cdot c_0} \approx 0 \quad \text{dB}$$

By (3.108) it follows from (3.107):

$$L_{WK}(f) = 10 \cdot \lg \sum 10^{\frac{L_{vj}(f) + L_{Sj} + \sigma'_j}{10}} \quad \text{dB} \tag{3.109}$$

In the derivation of (3.106) linearity was assumed, i.e. that the increase of alternating forces also results in a proportional increase of the oscillations. A non-linearity exists, for example, if there is play or a large damping in the transmission paths.

Basically, one distinguishes between two cases of vibration excitation:

(a) **Free oscillations**

Free oscillations occur when a vibratory system is excited to oscillate, e.g. by a shock, and is then left to oscillate on its own. In this case, the oscillating system always oscillates at its **natural frequencies**.

(b) **Forced oscillations**

Forced oscillations occur when energy, e.g. a periodic excitation force, is permanently or forcibly supplied to a vibratory system. In this case, the oscillating system always oscillates at the **excitation frequencies**.

It is particularly critical here if the excitation frequencies coincide with the natural frequencies of the oscillatory system. Depending on the damping, the amplitude of the oscillating system can assume very large values due to resonance excitation.

The influence of **resonance phenomena**, which in many cases are the cause of excessive noise due to structure-borne sound radiation, is contained in the sizes and $Z_{e,k}(f) h_{ü_{j,k}}(f)$.

Resonance phenomena can lead to relatively small forces having a decisive influence on the overall sound radiation, not through their strength but through their frequency position.

The natural frequencies correspond to the amplitude increases in the transfer function $hü_{j,k}$ or the amplitude dips in the mechanical input impedance $Z_{e,\ k}$. They can best be determined by measurement, e.g. by impact tests when the machine is at standstill.

With the help of (3.105), (3.106), (3.107) and (3.108), it is now possible to identify the design influence parameters for sound generation by mechanical vibrations. In principle, the following statements can be derived:

1. The sum of all forces is basically responsible for the structure-borne sound excitation. If the design remains unchanged, only the largest forces are decisive due to the quadratic or energetic addition (summation over "k").
2. For the structure-borne sound excitation of a construction, the mechanical input impedance (vibration resistance at the point of introduction) is decisive in addition to alternating forces.
3. In principle, all radiating surfaces are responsible for the structure-borne sound radiation, whereby in the sum over "j" only the partial surfaces are decisive for which the product $(v_j^2 \cdot S_j \cdot \sigma_j)$ or the sum $(L_{v,\ j} + L_{S,\ j} + \sigma_j)$ is greatest.
4. In the case of the radiated structure-borne sound power, only the frequencies that determine the level are important. Here, resonance-excited components are particularly critical.

The air vibrations excited by the structural vibrations, as far as their frequencies are in the audible range, are called structure-borne sound or secondary airborne sound.

These oscillations are excited at specific, usually localised areas and radiated both at the excitation site and at more distant locations in the structure.

The transmission into the structure takes place through longitudinal, shear, torsional and bending waves and can overcome larger distances, especially through the longitudinal waves.

While the input impedance Z_e plays an important role in the introduction of the disturbance into the structure, the transmission depends mainly on the structure-borne sound attenuation and damping of the structure.

Finally, structure-borne sound is radiated as secondary airborne sound at suitable sections of the structure surface. Decisive for the strength of the radiated sound power are above all

- the size of the radiating surfaces,
- the structure-borne sound velocity,
- the degree of radiation

of the vibrating structures.

Structure-borne sound radiation can be particularly efficient when plate-like elastic structures are excited to **bending vibrations**.

The bending vibrations can be fanned particularly well by transverse forces. However, such vibrations can also be initiated primarily by longitudinal waves. If such longitudinal waves hit corners and edges, which are always present in a structure, the longitudinal waves are converted into bending waves with correspondingly stronger radiation.

Structure-borne sound vibrations are basically generated in two forms [7]:

(a) **Force excitation**

If a structure, e.g. a machine, emits structure-borne sound or secondary airborne sound, certain alternating forces are effective on it as parts of the operating forces. Together with the reaction forces, they build up a closed force flow. All elastic elements in this force flow are then directly excited to vibrations. Such an excitation of the structure is called "force excitation". In Fig. 3.23, one can see a force flow of the gas forces and a force flow of the mass forces running over the suspension. Here, for example, the engine block in the force flow is force-excited.

(b) **Spring base point or velocity excitation**

Indirectly coupled structural areas that are not in direct force flow can also be excited to oscillate. In this case, it is a "velocity excitation". In mechanics, it is also called a spring-foot excitation. In the example shown in Fig. 3.23, the oil pan coupled to the bottom of the housing is velocity-excited because it is not in the force flow.

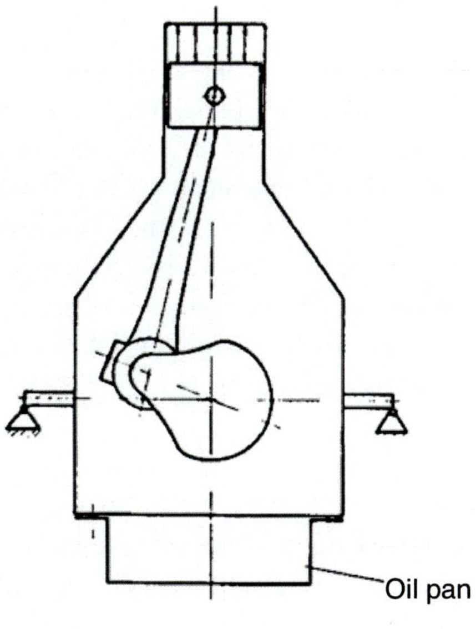

Fig. 3.23 Schematic representation of the force and speed excitation of an internal combustion engine

In principle, velocity-excited components are not responsible for the transmission of operating forces and are referred to as **non-load-bearing components**.

The differentiation of the type of excitation is above all decisive for the development of primary noise reduction measures, see Chap. 4.

3.2.1 Excitation of Structure-Borne Sound, Excitation or Alternating Force

In the case of force excitation of a structure by alternating forces, three excitation mechanisms can be distinguished:

- periodic excitation, e.g. unbalance forces
- Impulse excitation, e.g. brief impact forces and
- stochastic excitation, e.g. frictional forces or turbulent boundary layer flows

These alternating forces can occur individually or in any combination.

3.2.1.1 Periodic Excitation

This type of structure-borne sound excitation is relatively easy to detect. The time course of the forces is periodic, the associated amplitude spectrum is a line spectrum. Purely periodic excitation is primarily noise due to the effect of inertial forces. This type of excitation occurs in all rotating machine parts, e.g. pumps, turbines, electric engines, etc. The strength of the force excitation is determined by the size of the unbalance and the speed n of the rotor systems. In the case of pure unbalance, only the unbalance frequency f_U occurs in the spectrum:

$$f_U = \frac{n}{60} \text{ Hz} \quad n \text{ in } 1/\text{min} \tag{3.110}$$

If multiples of the unbalance frequency, the harmonics, also occur in the spectrum, then this is an indication that the unbalance forces – e.g. due to backlash – also cause shocks or that the rotating components (“rotors”) have different stiffnesses in the circumference.

Periodic structure-borne noise excitation can also be caused by oscillating masses, e.g. at non-uniformly translating gears, or by mass forces due to piston movements of an internal combustion engine. A modern high-speed industrial sewing machine is a typical example. Here, in addition to the oscillation frequency f_{Os}, the harmonics of the oscillation frequency also occur in $f_{Os,\,k}$ the excitation spectrum.

$$f_{Os,k} = (1 + k) \cdot f_{Os} \text{ Hz} \quad k = 1, 2, 3 \ldots \tag{3.111}$$

The oscillation frequency f_{Os} corresponds to the number of back-and-forth movements per second. The strength of the alternating force excitation is determined by the size of the oscillating masses and the oscillation frequency.

A special group of periodic excitation are the so-called magnetostrictive forces. The excitation is proportional to the square of the current strength, so that with an induction frequency of 50 Hz the excitation occurs with the fundamental frequency 100 Hz.

This type of structure-borne sound excitation plays a role above all in electric engines, generators and transformers. Due to the constancy of the induction frequency, the frequencies of the magnetostrictive forces are also constant and occur exactly at 100 Hz and multiples of 100 Hz.

3.2.1.2 Impulse or Shock Excitation

This is a short-term action of forces on a locally very limited structural area and a duration of action of fractions of a second. The impact may be a one-off or may only recur after longer intervals. However, it can also be a pulsating, intermittent impact process. In the first case, the single process alone is decisive. As an example, Fig. 3.24 shows the time course and the corresponding normalised frequency spectrum of a $|F(f)|\ /J_S$ half-cosine pulse. Here $|F(f)|$ is the magnitude of the spectral function and J_S is the pulse strength [7, 23–25]:

$$J_S = \int_{-t_{St}/2}^{+t_{St}/2} F(t) \cdot \mathrm{d}t = \frac{2}{\pi} \cdot F_{max} \cdot t_{St} \quad \mathrm{N\,s} \tag{3.112}$$

$$|\,F(f)\,| = \left|\int_{\frac{-t_{St}}{2}}^{\frac{+t_{St}}{2}} F_{max} \cdot \mathrm{e}^{-\mathrm{j}2\pi f t}\,\mathrm{d}t\right| = J_S \cdot \left|\frac{\cos(\pi \cdot f \cdot t_{St})}{1 - 4 \cdot f^2 \cdot t_{St}^2}\right| \quad \mathrm{N\,s} \tag{3.113}$$

It follows from this that the shorter the action time t_{St} for a blow or impact, the broader the spectral distribution of the excitation. If one assumes equal impulse strength J_S, then the shorter the action time t_{St}, the stronger the amplitude of the force excitation will be.

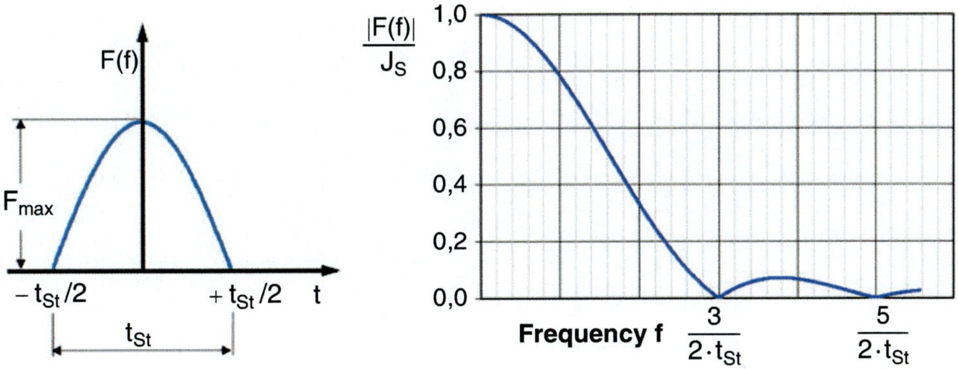

Fig. 3.24 Time and frequency diagram of a half-cosine pulse normalised by J_S

Intermittent pulse processes with periodic sequence

If the pulse-like fluctuations repeat themselves in shorter time intervals T, they are intermittent processes. Figure 3.25 shows the temporal and spectral course of intermittent half-cosine pulses. The frequency spectra of periodic pulse processes are basically line spectra, the distance between the individual amplitudes has the constant magnitude $\Delta f = 1/T$.

The spectrum of the $|G(f)|$ single pulse, see Fig. 3.24, determines the envelope of the line spectrum, the partial amplitudes $F_n(f_k)$ can be determined as follows [7, 25]:

$$\begin{aligned} |F_n(f_k)| &= \frac{2 \cdot |F(f_k)|}{T} = \frac{J_S}{T} \cdot \left| \frac{\cos(\pi \cdot f_k \cdot t_{St})}{1 - 4 \cdot f_k^2 \cdot t_{St}^2} \right| \\ &= \frac{4}{\pi} \cdot F_{max} \cdot \frac{t_{St}}{T} \cdot \left| \frac{\cos(\pi \cdot f_k \cdot t_{St})}{1 - 4 \cdot f_k^2 \cdot t_{St}^2} \right| \quad \mathrm{N} \end{aligned} \tag{3.114}$$

With

$$f_k = \frac{k+1}{T} \ \mathrm{Hz} \quad k = 0, 1, 2, 3, \ldots \tag{3.115}$$

It can also be seen here that the frequency range expands to the right to higher frequencies with smaller pulse duration t_{St} of the single pulse (Fig. 3.25). The magnitude of the partial amplitudes is determined by the pulse strength J_S and the time duration of the pulse train T.

Examples of this are impulse forces, which are generated e.g. by forging, punching, riveting, hammering, by rattling with free play in guides, in typewriters or also by roller bearing support forces or by tooth flank forces. Structure-borne sound can also be excited or generated as a result of intermittent pressure forces of hydraulic and gas-dynamic origin, see also Sect. 3.1.2.2.

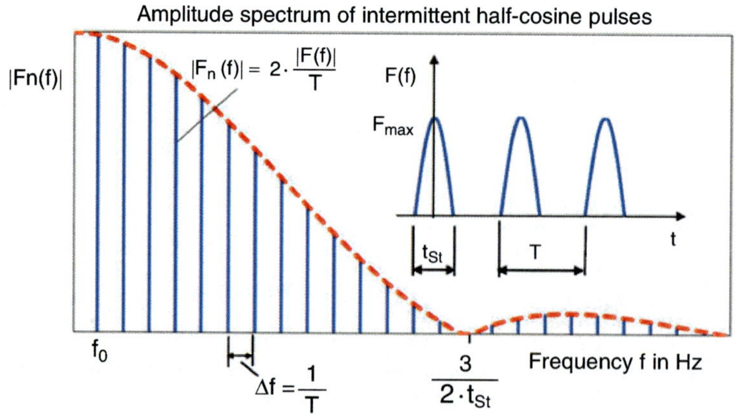

Fig. 3.25 Temporal and spectral course of intermittent half-cosine pulses

3.2.1.3 Stochastic Excitation

This type of excitation is probably the most common. It is random with the characteristics for stochastic time sequences, see Fig. 2.23. In particular, the frequency spectrum of the amplitude density is broadband. Stochastic structure-borne sound excitation occurs during rolling (unrolling) and sliding of two boundary surfaces relative to each other, which are pressed together by external forces. Due to the roughness of their surfaces, strongly varying local force peaks with a stochastic character occur in the boundary surfaces and lead to an equally stochastic structure-borne sound excitation of the coupled structure. Such excitations occur, for example, at the balls, rollers and needles of rolling bearings or at the tooth flanks of two meshing gears, and also at rolling wheels, such as the rolling noise of road and rail vehicles. All excitations increase very strongly with the increase in rolling and sliding speed, as well as with greater roughness.

The excitation mechanism is of a similar nature, as it occurs during machining, e.g. milling, turning, drilling, grinding, sawing or planing. Here the local, stochastically acting force peaks are even more intensive. The excitation spectrum and the noise itself are broadband. The excitation also increases strongly with the increase in machining speed, which is usually synonymous with an increase in excitation due to an increase in speed.

The stochastic structure-borne sound excitation of a structure can also be effected by a medium flowing against, around or through it. It does not matter whether the structure forming the obstacle is stationary and the medium is flowing or, conversely, whether the obstacle is moving in a stationary medium. As is well known, a system of vortices is formed at the back of the obstacles by detachment effects. These act back on the obstacle with changing forces. However, with the larger Reynolds numbers usually present, turbulent velocity fluctuations are superimposed on these vortices, leading to rapid decay of the vortices. This leads to random velocity fluctuations and consequently to stochastic alternating forces at the obstacle with a similar excitation of structure-borne sound. The excitation of structure-borne sound by vortex shedding increases very strongly with increasing inflow velocity. The structure-borne sound excitation of the body of an engine vehicle by the airstream is an apt example of this.

Of course, it is obvious that the three treated structure-borne sound excitations, the periodic excitation, the impact and shock excitation and the stochastic excitation, can not only interact separately, i.e. each separately, but also in superpositions. For example, during material processing, narrowband components can also occur in the broadband basic spectrum as a result of imbalance influences in the drive. When a wheel rolls on rails supported by sleepers, an intermittent impact noise is superimposed on the broadband noise of the rolling process. In the case of a fan, it is correspondingly the superposition of broadband flow noise and narrowband frequencies of the rotational sound and their harmonics.

Note

Finally, a less easily understood type of excitation is presented. It is based on the mechanisms of **self-excitation**, as they occur, for example, in the squealing of brakes, the screeching of wheels when rolling through tight curves, in short, in all sliding processes that take place under a shrill noise.

This excitation, which comes from feedback effects of an oscillating structure with a static energy store, is broadband with preference for the higher frequency ranges. In contrast to the usual excitation, where the structures oscillate only with excitation frequencies – **forced oscillations**[3] –, in self-excitation the structures often oscillate with their natural frequencies.

The same applies to **chatter vibrations**, such as those that can occur on metal and woodworking machines. This type of vibration is also caused by feedback effects in which several vibrating systems can be excited [26].

References

1. Sinambari, Gh.R.: Systematische Vorgehensweise bei der Geräuschminderung, Technische Lärmminderung bei Maschinen und Geräten – OTTI-Fachforum Regensburg, 2012
2. Sinambari, Gh.R., Kunz, F.: Primäre Lärmminderung durch akustische Schwachstellenanalyse. VDI-Bericht Nr. 1491 (1999)
3. Thorn, U., Wachsmuth, J.: Konstruktionsakustische Schwachstellenanalyse eines Straddle Carriers Typ Konecranes 54 DE und Erarbeiten von prinzipiellen Lärmminderungsmaßnahmen. VDI-Berichte 2118 (2010)
4. Strutt, J.W.: Lord Rayleigh: The Theory of Sound. Nachdruck, New York (1945)
5. Heckl, M.: Strömungsgeräusche. Fortschr. Ber. VDI-Z. Reihe 7, Nr. 20. VDI Verlag (1969)
6. Sinambari, Gh.R.: Ausströmgeräusche von Düsen und Ringdüsen in angeschlossenen Rohrleitungen, ihre Entstehung, Fortpflanzung und Abstrahlung. Dissertation, Universität Kaiserslautern (1981)
7. Sinambari, Gh.R., Sentpali, S.: Ingenieurakustik, 6. Aufl. Springer Vieweg, Wiesbaden (2020)
8. Heckl, M., Müller, H.A.: Taschenbuch der Technischen Akustik, 2. Aufl. Springer, Berlin (1994)
9. Dubbel: 21. Aufl. Springer (2005)
10. Morse, P.M., Ingard, K.U.: Theoretical Acoustics. Princeton University Press, Princeton (1986)
11. Sinambari, G.R., Thorn, U.: Strömungsinduzierte Schallentstehung in Rohrbündel-Wärmetauschern. Zeitschrift für Lärmbekämpfung. **47**(1) (2000)
12. VDI 3733: Noise at pipes (1996)
13. Henn, H., Rosenberg, H., Sinambari, G.R.: Akustische und gasdynamische Schwingungen in Gasdruck-Regelgeräten, ihre Entstehung und Fortpflanzung. gwf-gas/erdgas. **6** (1979)
14. Rosenberg, H., Henn, H., Sinambari, Gh.R., Fallen, M., Mischler, W.: Akustische und schwingungstechnische Vorgänge bei der Gasentspannung. DVGW-Schriftenreihe, Gas Nr. 32. ZfGW-Verlag, Frankfurt (1982)
15. Stadtmüller, M.: Abschätzung der strömungsbedingten Geräuschentwicklung von Ansaugöffnungen. Bachelorarbeit, FH Kaiserslautern (2014)

[3] In the case of forced oscillations, which usually occur during steady-state operation of a machine, the structures oscillate *only* with the frequency of the excitation, e.g. pulsation frequency of a compressor. In contrast, structures vibrate *only* at their natural frequencies when excited by free vibrations. Free oscillations occur when the structures are excited once, e.g. by a shock, and then left to themselves.

16. Sinambari, Gh.R., Heß, H., Stadtmüller, M.: Strömungsbedingte Geräuschentwicklung von Ansaugöffnungen. VDI-Berichte Nr. 2229 (2014)
17. Fritz, K.R., Stübner, B.: Schalldämmung und Abstrahlgrad von gasgefüllten Stahlrohren. Fortschritte der Akustik. VDE Verlag, Berlin (1980)
18. Cremer, L., Heckl, M.: Körperschall, 2. Aufl. Springer, Berlin (1996)
19. VDI 3731 Blatt 2: Characteristic noise emission values of technical sound sources; fans11/1990
20. Sinambari, Gh.R.: Einflussparameter bei körperschallbedingter Geräuschentwicklung einer Maschine, 37. Jahrestagung, DAGA 2011
21. Sinambari, Gh.R., Felk, G., Thorn, U.: Konstruktionsakustische Schwachstellenanalyse an einer Verpackungsmaschine. VDI-Berichte Nr. 2052 (2008)
22. Sinambari, Gh.R., Thorn, U., Tschöp, E.: “Konstruktionsakustik”, Seminarveranstaltung. IBS-Seminar, St. Martin/Pfalz, 2016
23. Kraak, W., Weißing, H.: Schallpegelmesstechnik. VEB Verlag, Berlin (1970)
24. Reichard, W.: Grundlagen der Technischen Akustik. Akademische Verlagsgesellschaft, Leipzig (1968)
25. Randall, R.B.: Frequency Analysis. Application of B & K Equipment (1979)
26. Sinambari, Gh.R., Walter, A., Thorn, U.: Entstehung und Vermeidung von Ratterschwingungen an einer Honmaschine. VDI-Berichte 2118 (2010)
27. Perez, P.C.: Untersuchung von Koronageräuschen an Hochspannungsfreileitungen. Diplomarbeit, FH Bingen (2006)

Measuring Technology

As already mentioned, the essential task of design acoustics is to avoid or reduce the noise development of machines responsible for disturbing sound events by suitable measures, preferably constructive ones. The necessary acoustic parameters for this can best be determined by measurements on existing machines or prototypes [1].

In this chapter, the main possibilities for the experimental determination of sound and vibration parameters are given which are necessary for the determination or description of primary and secondary noise reduction measures.

The fundamentals of sound and vibration measurement technology are standardised nationally and internationally. This is intended to ensure, among other things, that measurements taken at different locations by different people can be compared with each other. The following is a list of the currently valid standards that are used in sound and vibration measurement technology and acoustic weak point analysis. Since the standards are constantly updated due to new findings, it is important to check the topicality or validity of the standards when applying them.

4.1 National and International Standardisation

The basic principles, measurement systems and measurement methods for immission and emission measurement technology are regulated in numerous standards and regulations. Various types of standards and designations are compiled below [2].

G. R. Sinambari, *Design Acoustics*,
https://doi.org/10.1007/978-3-658-40183-2_4

National standardisation

DIN	German Institute for Standardisation e. V.
VDE	German Electrotechnical Association
VDI	Society of German Engineers
VDMA	German Engineering Federation (Verband Deutscher Maschinen- und Anlagenbau e. V.)
DKE	German Commission for Electrical Engineering Electronics Information Technology

European standardisation

CEN	Comité Européen de Normalisation (Engl.: European Committee for Standardisation)
CENELEC	Comité Européen de Normalisation Electrotechnique (Engl.: European Committee for Electrotechnical Standardisation)

International standardisation

ISO	International Standardisation Organisation
IEC	International Electrotechnical Committee

4.1.1 Emission Sound Pressure Level at the Workplace

Here, the noise exposure of the people concerned at a workplace is determined, which is caused exclusively by a machine (without the influence of room reflections and any interference levels that may be present). The associated standards are:

ISO 11201 (2010)
Acoustics – Noise emitted by machinery and equipment – Determination of emission sound pressure levels at a work station and at other specified positions in an essentially free field over a reflecting plane with negligible environmental corrections.

ISO 11202 (2010)
Acoustics – Noise emitted by machinery and equipment – Determination of emission sound pressure levels at a work station and at other specified positions applying approximate environmental corrections.

ISO 11203 (1995)
Acoustics – Noise emitted by machinery and equipment – Determination of emission sound pressure levels at a work station and at other specified positions from the sound power level.

ISO 11204 (2010)
Acoustics – Noise emitted by machinery and equipment – Determination of emission sound pressure levels at a work station and at other specified positions applying accurate environmental corrections.

ISO 11205 (2003)
Acoustics – Noise emitted by machinery and equipment – Engineering method for the determination of emission sound pressure levels in situ at the work station and at other specified positions using sound intensity.

4.1.2 Sound Power Level by Measuring the Sound Pressure Level

In design acoustics, the sound power level – as A-total and A-partial sound power level or its spectral distribution – is an essential acoustic parameter, both for the description of the ACTUAL state and for the verification of the achieved noise and sound reduction.

The design acoustic weak point analysis, an important tool in design acoustics, can only be realised by determining the partial and total sound power levels.

Numerous standards and regulations also exist for this purpose. In the following, only the essential main standards which are applied in design acoustics are given. The many machine-specific standards which have been published as subsequent sheets are not dealt with in detail here. However, it is recommended to consider the corresponding special standards and regulations depending on the application.

4.1.2.1 Basic Standards

DIN 45635 (1984-04)
Measurement of noise emitted by machines; airborne noise emission; enveloping surface method; basic method, divided into 3 grades of accuracy.

ISO 3740 (2000)
Acoustics – Determination of sound power levels of noise sources – Guidelines for the use of basic standards

4.1.2.2 Enveloping Surface Method

ISO 3744 (2010)
Acoustics – Determination of sound power levels and sound energy levels of noise sources using sound pressure – Engineering methods for an essentially free field over a reflecting plane.

ISO 3745 (2012+A1 2017)

Acoustics – Determination of sound power levels and sound energy levels of noise sources using sound pressure – Precision methods for anechoic rooms and hemi-anechoic rooms.

ISO 3746 (2010)

Acoustics – Determination of sound power levels and sound energy levels of noise sources using sound pressure – Survey method using an enveloping measurement surface over a reflecting plane.

Figure 4.1 summarises the essential requirements according to the above mentioned standards for the determination of the sound power level according to the enveloping surface method [2].

	Accuracy class 1 Precision method DIN EN ISO 3745	Accuracy class 2 Operating measurement DIN EN ISO 3744	Accuracy class 3 Overview method DIN EN ISO 3746
Measuring environment	anechoic room	outdoors or indoors	
Machine dimensions	Vol. ≤ 0.5% of the room	no restriction	
Noise type	any (broadband, narrowband, stationary, transient, pulsed, tonal)		
Measuring environment	$K_2 \leq 0.5$ dB	$K_2 \leq 4$ dB	$K_2 \leq 7$ dB
extraneous noise	$\Delta L \geq 10$ dB $K_1 \leq 0.4$ dB	$\Delta L \geq 6$ dB $K_1 \leq 1.3$ dB	$\Delta L \geq 3$ dB $K_1 \leq 3$ dB
Measurement area	Sphere / Hemisphere	Hemisphere / cuboid	Hemisphere / cuboid
Number of measuring points	≥ 10	≥ 9	≥ 4
Measurement equipment	Class 1	Class 1	Class 2
Accuracy, standard deviation σ_R	$\sigma_R \leq 1$ dB	$\sigma_R \leq 1.5$ dB	$K_2 \leq 5$ dB: $\sigma_R \leq 3$ dB 5 dB≤ K_2 ≤7 dB: $\sigma_R \leq 4$ dB

Fig. 4.1 Requirements for the determination of the sound power level according to the enveloping surface method

4.1.2.3 Reverberation Chamber Method

ISO 3741 (2010)

Acoustics – Determination of sound power levels and sound energy levels of noise sources using sound pressure – Precision methods for reverberation test rooms.

ISO 3743-1 (2010)

Acoustics – Determination of sound power levels and sound energy levels of noise sources using sound pressure – Engineering methods for small, movable sources in reverberant fields – Part 1: Comparison method for a hard-walled test room.

ISO 3743-2 (1994)

Acoustics – Determination of sound power levels of noise sources using sound pressure – Engineering methods for small, movable sources in reverberant fields – Part 2: Methods for special reverberation test rooms.

ISO 3747 (2010)

Acoustics – Determination of sound power levels and sound energy levels of noise sources using sound pressure – Engineering/survey methods for use in situ in a reverberant environment.

4.1.3 Sound Power Level by Measuring the Sound Intensity Level

ISO 9614-1 (1993)

Acoustics – Determination of sound power levels of noise sources using sound intensity – Part 1: Measurement at discrete points.

ISO 9614-2 (1996)

Acoustics – Determination of sound power levels of noise sources using sound intensity – Part 2: Measurement by scanning.

ISO 9614-3 (2002)

Acoustics – Determination of sound power levels of noise sources using sound intensity – Part 3: Precision method for measurement by scanning.

4.1.4 Sound Power Level by Measurement of the Structure-Borne Sound Level

DIN 45635-8 (1985-06)

Determination of airborne noise emitted by machines by, measurement of structure borne noise; basic requirements.

ISO/TS 7849-1 (2009-03)
Acoustics – Determination of airborne sound power levels emitted by machinery using vibration measurement – Part 1: Survey method using a fixed radiation factor.

ISO/TS 7849-2 (2009-03)
Acoustics – Determination of airborne sound power levels emitted by machinery using vibration measurement – Part 2: Engineering method including determination of the adequate radiation factor

4.2 Sound and Vibration Measurement Techniques

Various measurement methods, e.g. sound pressure, sound intensity or structure-borne sound measurements, can be used to determine the noise emission. The basic measurement techniques for determining the sound pressure level, the sound intensity level and the structure-borne sound level, as given in [5], are reproduced in abbreviated form below.

4.2.1 Sound Pressure Measurement

In acoustic measurement technology, the sound pressure level is usually measured using the so-called condenser microphones. Figure 4.2 shows a typical standard measuring microphone including the corresponding protective cap. The diaphragm is only a few μm thick and therefore mechanically very sensitive. If possible, the protective cap should not be removed. If this must nevertheless be done for special measurement tasks, e.g. in order to screw on a nose cone or a probe attachment, extreme care must be taken during conversion!

Condenser microphones are very sensitive and typically have omnidirectional characteristics up to high frequencies. The usable frequency range is limited downward by electrical and mechanical high-pass filtering (electrically by the series resistor, mechanically by capillary holes in the housing to compensate for the static air pressure in front of and behind the diaphragm). Upward, the usable frequency range is limited by the diaphragm's natural frequency. Each microphone has an individual frequency response, supplied by the manufacturers in the form of a calibration certificate, from which the usable frequency range of the microphone can be determined. Figure 4.3 shows the calibration certificate of a $\frac{1}{2}''$ condenser microphone.

The polarisation voltage can be dispensed with if a dielectric with permanent polarisation, a so-called electret foil, is inserted between the diaphragm and the counter-electrode. In this way, a permanently polarised measurement microphone (electret microphone) is obtained. Electret microphones can be manufactured inexpensively. Miniature

Fig. 4.2 Standard measuring microphone $\frac{1}{2}''$ Condenser microphone [3]

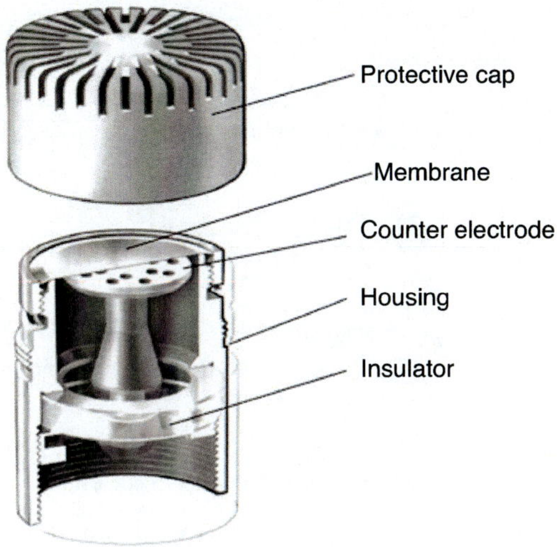

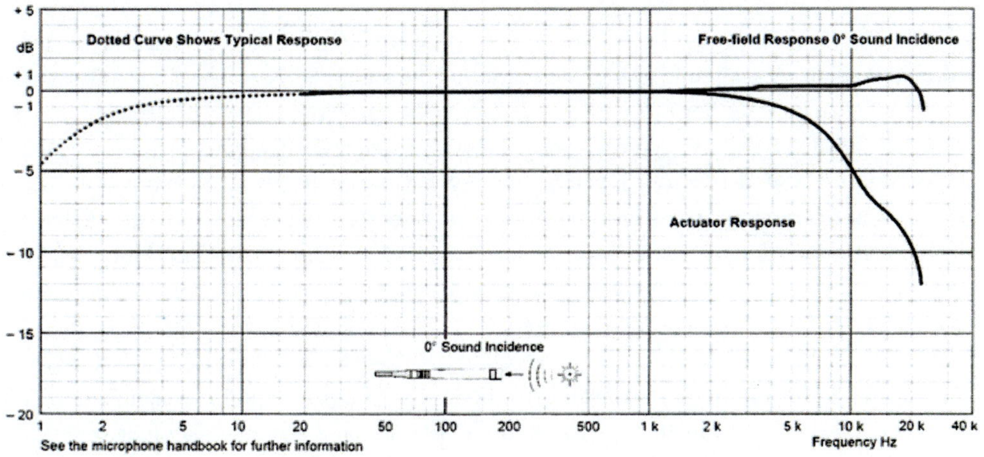

Fig. 4.3 Calibration certificate of a $\frac{1}{2}''$-condenser microphone; manufacturer: Brüel & Kjaer

microphones with dimensions of a few millimeters can also be produced in this way. However, compromises must then be made with regard to sensitivity and accuracy.

The signal voltages of the microphone capsule generated by sound pressure fluctuations are amplified by a preamplifier and fed for further processing to a sound level meter or analyser, which determines and displays the "sum level" or performs a frequency-dependent analysis and outputs a "spectrum". Figure 4.4 shows a schematic diagram of how a sound level meter works.

The pre-amplified voltage signal of the microphone, which is proportional to the sound pressure, is first frequency-weighted in the sound level meter. Depending on the type of sound measurement, A, B or C weighting is selected. If no frequency weighting is required,

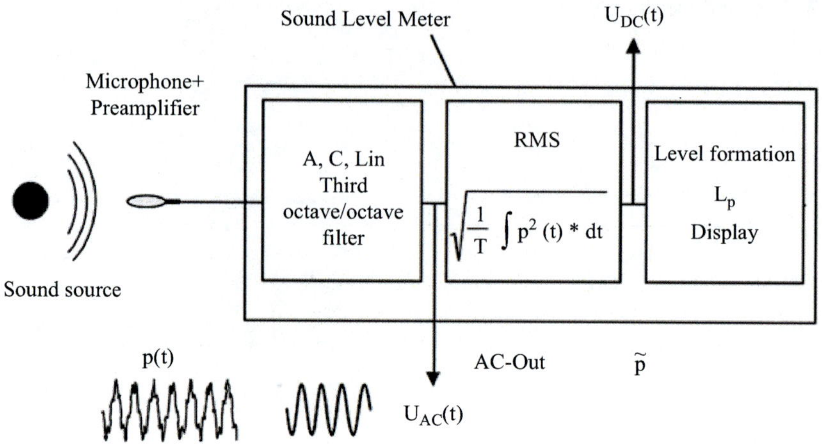

Fig. 4.4 Functional diagram of a sound level meter

LINEAR, FLAT or Z is set as the frequency weighting.[1] The frequency-weighted or filtered signals are available at the AC output of the sound level meter as an AC electrical voltage $U_{AC}(t)$ for further processing and/or recording. Older sound level meters can use external bandpass filters, typically third-octave or octave bandpass filters, to serially determine the sound pressure level in specific frequency bands. However, if the sound event is not stationary, the bandpass filters must operate in parallel and simultaneously, i.e. in real time. Modern sound level meters today have internal, parallel digital bandpass filters, which make them so-called real-time frequency analysers. From the frequency-weighted and/or filtered microphone signal, effective values, see (2.21), are formed in a next step in the sound level meter (also called RMS values). Depending on the type of sound measurement, the time constant for the RMS value formation is selected as 125 ms (FAST) or 1 s (SLOW) in accordance with the international standard. In older sound level meters, a DC voltage signal U_{DC} (t) proportional to the rms value can be tapped at the DC output, e.g. to feed a level recorder. Finally, with the obtained rms values, the corresponding sound pressure levels are formed and displayed in a further step. Furthermore, the corresponding average values such as the "equivalent continuous sound pressure level" L_{eq} and the "averaged maximum sound pressure level for consecutive intervals" L_{Teq}, maximum and peak levels L_{max} resp. L_{peak}, and level percentiles, e.g., the L_{95}, are formed and displayed. Depending on the settings made on the sound level meter, attention must be paid to an exact designation in the form of indices on the sound pressure level, e.g. L_{AFAeq}, L_{Cpeak}, L_{AFmax}. Figure 4.5 shows two commercially available precision handheld sound level meters.

[1] Depending on the manufacturer of the sound level meter, different designations are common here.

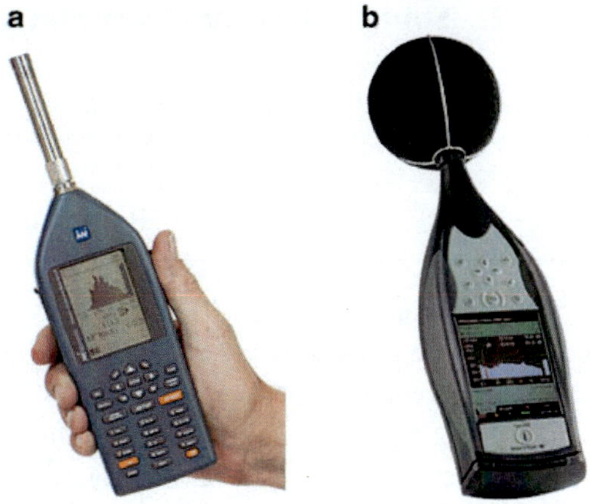

Fig. 4.5 Precision hand-held sound level meter; (**a**) Manufacturer: Norsonic-Tippkemper, Type 140; (**b**) Manufacturer: Brüel & Kjaer, Type 2250

For standard-compliant sound pressure level measurements (measurement of absolute sound pressure levels), sound level meters together with the associated measuring microphone must be calibrated regularly, e.g. by calibration offices. This includes checking the transmission characteristics over the entire frequency range. In order to check the correct functioning of the entire measurement chain during on-site measurements, the measurement chain is checked (calibrated) before and after the measurement using an airborne sound calibrator (Fig. 4.6) at a specified frequency (usually 1000 Hz) with a defined level, e.g. 94 dB or 114 dB.

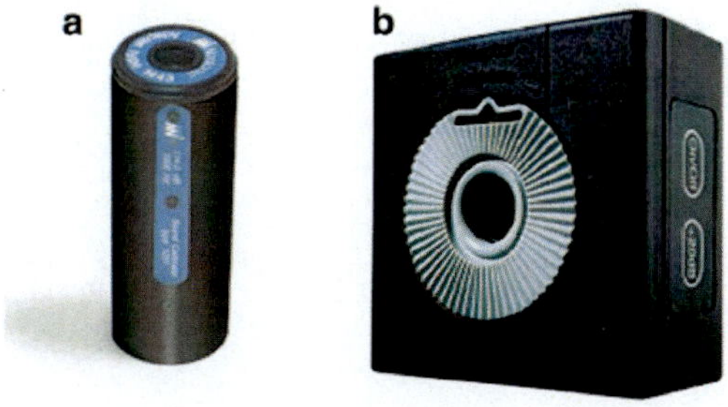

Fig. 4.6 Airborne sound calibrator; (**a**) Manufacturer: Norsonic-Tippkemper, Type 1251; (**b**) Manufacturer: Brüel & Kjaer, Type 4231

4.2.2 Sound Intensity Measurement

As described in Sect. 2.2.2, sound intensity is the time-averaged product of sound pressure and sound velocity:

$$\boldsymbol{I}(r) = p(r) \cdot \boldsymbol{v}(r) \tag{4.1}$$

Sound pressure can be measured relatively easily with a condenser microphone. Usually, in practice, the sound velocity is measured with the help of a so-called sound intensity probe, where two condenser microphones A and B are placed at a certain distance (Δr in practice usually 12 mm or 50 mm) from each other [4].

Theoretically, the sound velocity can be determined from the sound pressure as follows [5]:

$$\frac{\partial v}{\partial t} = -\frac{1}{\rho}\frac{\partial p}{\partial r} \tag{4.2}$$

respectively

$$\boldsymbol{v}(r) = -\frac{1}{\rho} \cdot \int \frac{\mathrm{d}p(r)}{\mathrm{d}r} \cdot \mathrm{d}t \tag{4.3}$$

If the two microphones of the sound intensity probe (Fig. 4.7) are arranged close to each other, the pressure gradient $\frac{\mathrm{d}p(r)}{\mathrm{d}r}$ can be replaced by $\Delta p/\Delta r$ in the first approximation.

With

$$\Delta p = p_{\mathrm{B}} - p_{\mathrm{A}}$$

it follows from (4.3):

$$\boldsymbol{v}(r) = -\frac{1}{\rho}\int \frac{\mathrm{d}p(r)}{\mathrm{d}r} \cdot \mathrm{d}t \approx -\frac{1}{\rho \cdot \Delta r}\int (p_{\mathrm{B}} - p_{\mathrm{A}}) \cdot \mathrm{d}t \tag{4.4}$$

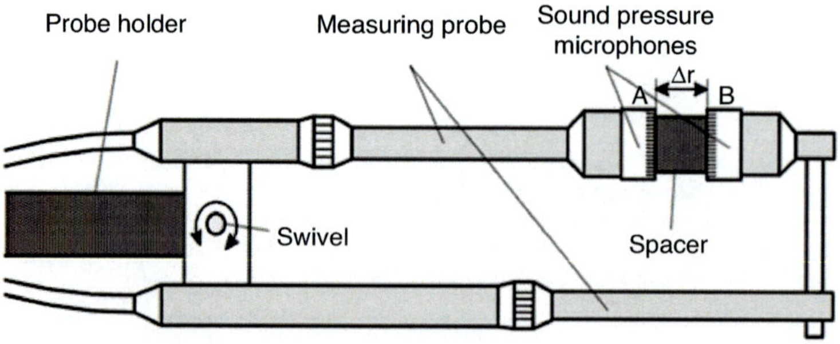

Fig. 4.7 Sound intensity probe [4]

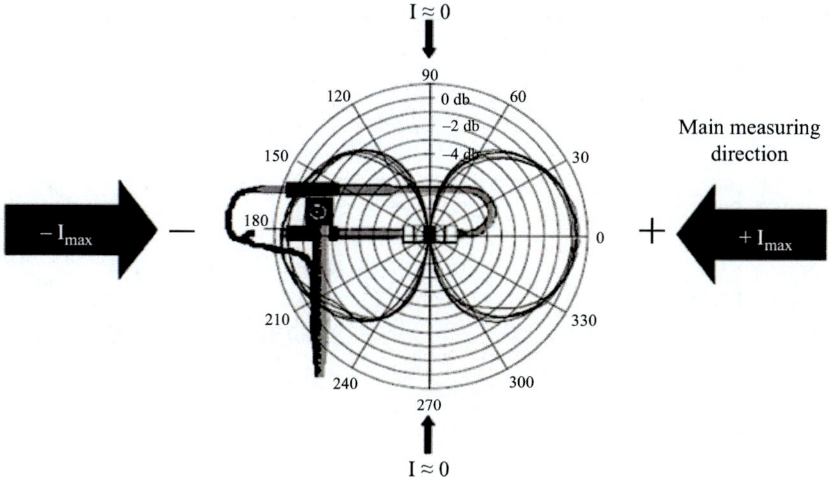

Fig. 4.8 Typical directional characteristic of an intensity probe [4]

The sound pressure in (4.1) can be determined from the mean value of the two sound pressures of microphones A and B:

$$p(r) = \frac{p_{\mathrm{A}} + p_{\mathrm{B}}}{2} \tag{4.5}$$

Using (4.4) and (4.5), the sound intensity according to (4.1) can be determined as follows:

$$\boldsymbol{I} = -\frac{p_{\mathrm{A}} + p_{\mathrm{B}}}{2 \cdot \rho \cdot \Delta r} \int (p_{\mathrm{B}} - p_{\mathrm{A}}) \cdot \mathrm{d}t \tag{4.6}$$

Equation (4.6) forms the basis of the sound intensity measurement technique by sound pressure measurements according to the so-called "two-microphone method". The calculated value of the sound intensity applies to the acoustic centre of the sound intensity probe, i.e. the centre between the two microphones. The mean sound pressure, which is calculated from the sound pressures of the two microphones, also refers to this point.

In practice, the microphone distances are realised by appropriate spacers. To determine the sound intensity, the sound pressures of the two microphones are measured and fed to a sound intensity analyser for further processing. The sound intensity analyser must have two input channels so that it can measure and process both microphone signals simultaneously and in phase. The sound intensity is calculated from the two microphone signals according to (4.6).

The sound intensity is a vectorial quantity and therefore directional. The sound intensity probes have a distinctive directional characteristic, see Fig. 4.8.

As can be seen from Fig. 4.8, a positive reading is obtained when sound is incident from the main direction of measurement (from the front) and a negative reading is obtained when sound is incident from behind. If the sound is incident perpendicular to the main measurement

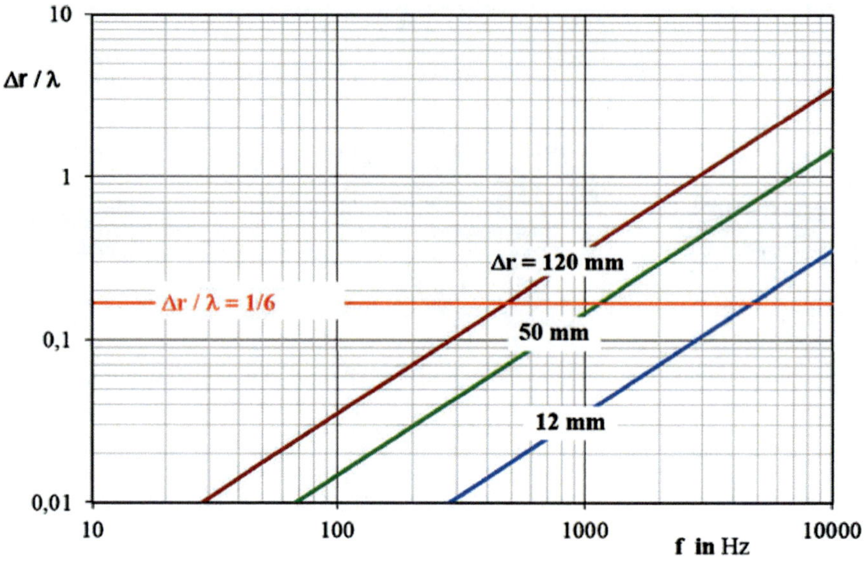

Fig. 4.9 $\Delta r/\lambda$ as a function of frequency for different microphone distances

direction, the sound intensity is approximately zero. This makes it possible to determine not only the amount of sound intensity $|\boldsymbol{I}|$, but also the direction of sound incidence. To determine the sound velocity or sound intensity, the microphone signals $p_A(t)$ and $p_B(t)$ of the two microphones must be measured simultaneously and evaluated in phase. The quality and accuracy of the sound intensity measurement depends, among other things, essentially on the phase-accurate measurement of the two sound pressures. For this reason, only special microphone pairs are used for sound intensity measurements, which are very precisely matched to each other with regard to their phase position. The same high requirements with regard to the so-called phase mismatch are also placed on the downstream analyser. The phase difference $\Delta\varphi$ of the two sound pressure signals can be determined as follows [5]:

$$\Delta\phi = 360\,° \cdot \frac{\Delta r}{\lambda} = 360\,° \cdot \frac{\Delta r \cdot f}{c} \tag{4.7}$$

In Fig. 4.9, sound propagation in air ($c = 340$ m/s),$\Delta r/\lambda$ as a function of frequency is shown graphically for three microphone distances commonly used in practice, 12, 50 and 120 mm, according to (4.7).

In Fig. 4.10, the sound propagation in the air ($c = 340$ m/s),$\Delta\varphi$ according to (4.7) is shown graphically as a function of frequency for three microphone distances commonly used in practice: 12, 50 and 120 mm.

Due to the analysis system, a small delay occurs between the two microphone channels A and B during the measurement, which causes a small phase change, the so-called phase mismatch. Depending on the microphone distance Δr used, a lower and an upper frequency limit within which sound intensity measurements are possible can be defined for the sound intensity measurement system.

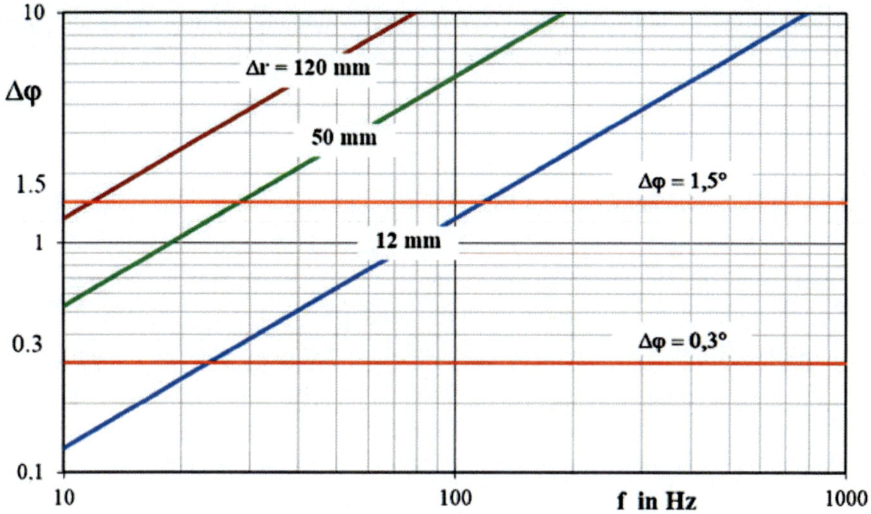

Fig. 4.10 As a $\Delta\varphi$ function of frequency for different microphone distances

Upper Frequency Limit: Error Due to Linear Pressure Gradient Approximation

At higher frequencies, an error occurs due to the simplification used with linear pressure gradient approximation, since the pressure gradient $\partial p/\partial r$ is only approximately described by the difference quotient $\Delta p/\Delta r$ from the pressure difference and the distance between the two microphones of the intensity probe. Figure 4.11 illustrates the problem.

If the wavelength λ is small compared to the microphone distance Δr, the pressure gradient is approximated very inaccurately. For a permissible deviation of ≤ 1 dB, the

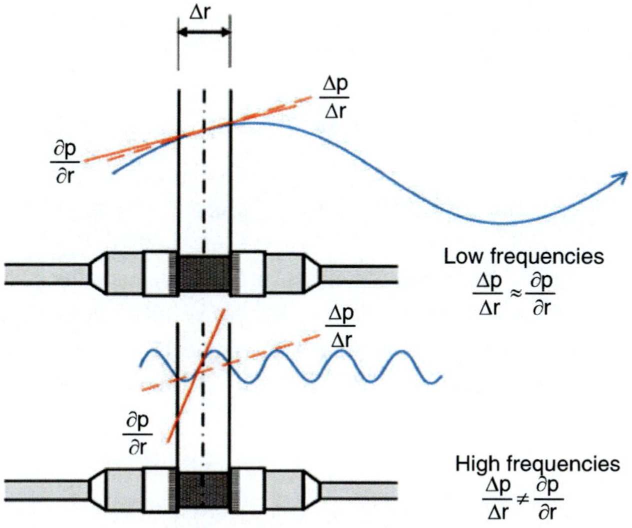

Fig. 4.11 Error due to linear pressure gradient approximation [6]

wavelength λ must be at least a factor of 6 larger than the microphone distance Δr, i.e.: $\Delta r/\lambda \leq 1/6$. From Fig. 4.9, the following upper frequency limits can be determined for the calculated microphone distances:

$$\Delta r = 120\ \text{mm} \quad f_o \approx 470\ \text{Hz} \quad f_{\text{o,third-oct}}: \quad \text{until} \quad 500\ \text{Hz}$$
$$\Delta r = 50\ \text{mm} \quad f_o \approx 1135\ \text{Hz} \quad f_{\text{o,third-oct}}: \quad \text{until} \quad 1.25\ \text{kHz}$$
$$\Delta r = 12\ \text{mm} \quad f_o \approx 4720\ \text{Hz} \quad f_{\text{o,third-oct}}: \quad \text{until} \quad 5\ \text{kHz}$$

If the wavelength λ is large compared to the microphone distance Δr, the linear approximation captures the slope of the sound pressure curve with good accuracy and the pressure gradient is well approximated. At low frequencies, therefore, the basic equation of the linear pressure gradient approximation is satisfied: $\partial p/\partial r \approx \Delta p/\Delta r$.

Therefore, a small spacer is required for measuring high frequencies.

Lower Frequency Limit: Error Due to Phase Mismatch

If the phase difference $\Delta\varphi$ to be measured is greater than the total phase mismatch (probe and analyser) of the sound intensity measurement system, the intensity can be reasonably calculated. For sound intensity analysers used in practice, the total phase mismatch is approximately $\leq \pm 0.3°$. It can be seen from Fig. 4.10 that the lower frequency limit of a sound intensity measurement system is determined by the smallest microphone spacing. This means that for a microphone spacing of $\Delta r = 12$ mm and a total phase mismatch of only 0.3° above $f = 24$ Hz, one obtains a useful signal that is larger than the noise of the measurement chain. To allow for a sufficient signal-to-noise ratio and thus keep the measurement error smaller than 1 dB, the phase difference $\Delta\varphi$ to be measured should be at least a factor of 5 larger than the total phase mismatch, i.e.: $\Delta\varphi \geq 1.5°$. From Fig. 4.10, the following lower frequency limits *valid under free-field conditions* can thus be read off for the calculated microphone distances:

$$\Delta r = 120\ \text{mm} \quad f_u \approx 12\ \text{Hz} \quad f_{\text{u,third-oct}}: \quad \text{above} \quad 25\ \text{Hz}$$
$$\Delta r = 50\ \text{mm} \quad f_u \approx 29\ \text{Hz} \quad f_{\text{u,third-oct}}: \quad \text{above} \quad 31.5\ \text{Hz}$$
$$\Delta r = 12\ \text{mm} \quad f_u \approx 120\ \text{Hz} \quad f_{\text{u,third-oct}}: \quad \text{above} \quad 125\ \text{Hz}$$

Therefore, a large spacer is required for measuring low frequencies.

If the sound hits the probe at an angle θ, the phase difference $\Delta\varphi$ to be measured is reduced, since not the full microphone distance Δr but only its projection $\Delta r \cdot \cos(\theta)$ is effective (see Fig. 4.12). Therefore, a sound intensity is measured that is lower by the factor $\cos(\theta)$. Since the sound pressure, in contrast to the sound velocity, is a scalar quantity, a difference between the sound pressure and intensity level arises depending on the angle of incidence.

At oblique sound incidence or in a reactive (near field) or diffuse sound field, the phase difference $\Delta\varphi$ to be measured is therefore reduced. As a result, the phase mismatch is more

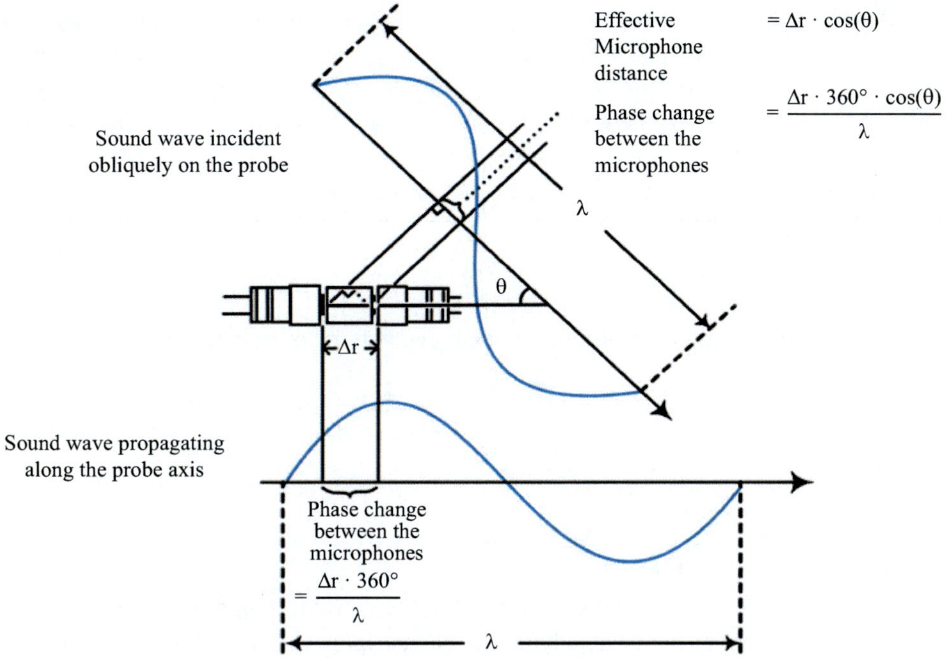

Fig. 4.12 Phase difference at oblique sound incidence [6]

significant there. If the sound does not propagate in a free field along the probe axis, this will affect the lower frequency limit. The measurable frequency range is reduced because the lower frequency limit shifts towards higher frequencies.

In a reactive or diffuse sound field, the difference $L_p - L_I$ between the sound pressure level and the sound intensity level is referred to in this context as so-called "pressure intensity index" F_{pI}. The pressure intensity index characterises the "reactivity" of a sound field and is very important for assessing the accuracy of a sound intensity measurement. It describes the phase change between the two microphone signals caused by the reactivity of the sound field. The pressure intensity index thus allows conclusions to be drawn about the phase difference between the microphone signals and thus makes it possible to assess whether the phase mismatch causes an inaccurate measurement in relation to the determined phase difference.

The phase mismatch of the entire sound intensity measurement system can also be quantified. It is described by the so-called "cross-field suppression" δ_{pIo}, also referred to as "pressure-residual intensity distance" in current standardisation [7]. If the same sound signal is incident on both microphones perpendicular to the probe axis, no sound intensity should ideally be measurable. However, the unavoidable phase mismatch causes an "apparent intensity" which is comparable to an inherent noise of the measuring system. The difference between the sound pressure level detected here and the (apparent) intensity level is defined as the cross-field suppression δ_{pIo}.

As described earlier, the phase difference $\Delta\varphi$ to be measured should be at least a factor of 5 larger than the total phase mismatch if the sound intensity level is to be determined with an accuracy of 1 dB. This is consistent with the requirement that the pressure intensity index F_{pl} must be at least 7 dB above the cross-field suppression δ_{plo}. Subtracting 7 dB from the cross-field rejection gives the usable device dynamic range for this accuracy. In current standardisation, the device dynamic range is referred to as the working range L_d [7]:

$$L_d = \delta_{plo} - K \quad \text{dB} \tag{4.8}$$

K describes the systematic error measure and, depending on the desired accuracy, is given by

$K = 7$ dB	Accuracy: ±1.0 dB
$K = 10$ dB	Accuracy: ±0.5 dB

The cross-field suppression δ_{plo} is determined with a special coupler using a sound intensity calibrator (see Fig. 4.13).

If the cross-field suppression δ_{plo} is known, the working range of the sound intensity measurement system or the device dynamics can be specified for each spacer as a function of frequency. Figure 4.14 shows the usable frequency range at a required measurement accuracy of 1 dB ($K = 7$ dB) as a function of pressure intensity index F_{pl}, microphone spacing Δr and total phase mismatch.

As a rule, in many practical applications the distance piece $\Delta r = 12$ mm is used for measurement. With a total phase mismatch of the sound intensity measuring system of, ≤0.3 depending on the pressure intensity index F_{pl} (reactivity of the sound field), with a desired accuracy of ±1.0 dB, a usable frequency range of:

$$F_{pl} = 0 \text{ dB} \qquad \text{(open field conditions)} \qquad 125\text{Hz} \le f_{\text{third}-\text{oct}} < 5 \text{ kHz}$$

$$F_{pl} = 10 \text{ dB} \qquad \text{(reactive sound field)} \qquad 1.25\text{kHz} \le f_{\text{third}-\text{oct}} < 5 \text{ kHz}$$

can be achieved. In reactive sound fields, a larger spacer must therefore be used for measuring low frequencies!

Similar to sound pressure measurements, the microphones of the sound intensity probe must also be calibrated before each measurement. In addition, the cross-field suppression must be checked at regular intervals. With the aid of an airborne sound calibrator, both microphones can be checked successively or, with the aid of an appropriate coupler, simultaneously for compliance with an appropriate sound pressure calibration level at a specified frequency, but the phase position of both microphone signals, which is essential

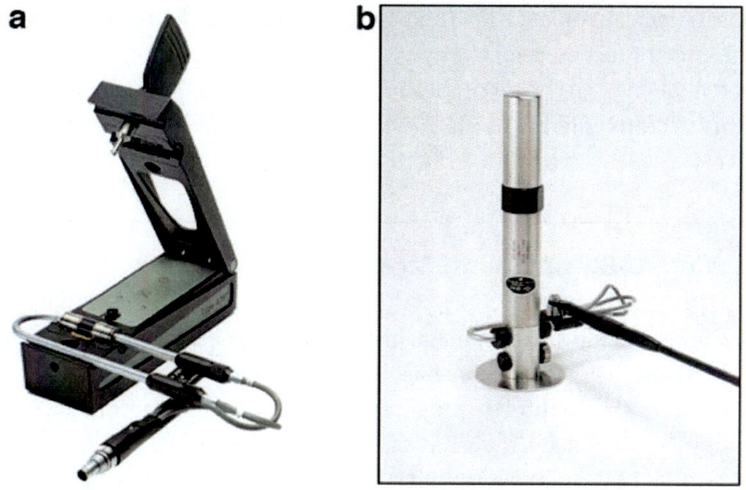

Fig. 4.13 Sound intensity calibrators, (**a**) Manufacturer: Brüel & Kjaer, Type 4297; (**b**) Manufacturer: Brüel & Kjaer, Type 3516

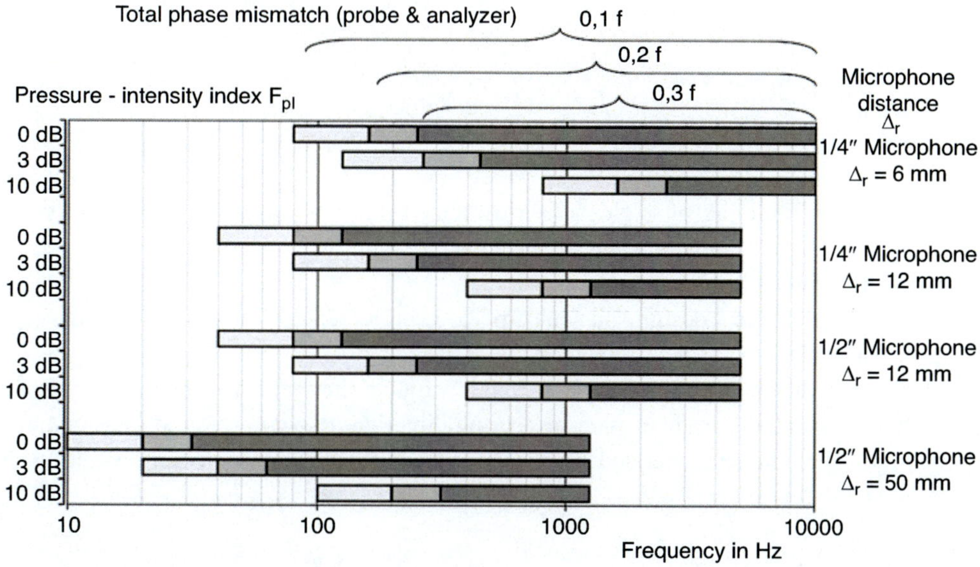

Fig. 4.14 Usable frequency range at a required measurement accuracy of 1 dB as a function of pressure intensity index, microphone spacing and phase mismatch. (Modified after [6])

for sound intensity measurements, can only be checked with appropriate sound intensity calibrators.

Even more critical than with sound pressure measurements are flow-induced pressure fluctuations, e.g. as a result of wind or cooling air flows, at the probe microphones during sound intensity measurements. If currents occur in the measurement area, a corresponding probe wind screen must be used.

Since the frequency spectrum of the sound source (e.g. a fan), which is caused by wind or flow noise, drops sharply with increasing frequency, the low-frequency intensity measurements (usually below 200 Hz) are generally most affected by this.

4.2.3 Structure-Borne Sound Measurement

In structure-borne sound measurement, the vibration accelerations of the structures are primarily measured. This is usually done by so-called piezoelectric accelerometers. They generate an electric charge directly proportional to the vibration acceleration a of the object on which they are mounted. Their advantageous properties are a relatively small size, a wide frequency and dynamic range, and a long, maintenance-free service life. In addition, signals proportional to velocity v and displacement s can be derived from their acceleration-proportional output signal in a relatively simple manner by electrical integration:

$$v = \int a(t) \cdot \mathrm{d}t \rightarrow v = \frac{a}{\omega} = \frac{a}{2\pi f} \tag{4.9}$$

$$s = \int v(t) \cdot \mathrm{d}t \rightarrow s = \frac{a}{\omega^2} = \frac{a}{(2\pi f)^2} \tag{4.10}$$

The operating principle of a piezoelectric accelerometer is based on the property of a piezoelectric material which, when subjected to external forces, generates a charge proportional to the applied forces (see Fig. 4.15).

Depending on their design, piezoelectric accelerometers are distinguished between a compression type and a shear type (see Fig. 4.16) [8].

If the transducers are constructed in such a way that they react to tensile and compressive forces essentially in their main measuring direction, they are referred to as compression type. The charges generated by shear forces in this type, which can be caused, for

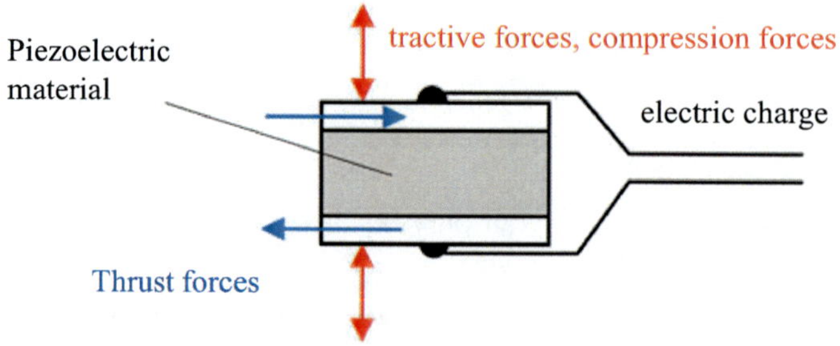

Fig. 4.15 Operating principle of a piezoelectric accelerometer

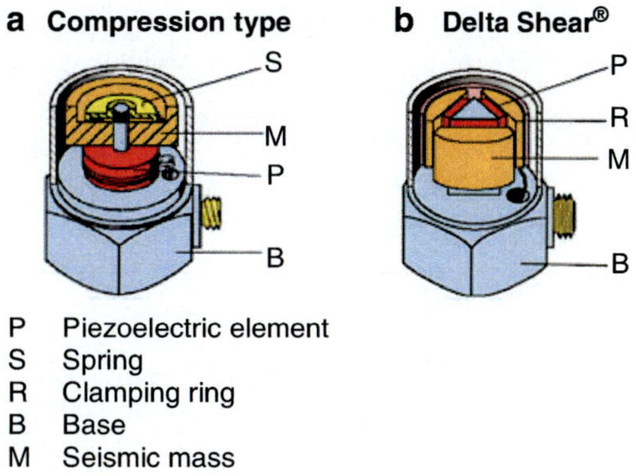

Fig. 4.16 Section through two accelerometer types (B&K)

example, by horizontally acting accelerations, are much lower compared to the charges generated in the main measuring direction, in this case perpendicular to the surface of the structure. In the shear type, the vibrating mass in the main measurement direction causes a shear force on the piezoelectric element. With this type, the charges generated by tensile and compressive forces, which can be caused, for example, by horizontally acting accelerations, are correspondingly negligible compared to the charges generated in the main direction of measurement.

Figure 4.17 shows a typical frequency response of a piezoelectric accelerometer as a function of the frequency ratio (vibration frequency to natural frequency) of the accelerometer. A distinction is made here between the main sensitivity in the direction of the mounting axis (main measurement direction) and the sensitivity transverse to the main measurement direction, the so-called cross-sensitivity. From this it can be seen that the sensitivity in the main measuring direction is significantly higher than in the transverse direction. The transmission behavior is clearly influenced by the resonant frequency of the spring-mass system of the accelerometer. Measurements in the frequency range of the transducer's natural frequency do not provide a true picture of the actual vibrations at the measuring point, but falsify the measurement result. For vibration measurements, it must therefore be borne in mind that the usable frequency range of an accelerometer ends at approx. 1/3 of the resonant or natural frequency of the accelerometer. If the upper frequency limit is set equal to 1/3 of the resonant frequency of the transducer, it can be assumed that the vibration components measured at the edge of the frequency limit were recorded with a measurement error of less than approx. +12% (accuracy <1 dB) [8]. The resonant frequency of the transducers can usually be taken from the calibration certificates of the manufacturers.

The charge signal generated by the accelerometer is fed via a special low-noise charge cable to a so-called charge amplifier, which outputs and amplifies a voltage proportional to the electrical charge at the input. Figure 4.18 shows an accelerometer and an integrating charge amplifier.

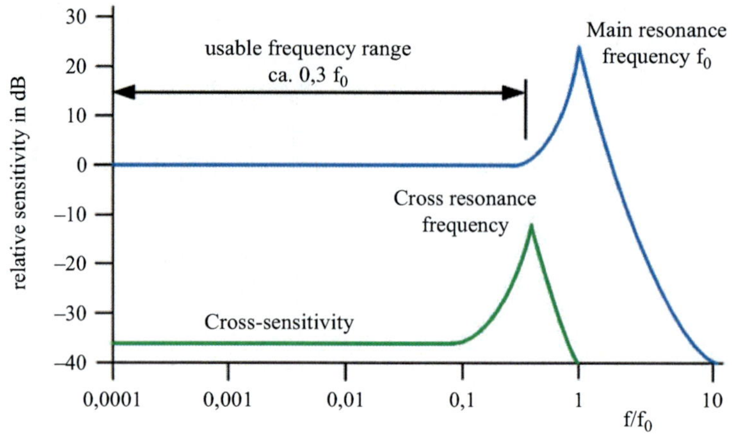

Fig. 4.17 Typical frequency response of a piezoelectric accelerometer

Fig. 4.18 Piezoelectric accelerometer, low-noise charge cable and integrating charge amplifier [9]

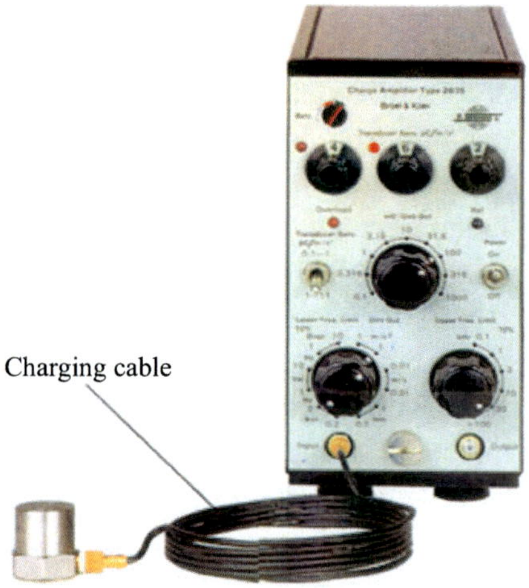

If the acceleration-proportional signal is to be converted into a velocity-proportional signal, the acceleration signal must be integrated. The integration is either carried out in analogue form directly in the charge amplifier or can subsequently be carried out by calculation in the computer or analyser. To obtain a displacement-proportional signal, the acceleration signal must be integrated twice accordingly.

In order to keep interfering signal components away from the downstream analysis unit, e.g. in order to be able to make the best possible use of the dynamic range of the analysis unit or the recording device or to mask out interfering signal components outside the linear transducer frequency range, the vibration signal can be high-pass and low-pass filtered. In the analysis unit, the signal is bandpass-filtered, e.g. via third-octave or octave-band filters, or fed to the detection unit as a broadband signal. Depending on the measurement task, the effective value, the peak value or the peak-to-peak value are formed or determined there and displayed.

Since this transducer system requires one charge amplifier per accelerometer, multi-channel measurements in particular often become a complex and confusing measurement setup. For this reason, in addition to the conventional charge types, accelerometers with integrated charge amplifiers, which are supplied by means of a constant direct current via the signal line, are also common, especially in automotive and test bench acoustics. Depending on the manufacturer, they are referred to as ICP® or DeltaTron® transducers, for example. These transducers can be connected via relatively long (up to several 100 m) cables directly to recording or analysis units with a corresponding built-in power supply. The disadvantage, however, is that these transducers can only be used up to a temperature range of approx. 125 °C, as their operating temperature is limited by the integrated electronics. With conventional charge types, the application range is usually around 250 °C [9].

In order to record the structural vibrations, the accelerometers must be connected or fastened to the structure in a non-positive way as ideal as possible. The ideal type of coupling is the use of a threaded bolt with a flat and smooth support to which a thin layer of grease is applied. However, since coupling via a threaded pin requires the insertion of a threaded hole in the measured object, this type of fastening is often not possible in practice. However, there are several alternative coupling options that are applied depending on the frequency range of interest. It should be noted here that the accelerometer coupling represents another spring-mass system that shifts the natural frequency of the overall arrangement towards lower frequencies. The natural frequency of this system is very strongly dependent on the stiffness of the coupling and is the lower, the softer the coupling is.

Figure 4.19 shows some of the fastening types that are most frequently used in practice. The natural frequencies given therein are only intended for orientation. They represent the relative displacement of the natural frequency of the overall arrangement compared to the natural frequency of a universal accelerometer (natural frequency: approx. 32 kHz) mounted "ideally" smooth and level under calibration conditions. However, the seismic transducer in the lower right corner of the graph has a natural frequency of only 4.5 kHz under ideal coupling conditions. It can be seen that the natural frequency is only slightly lower than the optimum value when using adhesive wax, e.g. beeswax. However, this method is limited to a temperature of approx. 40°C. The natural frequency is already significantly reduced when fastening by means of a permanent magnet. The lowest natural frequency is achieved when using a probe.

It is recalled that the upper useful frequency range also ends here again already at approx. 1/3 of the reduced natural frequency. The mounting method must therefore be selected according to the measurement task.

Where measurements are to be limited to low frequencies up to about 1 kHz, but at the same time very large high-frequency accelerations are present, the influences of high-frequency vibrations can be eliminated with the aid of a mechanical filter. Applications

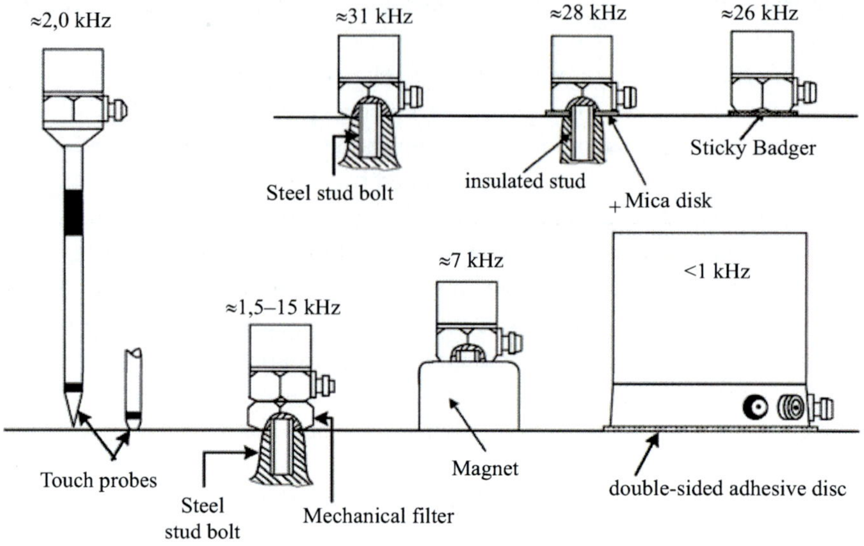

Fig. 4.19 Different mounting types of accelerometers [10]

include measurements for chassis tuning or on wheelset bearings of rail vehicles. The filters consist of an elastic material, usually rubber, bonded between two discs. As screw bases, they are inserted between the structure and the transducer. Their task is to reduce the passage of disturbing high-frequency vibrations in order to prevent overdriving of the transducer and the subsequent electronics [9].

A large number of accelerometers of different sensitivities for very different applications with different masses are offered on the market. The range extends from miniature transducers with weights <1 g to seismic transducers of approx. 500 g. Basically, the resonant frequency of the transducer decreases with its mass. The lighter a transducer is, the higher its resonant frequency. Therefore, if high frequencies are to be measured, small transducers must be used. On the other hand, smaller transducers have a lower sensitivity, since a high sensitivity usually requires a correspondingly large piezoelectric system, i.e. a large, heavy transducer. This means that the transducers must always be used in an application-oriented manner. One must therefore select the appropriate accelerometer depending on the frequency spectrum and vibration amplitude and the type of structure on which the vibrations are to be measured.

Depending on their design, accelerometers react differently to environmental influences, such as temperature, humidity, radioactive radiation, magnetic fields, etc. Modern accelerometers and accelerometer cables are designed in such a way that they react relatively insensitively to the above-mentioned environmental influences. For special applications, e.g. vibration measurements at high temperatures, there are special accelerometers available with and without external cooling. If, for example, the accelerometer is to be mounted on a surface with temperatures >250 °C, a cooling plate and a

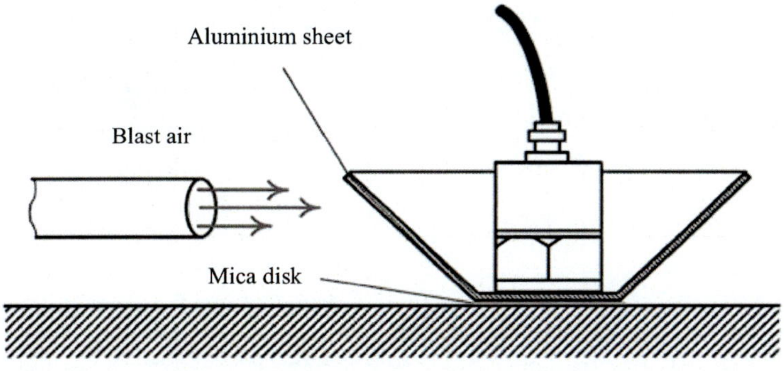

Fig. 4.20 Air cooling of an accelerometer [10]

mica disc can be placed between the transducer base and the structure. This allows the temperature of the transducer base to be maintained at 250°C for surface temperatures of 350–400 °C. Heat dissipation can be further aided by the supply of cooling air (see Fig. 4.20).

Another method with a water-cooled transducer base (forced cooling) is shown in Fig. 4.21. This allows vibration measurements to be made at surface temperatures of up to 600 °C.

So-called ground loops are a frequent interference signal in vibration measurements. Ground loops occur when the accelerometer and the measuring instruments are individually grounded. If a measurement chain is earthed at more than one point due to the contact of the accelerometer with the machine under investigation and, for example, the operation of the analyser on the AC mains, earth loop currents can flow off via the shielding sheath of the transducer cable. Figure 4.22 schematically shows the formation of earth loops.

If ground loops occur during a vibration measurement, the measurement signals are distorted by superimposing an interference signal with the frequency of the electrical mains frequency (e.g. 50 Hz in Europe, 60 Hz in the USA, German railway current: $16\frac{2}{3}$ Hz). In order to avoid earth loops, care must always be taken to ensure that the measurement chain is only earthed at one point during vibration measurements. For this purpose, the accelerometers can be electrically insulated from the structure, e.g. by means of an insulating grub screw and a mica disk, or the measuring devices used can be completely battery-powered.

To determine the structural vibrations, the vibration acceleration is measured at grid measurement points distributed as uniformly as possible over the area of interest (see Fig. 4.23). The number of grid measurement points required depends on the desired accuracy of the measurement result for the mean fast level. For an accuracy of <1 dB, the following number of grid measurement points can be assumed as a function of the structure area *S:*

From the acceleration measured at each grid measuring point *j*, the corresponding velocity level can be determined as follows:

Fig. 4.21 Triaxial accelerometer with water-cooled transducer base; manufacturer: Brüel & Kjaer, Type 4326 with Cooling Unit UA 3014

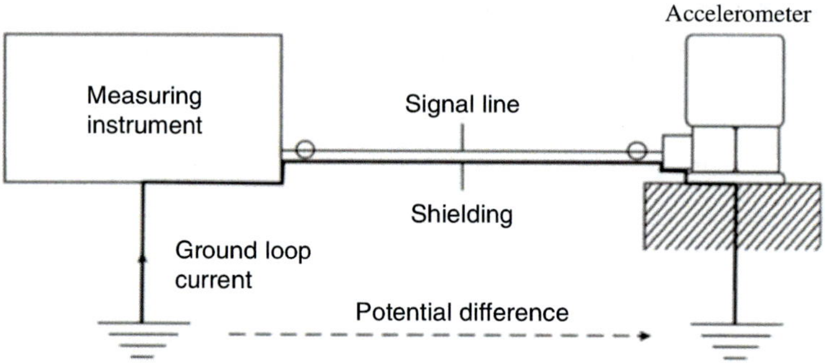

Fig. 4.22 Schematic representation of the formation of ground loops [9]

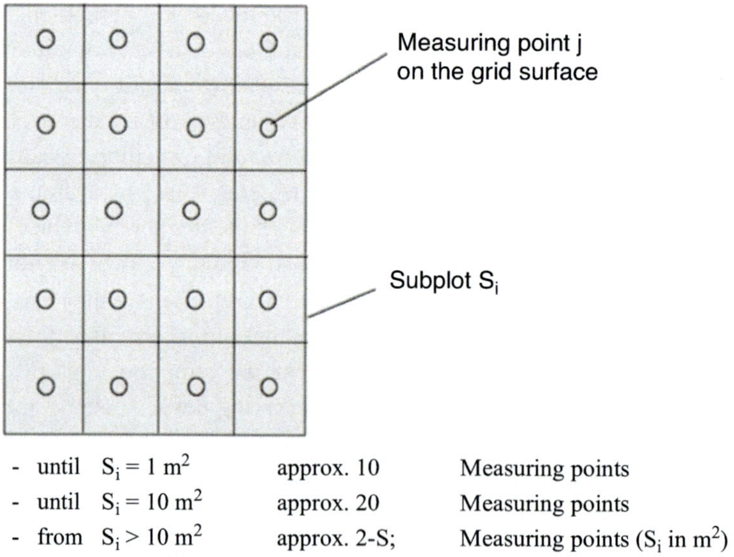

Fig. 4.23 Schematic arrangement of the measuring points for determining the average fast level

$$L_{v,j} = L_{a,j} + 20\lg\frac{a_0}{v_0 \cdot \omega} \quad (4.11)$$

with

$$v_0 = 5 \cdot 10^{-8}\,\mathrm{m/s}$$
$$a_0 = \text{z.B.}\quad 10^{-6}\,\mathrm{m/s^2} \quad (\text{since “freely selectable”})$$
$$\omega = 2\pi f$$

By energetically averaging the velocity levels of all grid measurement points, the average velocity level $\overline{L_{v,i}}$ of the subarea S_i *is* obtained:

$$\overline{L_{v,i}} = 10 \cdot \lg\left[\frac{1}{n}\sum_{j=1}^{n} 10^{0.1 \cdot L_{v,j}}\right] \quad (4.12)$$

The number of measurement points shall be doubled until the average velocity level $\overline{L_{v,i}}$ remains constant within a range of 1 dB.

In the following, the sources of error in the determination of structural vibrations as a result of transducer feedback and spurious vibrations are given.

(a) **Influence of the mass of the accelerometer**

Assuming that the impedance of the accelerometer can be considered as a compact mass, the influence of the mass m_A of the accelerometer, can be given by the correction level K_m, in case of mounting on plate-like structures, which often occurs in practice, as follows [5]:

$$K_m \approx 20 \cdot \log\left(\sqrt{1 + \left(\frac{m_A}{m_{b,Pl}}\right)^2}\right) \quad \mathrm{dB} \quad (4.13)$$

with

m_A	Mass of the transducer in kg = dyn. mass of the transducer
$m_{b,Pl}$	Dynamic mass of the plate, see Eq. (2.39).

with

$$m_{b,Pl} = \frac{8\sqrt{B' \cdot \rho \cdot h}}{\omega} = \frac{0.36 \cdot h^2 \cdot \sqrt{E \cdot \rho}}{f} \quad \mathrm{kg} \quad (4.14)$$

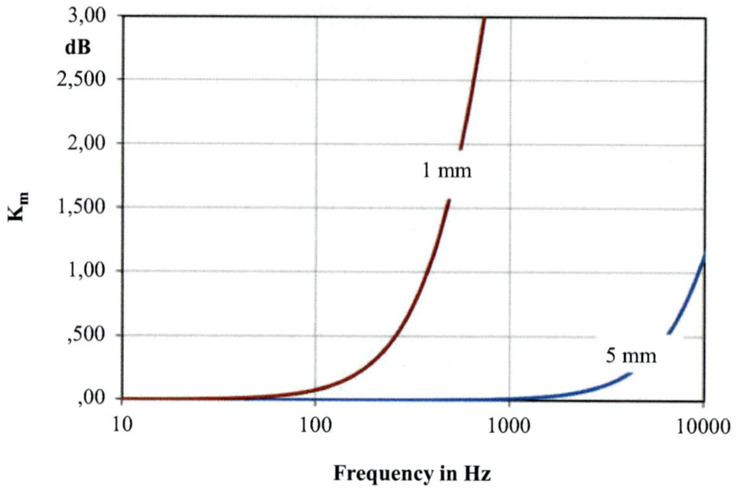

Fig. 4.24 Correction level K_m for two steel plates of thicknesses $h = 1$ mm and $h = 5$ mm ($m_A = 0.02$ kg)

For rigid coupling of the transducer to plate-like structures, the correction level K_m can then be determined as follows:

$$K_m \approx 20 \cdot \log\left(\sqrt{1+\left(\frac{m_A}{m_{b,Pl}}\right)^2}\right) = 20 \cdot \log\left[\sqrt{1+\left(\frac{m_A \cdot f}{0.36 \cdot h^2 \cdot \sqrt{E \cdot \rho}}\right)^2}\right] \quad \text{dB} \qquad (4.15)$$

E	Modulus of elasticity of the plate in N/m^2
ρ	Density of the plate in kg/m^3
h	Slab thickness in m

For the rigid coupling of an accelerometer $m_A = 0.02$ kg to a steel plate of thickness $h = 1$ mm or $h = 5$ mm, (4.15) is shown graphically in Fig. 4.24. Here it can be seen that the correction level K_m is very dependent on the impedance or dynamic mass of the structure. For $m_A \ll m_{b,Pl}$ the influence of the transducer on the measurement signal can be neglected. However, for measurements on light or soft structures, the feedback effect of the transducer mass on the vibrating structure must be taken into account!

(b) **Influence of interfering vibrations/external structure-borne sound**

Similar to the airborne sound measurement, the influence of extraneous or spurious vibrations can be taken into account. Analogous to (2.101) it follows for the correction level K_v due to disturbing or extraneous vibrations:

$$K_v = -10 \cdot \log\left(1 - 10^{-0.1 \cdot \Delta L_v}\right) \quad \text{dB} \tag{4.16}$$

with

$$\Delta L_v = \overline{L'_v} - \overline{L''_v} \tag{4.17}$$

$\overline{L'_v}$	Measured average fast level in dB with machine switched on with extraneous vibrations
$\overline{L''_v}$	Measured average fast level in dB with machine switched off (only external vibrations!)

Using the correction levels K_m and K_v, according to (4.15) and (4.16), the corrected average velocity level $\overline{L_{v,i}}$ of subarea *i is* obtained by applying (4.12):

$$\overline{L_{v,i}} = \overline{L'_{v,i}} - K_{m,i} - K_{v,i} \tag{4.18}$$

$\overline{L'_{v,i}}$	Measured average fast level in dB in the presence of spurious oscillations. The correction level $K_{m,\mathrm{i}}$ takes into account the influence of the transducer mass

With the help of the measurement techniques described in Sect. 4.2, it is possible to experimentally determine the necessary emission characteristics for a design acoustic weak point analysis of a machine, see Chap. 3. As already mentioned, the measurements can best be carried out on existing machines or prototypes.

Some methods for the determination of emission values are given below.

4.3 Direct Determination of the A-total Sound Power Level

An essential task of design acoustics or primary noise reduction is to reduce the noise emission of a machine. To do this, one must first determine the current noise emission of the machine, e.g. by measurements: Description of the "ACTUAL state".

The acoustic "ACTUAL state" of a machine can best be described by direct measurement of the A-total sound power level, as a frequency spectrum, narrowband, third-octave or octave spectrum or their sum over the entire frequency range. This can be determined experimentally using various methods [11].

4.3.1 By Sound Pressure Measurement

Using the enveloping surface method, the total sound power level can be determined, e.g. according to DIN 45635, Part 1, or DIN EN ISO 3744 or 3746, by sound pressure measurements as follows, see Sect. 2.6:

$$\left.\begin{aligned} L_{WA}(f) &= \overline{L_{pA}(f)} + L_S - K_1(f) - K_2(f) \quad \mathrm{dB(A)} \\ L_{WA,\mathrm{total}} &= 10\lg\left[\sum_{f_{\min}}^{f_{\max}} 10^{\frac{L_{WA}(f)}{10}}\right] \quad \mathrm{dB(A)} \end{aligned}\right\} \tag{4.19}$$

$\overline{L_{pA}(f)}$	Average A-sound pressure level on the enveloping surface in dB(A). The A-weighting can also be taken into account subsequently by the attenuation level of the A-weighting, according to Tables 2.2 and 2.4, see (2.84), (2.85)
L_S	Enveloping surface measure in dB
$K_1(f)$	Correction level for interference level in dB
$K_2(f)$	Ambient correction level in dB
f	Centre frequency of a frequency band, e.g. third octave or octave, or discrete frequency in narrowband analysis

Figure 4.25 shows an example of the cuboidal arrangement of an enveloping surface. The enveloping surface is oriented to the sound source shape. In [15, 16] corresponding enveloping surfaces including number and position of measuring points are defined for different sound source shapes.

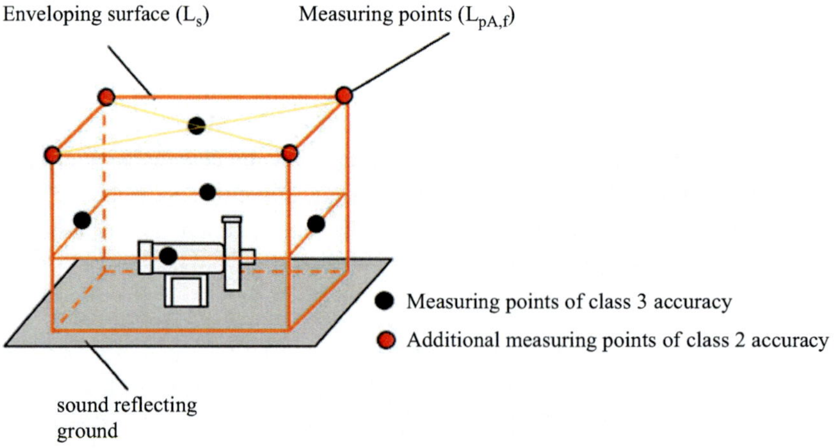

Fig. 4.25 Determination of the total sound power level, enveloping surface method according to DIN 45635, Part 1, or DIN EN ISO 3744 or 3746

4.3.2 By Sound Intensity Measurements

The A-total sound power level can also be determined by means of sound intensity measurement according to DIN EN ISO 9614, Part 2, by continuous sampling.

$$L_{WA,i}(f) = L_{IA,i}(f) + L_{S,i} \quad \text{dB(A)} \tag{4.20}$$

With

$L_{IA,\,i}(f)$	A total sound intensity level at frequency *f of* subarea *i*
$L_{S,\mathrm{i}}$	Area measure of the partial area *i*

The A-total sound power level is obtained by energetic addition over all partial areas

$$L_{WA} = 10\lg\left[\sum_{i=1}^{N} 10^{\frac{L_{IA,i}}{10}}\right] \quad \text{dB(A)} \tag{4.21}$$

N	Number of subplots

The A-sound intensity level $L_{IA,\,i}$ can be determined according to ISO 9614, Part 2, by continuous sampling, see Fig. 4.26.

By energetic addition over all frequencies the A-total sound intensity level of the partial area *i is* obtained.

$$L_{IA,i} = 10\lg\left[\sum_{f_{\min}}^{f_{\max}} 10^{\frac{L_{IA,i}(f)}{10}}\right] \quad \text{dB(A)} \tag{4.22}$$

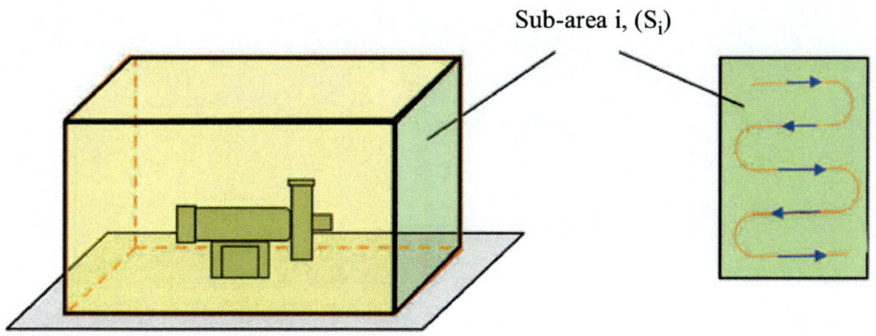

Fig. 4.26 Determination of the total sound power level by sound intensity measurements according to ISO 9614, Part 2

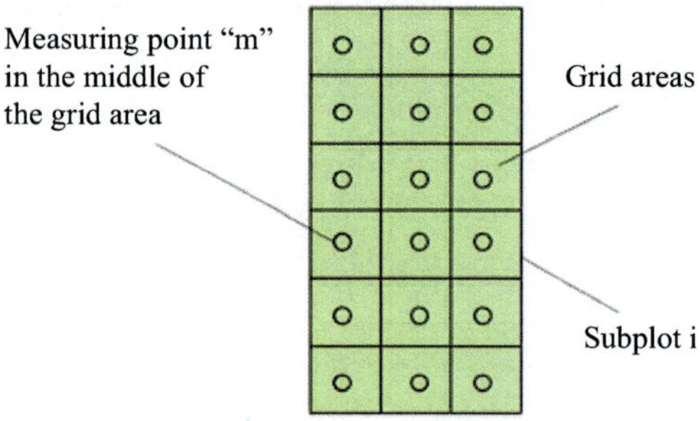

Fig. 4.27 Decomposition of subarea *i* into square grid areas

The A-sound intensity level $L_{IA,\ i}$ can also be determined by measurements at discrete points according to ISO 9614, Part 1. For this purpose, the partial area *i*, see Fig. 4.26, is divided into small square grid areas with edge lengths of approx. 3 to 5 cm, see Fig. 4.27. For each grid area – measuring point "m" in the centre of the grid area – $L_{IA,\ m,\ i}(f)$, the A-sound intensity level is measured at the frequency *f.*

By energetic addition of the sound intensity levels of all measuring points on the grid area, the total sound intensity level of the partial area *i* at the frequency *f is* obtained:

$$L_{IA,i}(f) = 10\lg\left[\sum_{m=1}^{M} 10^{\frac{L_{IA,m,i}(f)}{10}}\right] \quad \mathrm{dB(A)} \tag{4.23}$$

The determination of the A-total sound power level according to ISO 9614, part 1, by measurements at discrete points is, of course, much more complex than the corresponding measurements according to ISO 9614, part 2, by continuous sampling.

Measurements at discrete points, however, have the advantage of providing much more detailed information about the noise emission of the investigated subarea. It is also possible to use graphical representations, e.g. intensity mapping, to illustrate the sound radiation of the subarea (Fig. 4.28).

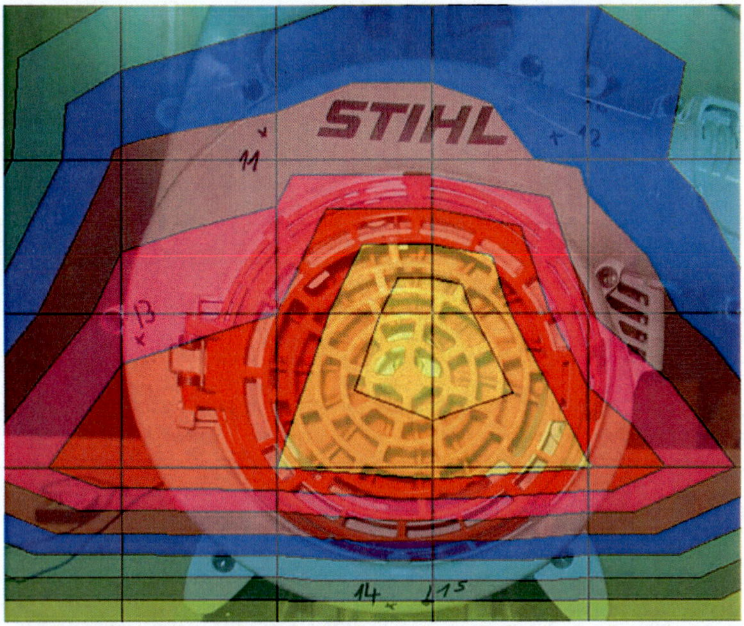

Fig. 4.28 Intensity mapping of a blowing device [5, 12]

4.4 Indirect Determination of the A-total Sound Power Level

The total noise development of a machine can also be determined indirectly via its partial sound power level. For this purpose, based on a design analysis, the possible components are first determined which are eligible for the airborne and structure-borne noise radiation or for the total noise development of the machine.

Subsequently, the partial sound power levels are determined quantitatively for the previously qualitatively determined components by means of experimental investigations.

An important task here is the quantitative separation between airborne and structure-borne sound radiation of the machine, see Sect. 4.1.1.

4.4.1 Partial Airborne Sound Power Level

This includes all opening surfaces, slots and leaks which are responsible for the airborne sound radiation of the machine. The associated partial airborne sound power level can be determined, among other things, with the aid of the sound intensity measurement technique according to ISO 9614, Part 1 or Part 2, analogous to (4.20):

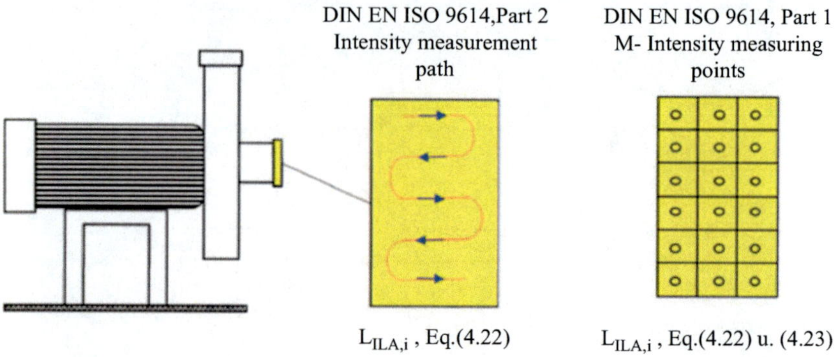

Fig. 4.29 Determination of the partial airborne sound power levels using the example of the intake opening of a fan

$$L_{WLA,i} = L_{ILA,i} + L_{SL,i} \quad \text{dB(A)} \tag{4.24}$$

L	The index L means that it is an airborne sound source

The A-part sound intensity level $L_{ILA,\,i}$ can be determined according to ISO 9614, Part 1 (discrete measuring points) or Part 2 (continuous sampling). Figure 4.29 shows a schematic representation of the above measurement methods using the example of the intake opening of a fan.

It is pointed out that it is not always possible to measure the partial sound power levels directly. This is due, for example, to the fact that the sound radiation of the neighbouring components is too high or that the sound intensity measurements are not possible due to flow-related pressure fluctuations.

In order to be able to determine the partial sound power levels nevertheless, it is sometimes unavoidable to acoustically mask out the sound radiation of the machine which does not belong to the partial sound power levels sought. The determination of these emission characteristics requires not only metrological knowledge but also special practical experience [13, 14].

4.4.2 Partial Structure-borne Sound Power Level by Sound Intensity Measurements

This includes sound-emitting machine housings or their sub-areas. This also includes all footpoint or speed-excited components such as foundations, control cabinets, buildings, etc., which are connected to the machine. The associated partial structure-borne sound power level can also be determined by means of the sound intensity measurement technique according to ISO 9614, Part 1 or Part 2, analogous to (4.20):

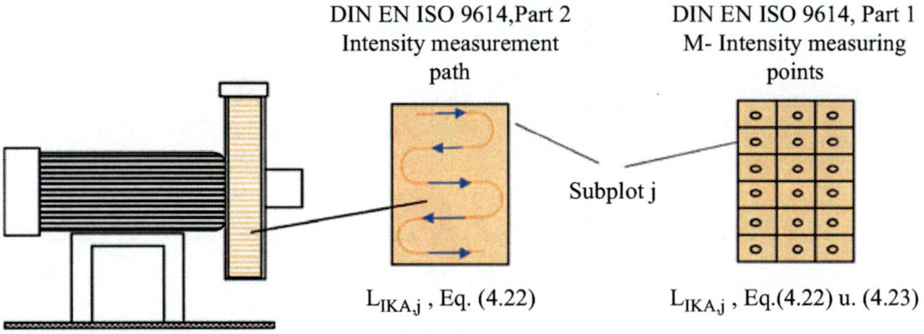

Fig. 4.30 Determination of partial body sound power levels by sound intensity measurements

$$L_{WKA,j} = L_{IKA,j} + L_{SK,j} \quad \text{dB(A)} \tag{4.25}$$

K	The index K means that it is a source of structure-borne sound

The A-part sound intensity level $L_{IKA,\,j}$ can be determined according to ISO 9614, Part 1 (discrete measuring points) or Part 2 (continuous sampling). Figure 4.30 schematically shows the sound intensity measurement method using the example of the sound radiation of the partial area of a fan.

4.4.3 Partial Structure-Borne Sound Power Level by Structure-Borne Sound Measurements

With the aid of structure-borne sound measurements, it is also possible to specifically determine the partial structure-borne sound power level. The advantage here is that the sound radiation of the adjacent components has less influence on the measurements than with sound intensity measurements. The disadvantage, however, is that the radiation level must often be roughly estimated.

The A-part structure-borne sound power level $L_{WAK,\,j}$, of the partial area "j" can be determined according to DIN 45635, Part 8, ISO/TS (pre-standard) 7849-1 and 2 (2009-03), as follows.

$$L_{WAK,j} \approx L_{vA,j} + L_{SK,j} + \sigma'_j \quad \text{dB(A)}; \qquad (\rho \cdot c \approx \rho_0 \cdot c_0) \tag{4.26}$$

$$L_{vA,j} = 20 \cdot \lg \frac{v_{A,j}}{v_0} \quad \text{dB(A)} \tag{4.27}$$

$$L_{SK,j} = 10 \cdot \lg \frac{S_{K,j}}{S_0} \quad \text{dB} \tag{4.28}$$

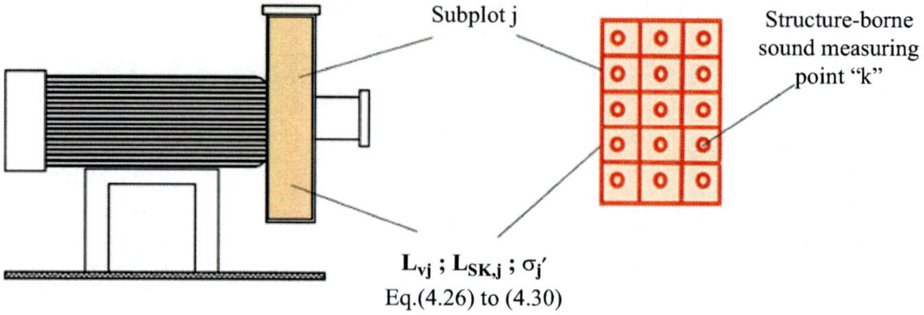

Fig. 4.31 Determination of partial structure-borne sound power levels by structure-borne sound measurements

$$\sigma_j' = 10 \cdot \lg \sigma_j \quad \text{dB} \tag{4.29}$$

$L_{WAK,\,j}$	A-weighted structure-borne sound power level of the partial area j
$L_{vA,\,j}$	Mean A-weighted fast level of the subarea j
$L_{SK,\,j}$	Area measure of the partial area j
σ_j'	Radiation coefficient of the sound-radiating surface j. For plate-like structures it can be estimated according to σ' (5.81)
$v_{A,\,j}$	A-weighted velocity of the partial surface j in m/s
$S_{K,\,j}$	Sound-radiating partial area j in m^2
$v_0 = 5 \cdot 10^{-8}$ m/s	Quick reference
$S_0 = 1$ m^2	Reference area

Figure 4.31 schematically illustrates the structure-borne sound measurement method using the example of the sound radiation of the partial surface of a fan.

The mean A-weighted fast level of subarea j can be determined as follows:

$$L_{vA,j}(f) = 10 \cdot \lg\left(\frac{1}{n} \cdot \sum_{n=1}^{n} 10^{0.1 \cdot L_{vA,n}(f)}\right) \quad \text{dB(A)} \tag{4.30}$$

n	Number of measuring points on the subplot j
$L_{vA,\,j}(f)$	Average A-velocity level of the partial area j at the frequency f

$$L_{vA,j} = 10 \cdot \lg\left[\sum_{f_{\min}}^{f_{\max}} 10^{\frac{L_{vA,j}(f)}{10}}\right] \quad \text{dB(A)} \tag{4.31}$$

With the experimentally determined emission characteristics:

- $\mathbf{L_{WA,\ total}}$ $\mathbf{L_{WA,\ total}(f)}$
- $\mathbf{L_{WAL}}$ $\mathbf{L_{WAL}(f)}$
- $\mathbf{L_{WAK}}$ $\mathbf{L_{WAK}(f)}$

it is now possible, as described in Sect. 4.1.1, to describe the sound generation with the aid of a design analysis and to carry out an acoustic weak point analysis for the machine under investigation. Here, among other things, the separation of airborne and structure-borne sound radiation can be carried out in order to plan the noise reduction quantitatively.

In addition, it is possible to define priorities on the basis of a priority plan,

- **on which component,**
- **at what frequency,**
- **in what order,**
- **and to what extent**

noise reduction must be carried out in order to achieve the desired overall noise reduction.

For special questions, further investigations are to be carried out, e.g:

- Determination of critical natural frequencies and natural modes
- Measurement of force-time curves
- Determination of the input and transmission impedances.

In addition to the measurement techniques described so far, other powerful measurement and analysis techniques are available today, which will not be discussed in detail here. Some of the special methods used in design acoustics are given below:

- **Finite element method** for the theoretical determination and representation of natural frequencies and natural modes
- **Modal analysis for** the experimental determination and representation of natural frequencies and natural modes
- **CFD-Anylysis** (Computational Fluid Dynamics) for simulation and calculation of fluidic problems
- **Laser scanning vibrometer** for non-contact vibration measurements. Among other things, the operating and natural vibrations of a machine can be clearly displayed with the machine or structural images. In conjunction with a so-called **derotator**, measurements can also be made on rotating components.
- **Rotational vibrometer** for non-contact measurement of torsional vibrations
- **Acoustic camera** for non-contact measurement of the sound emission of a machine or system with the aid of a microphone array. The localisation of the emission sources shall be facilitated by colored representation of the sound radiation.

All described measurement and analysis techniques are only aids for the description of the sound and vibration generation and serve as a decision basis for the development engineer in the field of design acoustics. Which measurement and/or analysis technique is used to solve the sound and vibration problem depends on the problem and should be determined on a case-by-case, task-oriented basis.

What is certain is that measurement data alone can never solve the problem!

A precise problem analysis in advance of the planned measurements, e.g. by means of a design analysis with the support of the machine designer, often helps to optimise the effort and costs of the planned investigations so that only those measurements need to be carried out which are actually necessary for solving the problem.

References

1. Sinambari, Gh.R., Kunz, F.: Primäre Lärmminderung durch akustische Schwachstellenanalyse. VDI-Bericht Nr. 1491 (1999)
2. Thorn, U., Tschöp, E.: IBS – Aufbauseminar-Konstruktionsakustik II, Messtechnische Ermittlung der Geräuschemission einer Maschine. FH Bingen (2012)
3. Brüel & Kjaer [Hrsg.]: Microphone Handbook. Technical Documentation (1996)
4. Brüel & Kjaer [Hrsg.]: Schallintensität in der Bauakustik. Informationsbroschüre (2001)
5. Sinambari, Gh.R., Sentpali, S.: Ingenieurakustik, 5. Aufl. Springer Vieweg, Wiesbaden (2014)
6. Brüel & Kjaer [Hrsg.]: Schallintensität. Informationsbroschüre (1991)
7. ISO 9614-2: Acoustics – Determination of sound power levels of noise sources using sound intensity – Part 2: Measurement by scanning,1996
8. Brüel & Kjaer [Hrsg.]: Schwingungsmessung. Informationsbroschüre, undatiert
9. Brüel & Kjaer [Hrsg.]: Schwingungsmessung und -analyse. Informationsbroschüre, undatiert
10. Brüel & Kjaer [Hrsg.]: Beschleunigungsaufnehmer. Datenblatt und Beschreibung BRÜEL & KJAER Messgeräte, Informationsbroschüre, undatiert
11. Sinambari, Gh.R., Tschöp, E.: "Konstruktionsakustik", Seminarveranstaltung. IBS-Seminar, St. Martin/Pfalz, 2016
12. Untersuchung zur Geräuschabstrahlung eines Blasgerätes. Interne Dokumentation der Fa. IBS Ingenieurbüro für Schall- und Schwingungstechnik GmbH (nicht veröffentlicht), 2003
13. Thorn, U., Wachsmuth, J.: Konstruktionsakustische Schwachstellenanalyse eines Straddle Carriers Typ Konecranes 54 DE und Erarbeiten von prinzipiellen Lärmminderungsmaßnahmen. VDI-Berichte 2118 (2010)
14. Sinambari, Gh.R., Felk, G., Thorn, U.: Konstruktionsakustische Schwachstellenanalyse an einer Verpackungsmaschine. VDI-Berichte Nr. 2052 (2008)
15. DIN 45635-1: Measurement of noise emitted by machines; airborne noise emission; enveloping surface method; basic method, divided into 3 grades of accuracy, 1984-4
16. ISO 3744: Acoustics – Determination of sound power levels and sound energy levels of noise sources using sound pressure – Engineering methods for an essentially free field over a reflecting plane, 2010

Noise Reduction

Noise reduction is always associated with the reduction of people's exposure to noise. This is due to the fact that the term noise is understood to mean the unpleasant part of the subjective perception of sound events by the human ear.

The task of design acoustics is to avoid or reduce the noise generated by machines by suitable measures, preferably design measures [1, 2].

Depending on the task and the possibilities, various noise reduction measures can be considered (Fig. 5.1). In this chapter, some essential possibilities of primary and secondary noise reduction, as they are applied today in technical acoustics and noise control, are shown.

Organizational and personal noise protection is often used in practice in noise protection at the workplace. It is usually suitable as a short-term and quick measure and is not the subject of this book.

Active noise protection, noise reduction by counter- or anti-noise, is not suitable for noise reduction on machines due to the spatial distribution of the sound field or is only possible with very high effort and is therefore not discussed further here. However, it is pointed out that for special applications, e.g. planar sound propagation in pipelines and punctual noise reduction, as in the case of headphones, market-ready products are also available [3–5]. Among other things, special anti-noise solutions for household appliances, vehicles and jet aircraft are currently being developed or tested, which may be used in the future.

In the following, the technical possibilities of noise protection, here especially the primary and secondary measures, are discussed.

G. R. Sinambari, *Design Acoustics*,
https://doi.org/10.1007/978-3-658-40183-2_5

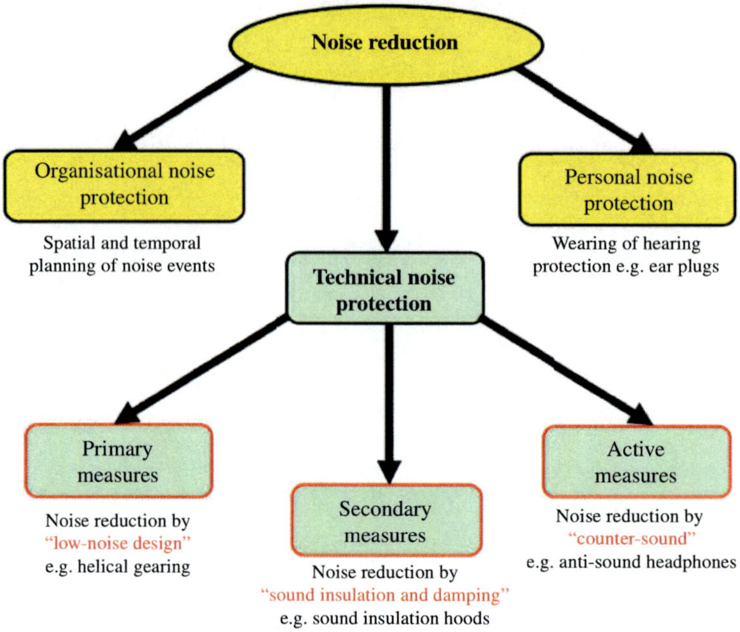

Fig. 5.1 Various noise reduction measures

5.1 Primary Noise Reduction

Primary noise reduction means avoiding or reducing the **generation of noise** at the source through constructive measures. For this purpose, however, it is necessary to understand the mechanisms involved in the generation, transmission and radiation of the noise. Among other things, it is pointed out that noise reduction can only be achieved if the **loudest or level-determining sources** are designed to be low-noise.

Loudest sources may include:

- Individual machines within a plant that determine the overall level,
- Components of a machine with the highest sound power level,
- Certain work processes of a production,
- Individual frequency components or frequency ranges that determine the overall level.

The **total level** can be very different depending on the task, e.g:

- A total sound power level,
- A-total sound pressure level, workplace or neighbourhood noise,
- Rating Level,
- Guarantee values.

It follows from the above that, before noise reduction measures can be defined or provided for, it is first necessary to describe the noise generation and identify the main sources of noise by means of suitable investigations, e.g. with the aid of a noise weak point analysis [6].

Furthermore, the quantitative separation between airborne and structure-borne noise of a machine or a complete plant is necessary for the development of suitable noise reduction measures.

The A-sound power level or its spectral distribution, e.g. one-third octave spectra, is decisive as a parameter for the objective quantification of sound radiation.

In principle, the implementation of noise reduction measures should only be started if the following questions have been clarified first:

- **What** is to be reduced in terms of noise? This implies the localization of the sources responsible for the overall noise development.
- **How** is noise reduction to be achieved? For this purpose, the sound generation mechanisms must be described and it must be determined whether airborne or structure-borne sound is decisive for the noise development. Furthermore, the level-determining frequency components and the sound transmission are to be determined.
- **How much** noise reduction is necessary? In order to plan noise reduction in a targeted manner, the noise reduction measures for airborne and structure-borne noise, both as a total value and in level-determining frequency components, must be quantified.

All these questions can be answered by an acoustic weak point analysis.

5.1.1 Acoustic Weak Point Analysis

An acoustic weak point analysis can be sensibly carried out on existing machines/plants or on a prototype by means of experimental investigations.

For machines in the planning phase, a rough acoustic weak point analysis can be carried out after the functional and design drawings are available. Here, the decisive alternating forces, the critical introduction points and the main sound-emitting components are localised on the basis of drawings. By estimating the expected sound power level on the basis of empirical values, measurements on similar machines or with the aid of numerical simulation methods, critical components are to be located. Based on the results, primary and secondary noise reduction measures can then be planned as a first step towards noise optimization.

After the construction of a prototype, it is then possible to carry out further noise optimisation or reduction with the aid of a precise acoustic weak point analysis.

For the quantitative separation between airborne and structure-borne noise development of a machine or a plant, the partial sound power levels in relation to the total sound power level must be determined within the framework of the acoustic weak point analysis

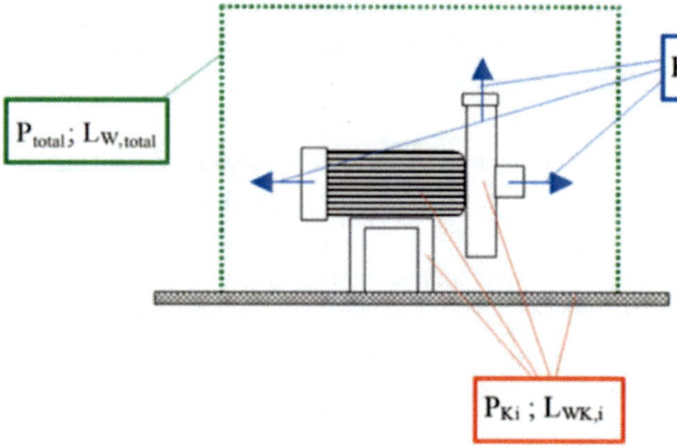

Fig. 5.2 Schematic representation of the sound radiation of a machine

[6, 7]. Figure 5.2 schematically shows the relationship between the various emission values of a machine.

The following relationships exist between the emission parameters given in Fig. 5.2:

$$P_{\text{total}}(f) = \eta_{\text{A}}(f) \cdot P_{\text{M}} \quad \text{W} \tag{5.1}$$

$$P_{\text{total}}(f) = P_{\text{L}}(f) + P_{\text{K}}(f) \quad \text{W} \tag{5.2}$$

$$P_{\text{L}}(f) = \sum_{i=1}^{N} P_{L,i}(f) \quad \text{W} \tag{5.3}$$

$$P_{\text{K}}(f) = \sum_{j=1}^{M} P_{\text{K},j}(f) \quad \text{W} \tag{5.4}$$

Meaning herein:

f	Frequency
η_A	Acoustic efficiency of the machine
P_M	Mechanical or electrical power of the machine
$P_L(f)$; $P_{L,i}(f)$	Total airborne sound power or airborne sound power of the *i-th* opening
$P_K(f)$; $P_{K,j}(f)$	Total structure-borne sound power or structure-borne sound power of the *j-th* component

The acoustic efficiency indicates what proportion of the mechanical power of a machine is converted into sound power. It is essentially dependent on the type of machine, e.g. pump, gearbox, press, automatic punching press, etc. It is particularly suitable for comparing different machines with each other as a function of the converted mechanical power or for evaluating or quantifying the effect of constructive noise reduction measures on a machine.

Acoustic efficiency is usually a constant, i.e. an increase in mechanical power will also cause a proportional increase in acoustic power. A particular increase in sound power at certain power levels and speeds indicates a possible misbehaviour or resonance excitation of the machine [6].

If one takes into account that a sound power level of 120 dB corresponds energetically to 1 W of power, it is understandable that the acoustic efficiency, which can be determined by measurements, can assume very low values, approx. 10^{-9} to 10^{-2}, depending on the machine.

5.1.1.1 Separation of Airborne and Structure-Borne Sound

For the development of constructive noise reduction measures, it is necessary to determine the airborne and structure-borne sound radiation of a machine or system or its share in the total sound radiation in a quantitative way. For this purpose, it is useful to indicate the total sound power level $L_{W,\text{ total}}$ of a machine or a complete plant both directly and indirectly via the sum of the airborne and structure-borne sound power levels (L_{WL} and L_{WK}). For this purpose, (5.2), (5.3) and (5.4) are given in level notation:

$$L_{W,\text{total}} = 10 \cdot \log\left(10^{L_{WL}/10} + 10^{L_{WK}/10}\right) \quad \text{dB} \tag{5.5}$$

$$L_{WK} = 10 \cdot \log\left(\sum_{j=1}^{M} 10^{L_{WK,j}/10}\right) \quad \text{dB} \tag{5.6}$$

$$L_{WL} = 10 \cdot \log\left(\sum_{i=1}^{N} 10^{L_{WL,i}/10}\right) \quad \text{dB} \tag{5.7}$$

L_{WL} $L_{WL,\ i}$	Total and partial airborne sound power level of the opening areas $S_{L,\ i}$
L_{WK} $L_{WK,\ j}$	Total and partial structure-borne sound power level of the radiating surfaces $S_{K,\ j}$
i,j	Number of opening surfaces (i) or surfaces radiating structure-borne sound (j)

The partial and total sound power levels can best be determined by measurements, see Chap. 4. In the case of machines in the planning stage, as already mentioned, a rough acoustic weak point analysis can be carried out after the functional and construction drawings have become available. An exact weak point analysis can only be carried out after the construction of a prototype.

The total acoustic energy or acoustic power of a machine is a constant for constant operating parameters such as power, speed, feed rate, number of cycles, etc., i.e. if the operating conditions remain unchanged, the noise generated by the machine is also constant. It is therefore essential that the measured data for acoustic energy balancing are all obtained under the same operating conditions.

The more precisely the data are measured and the more stable the operating parameters could be kept during the measurements, the more precise the acoustic energy balance will be. Since it is usually not possible to record all measurement data simultaneously, at least one acoustic reference measurement point, airborne and/or structure-borne sound, should be measured at the same time for control purposes.

A criterion for the accuracy of the acoustic balance and the measured sound power levels is how well the directly measured total sound power level and the total sound power level indirectly calculated via the partial sound power levels according to (5.5) agree.

The more precisely the measurement data are determined, the more accurately noise reduction can be planned and appropriate measures can be defined. For meaningful noise reduction planning, the following accuracies should be aimed for [6–8]:

A total sound power level:	±1 dB(A)
Partial sound power level in frequency bands, e.g. in third octaves:	±2–5 dB

Practical experience has shown that, with appropriate care, it is possible to achieve the specified accuracy in the measurement data. Furthermore, a noise reduction of >1 dB(A) can be planned with the emission characteristics obtained in this way.

It makes sense to determine the partial and total sound power levels in frequency bands, i.e. narrowband, in thirds or octaves. The narrower the bandwidth of the data, the more precisely the causes of the noise development can be analysed or described and thus targeted design measures can be developed. However, the measurement and analysis effort is considerably higher the narrower the bandwidth of the measurements. In most practical applications, it is sufficient if the measurements and analyses are carried out in one-third octave bandwidth.

If the measurement data are available with the above accuracy, then the following questions can be answered for a desired noise reduction $\Delta L_{\text{target}} > 1$ dB(A):

1. The airborne sound is decisive for the total noise development:

$$L_{WAL} - L_{WAK} < \Delta L_{\text{target}} \tag{5.8}$$

2. The structure-borne sound is decisive for the total noise development:

$$L_{WAK} - L_{WAL} > \Delta L_{\text{target}} \tag{5.9}$$

3. Both airborne sound and structure-borne sound are decisive for the total noise development:

$$| L_{WAK} - L_{WAL} | < \Delta L_{\text{target}} \tag{5.10}$$

With the help of (5.6) and (5.7) it is then possible to determine the relevant frequency components responsible for the A-total sound power level.

Based on the data obtained in this way, specific noise reduction measures can be defined for airborne and structure-borne noise sources (ΔL_{L}; ΔL_{K}).

From this, noise reduction measures for the level-determining frequency components can also be determined with the help of (5.6) and (5.7) ($\Delta L_{\text{L},\,i}$ and $\Delta L_{\text{K},\,j}$).

The noise reduction measures determined in this way can also be used to determine the expected reduction in the A total sound power level:

$$L_{\text{WA (with measures)}} = 10 \cdot \log\left(10^{(L_{\text{WAL}} - \Delta L_{\text{L}})/10} + 10^{(L_{\text{WAK}} - \Delta L_{\text{K}})/10}\right) \quad \text{dB(A)} \tag{5.11}$$

This can be used to check whether the desired noise reduction has been achieved:

$$L_{\text{WA(without measures)}} - L_{\text{WA(with measures)}} = \Delta L_{\text{is}} \quad \text{dB(A)} \tag{5.12}$$

Noise mitigation planning can be considered successful if:

$$\Delta L_{\text{is}} \geq \Delta L_{\text{target}} \tag{5.13}$$

In summary, it follows that the following questions can be answered with the help of the acoustic weak point analysis:

(a)	Is the total sound radiation determined by airborne or structure-borne sound radiation?
(b)	Which component is mainly responsible for the total, airborne or structure-borne sound power level?
(c)	Which frequency components are responsible for the noise?
(d)	What are the necessary noise reduction measures?

Together with an acoustic design analysis based on drawings and operating data as well as the description of the sound generation mechanisms, see Chap. 3, it is then possible to determine the **acoustic weak points of** a machine or even a complete plant.

Figure 5.3 summarises the main tasks of the acoustic weak point analysis.

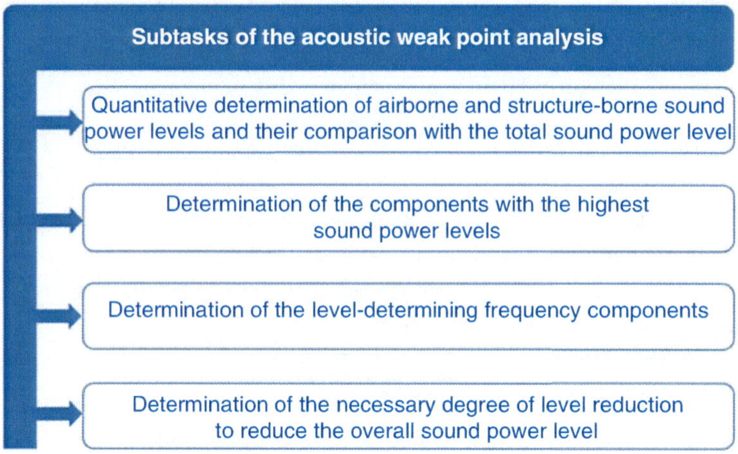

Fig. 5.3 Acoustic weak point analysis

5.1.1.2 Data Acquisition for the Energy Balance Sheet

All emission parameters required for the acoustic weak point analysis can be determined by measurements, see Chap. 4.

Figure 5.4 shows the normative determination of the emission parameters, i.e. the data acquisition for the energy balance in a flow chart. These data form the basis of the acoustic weak point analysis, if they have been determined with the required accuracy.

Figure 5.5 shows a schematic diagram of the procedure for the acoustic weak point analysis and noise reduction planning [7].

With the help of the noise reduction measures determined in this way for the acoustic weak points, it is now possible to develop suitable noise reduction measures. In principle, the following sequence should be followed:

- **Sound generation**
- **Sound transmission**
- **Sound radiation**

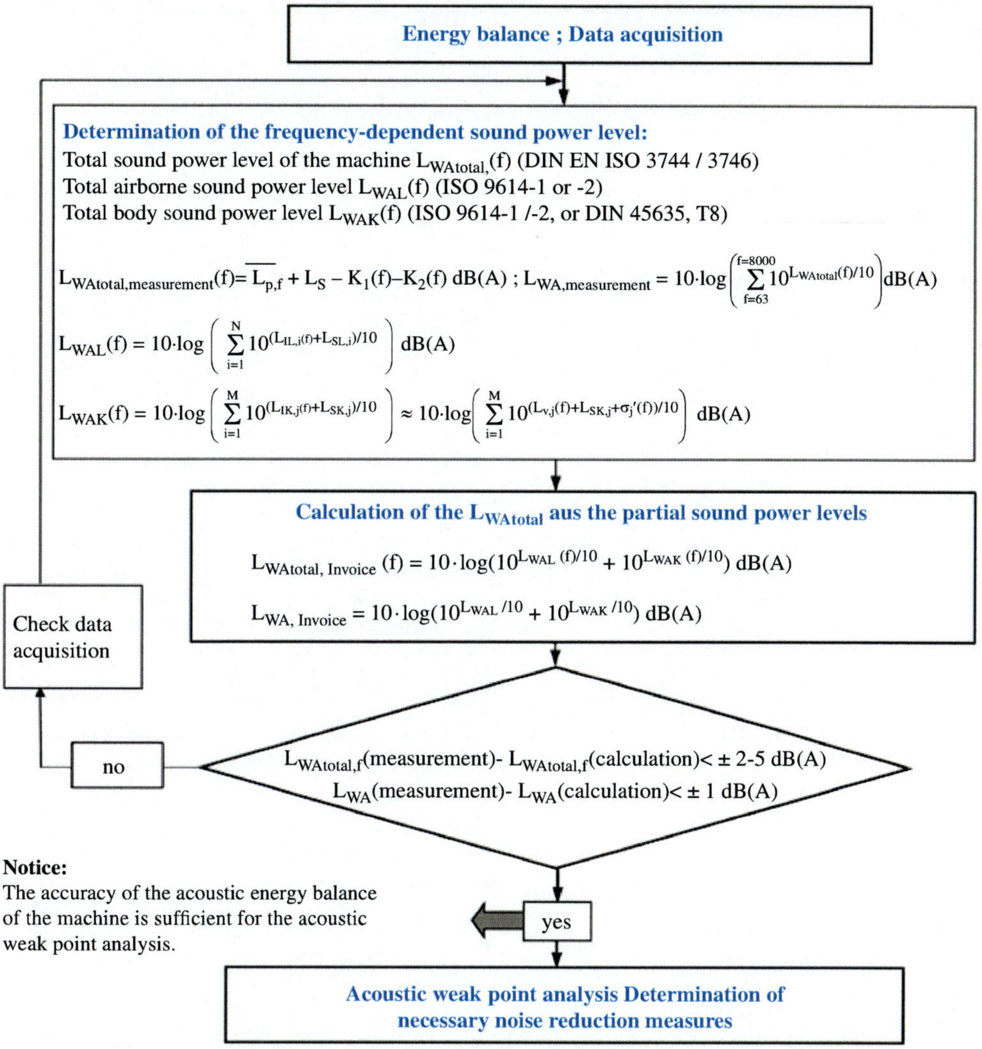

Fig. 5.4 Determination of the emission parameters for the acoustic weak point analysis

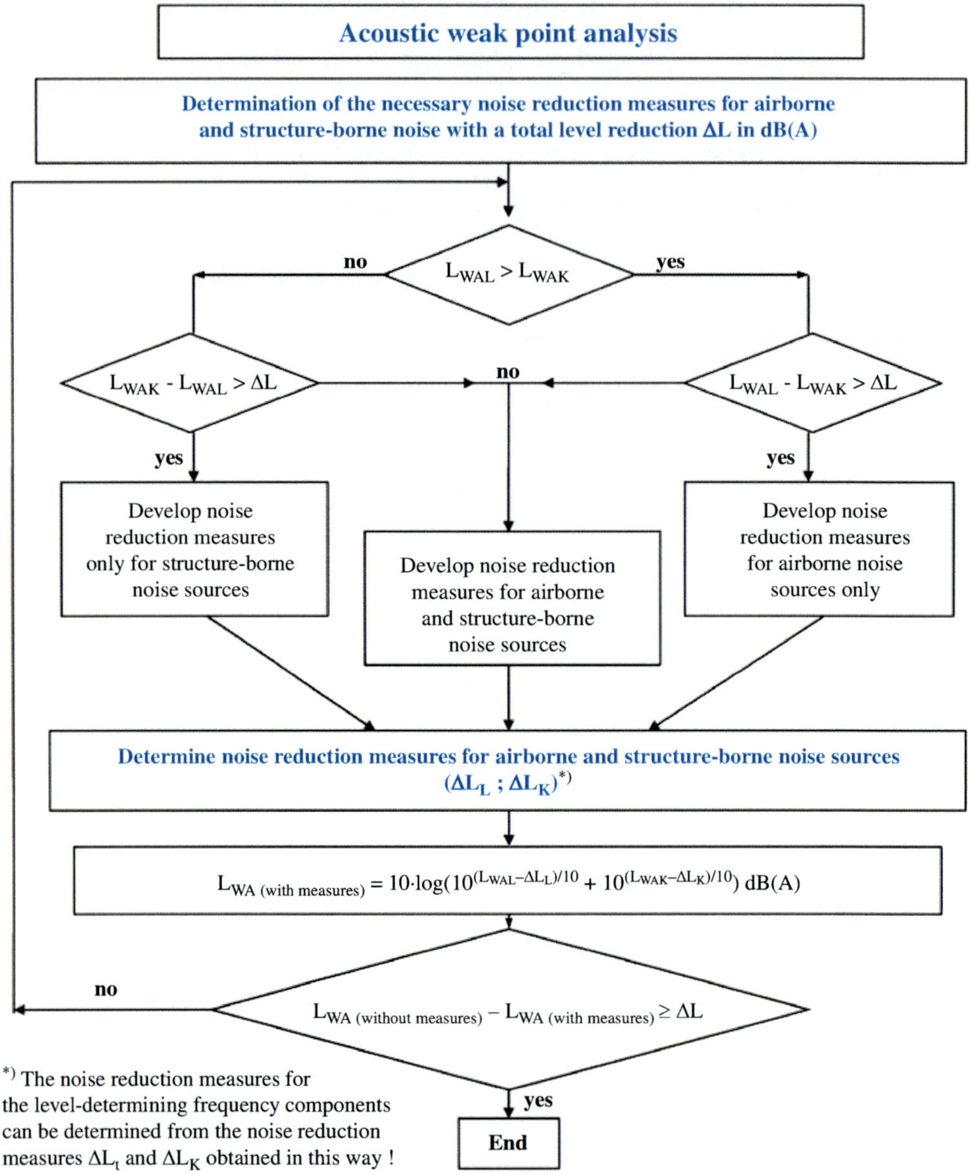

Fig. 5.5 Acoustic weak point analysis, noise reduction planning

5.1.2 Basic Measures for Noise Reduction

Basic options for primary noise reduction are explained below.

5.1.2.1 Influencing the Generation of Sound

(a) **Basic measures**

Reducing the forces that occur during the operation of a machine is the first possibility of primary noise reduction measures. Here are a few examples:

- Optimum clearance adjustment and correct lubrication for bearings, guides and gearboxes. If the clearance is too large, the excitation forces caused by shocks result in pulse-like noises.
- Reduction of surface roughness on the contact surfaces of moving parts, which may be caused by dirt, corrosion, transport damage, etc.
- Avoidance of impact points, e.g. when joining running surfaces or rails. Power transmission with as little impact as possible, e.g. for camshafts and cam gears.
- Avoidance or reduction of shocks or impacts with machine parts moving relative to each other, low drop heights
- Reduction of speeds, lowering of speeds
- Avoidance of cavitation in hydraulic systems or fluidic installations
- Optimization of the flow guidance, especially at the points with the highest flow velocity
- Avoidance of resonance, cutting and whistling sounds
- Acoustic sealing of openings and leaks.

(b) **Change in the force-time curve**

Another way to influence forces is to change the force spectrum. Very often impulsive forces cause structure-borne noise in mechanical systems. A change in the time course of the force can achieve a significant noise reduction with the same impulse strength. The flatter the rise and/or fall of the force, the less forces are generated in the medium and high frequency range. As already explained in Sect. 3.2.1.2, the longer the impact time t_{St}, the earlier the force spectrum falls, see Fig. 3.24.

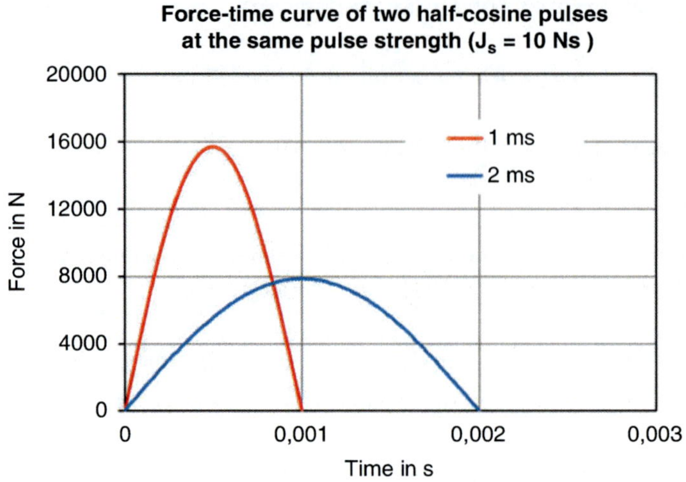

Fig. 5.6 Force-time history of two idealised half-cosine pulses at the same pulse strength ($J_S = 10$ N s) and different impact times t_{St}

Figure 5.6 shows an example of the force-time curves of two half-cosine impulses with different action times t_{St}. For better illustration, the time histories are shifted by π/2 in each case.

At this:

$$F(t) = F_{max} \cdot \cos\left(\frac{\pi}{t_{St}} \cdot t - \frac{\pi}{2}\right) \quad \text{N} \tag{5.14}$$

The momentum, the area underneath the diagrams, is obtained according to (3.112):

$$J_S = \int_0^{t_{St}} F_{max} \cdot \cos\left(\frac{\pi}{t_{St}} \cdot t - \frac{\pi}{2}\right) \cdot \mathrm{d}t = \frac{2}{\pi} \cdot F_{max} \cdot t_{St} \quad \text{N s} \tag{5.15}$$

At constant momentum, F_{max} can be determined as follows:

$$F_{max} = \frac{\pi \cdot J_S}{2 \cdot t_{St}} \quad \text{N} \tag{5.16}$$

With $J_S = 10$ N s you get the maximum force:

$$t_{St} = 10^{-3}\text{s} \quad (1 \text{ ms}) \Rightarrow F_{max} \approx 15708 \text{ N}$$

$$t_{St} = 2 \cdot 10^{-3}\text{s} \quad (2\text{ ms}) \Rightarrow F_{max} \approx 7854\text{ N}$$

The normalised frequency spectra of the force-time histories given in Fig. 5.6 can be determined, according to (3.113), as follows [9, 10]:

$$\frac{|F(f)|}{J_S} = \left| \frac{\cos(\pi \cdot f \cdot t_{St})}{1 - 4 \cdot f^2 \cdot t_{St}^2} \right| \tag{5.17}$$

Figure 5.7 shows the normalised frequency spectra of the above-mentioned force-time curves. As can be clearly seen from this, the frequency spectrum of the excitation force is shifted to the left or to lower frequencies by stretching the force-time curve from 1 ms to 2 ms with the same pulse strength. As a result, significantly less excitation force is available at higher frequencies. Consequently, the force curves for pulse-like excitations should be stretched as much as possible, i.e. they should run as smoothly and steadily as possible.

Examples include:

- Helical gearing instead of degree gearing,
- gradual valve opening instead of sudden opening,
- Cleaning drum with rubber pad,
- Rubber or plastic hammer instead of steel hammer,
- Elastic intermediate layer for parts that collide with each other.

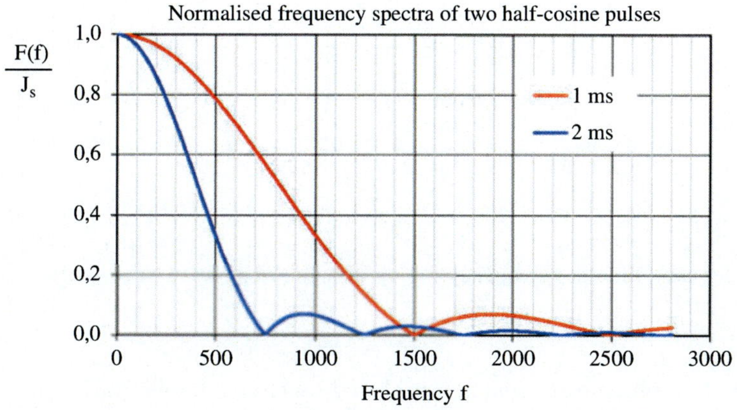

Fig. 5.7 Normalised frequency spectra of two semi-cosine pulses with different peak times

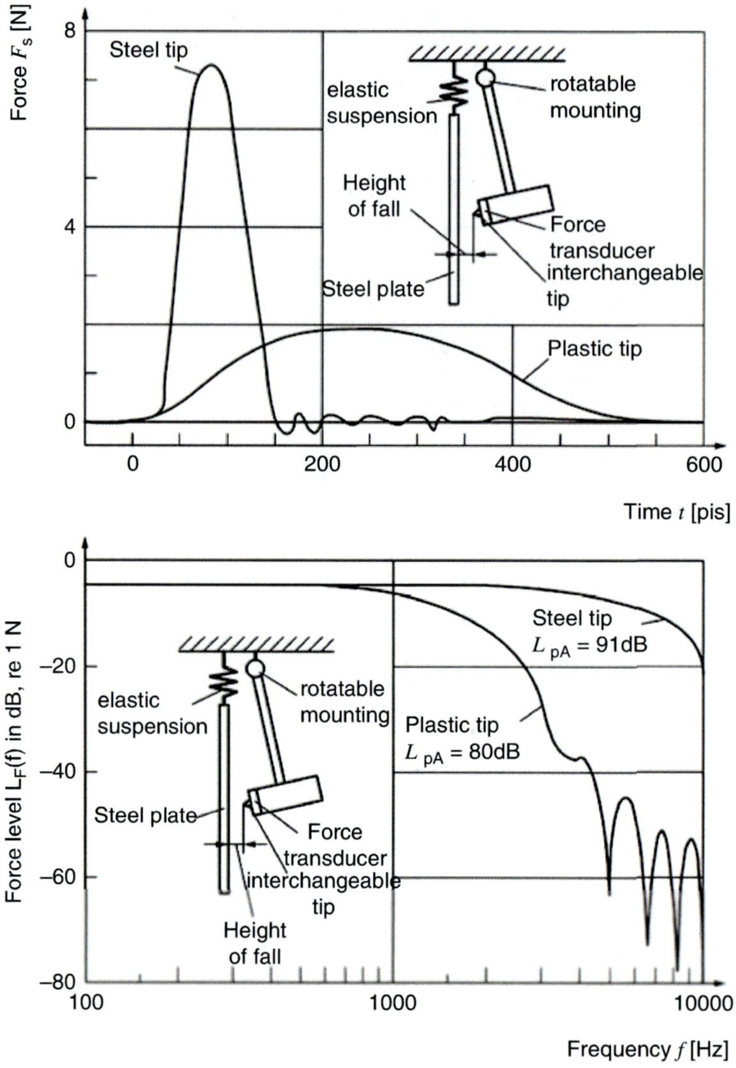

Fig. 5.8 Measured time and frequency curves of hammer blows with plastic and steel tips at the same pulse strength [1, 11]

Figure 5.8 shows the noise reduction achieved by stretching the force-time curve using a model as an example [11]. The experiment confirms that stretching the force-time curve also significantly reduces the airborne noise level for the same impulse strength and structure due to the absence of excitation forces in the high-frequency range.

5.1.2.2 Influence of Insulation on Structure-Borne Sound Transmission

Physically, the improvement in structure-borne sound transmission is achieved by increasing the mechanical input impedance through reflection of structure-borne sound at the point of introduction. This means that part of the structure-borne sound energy is reflected back at the point of introduction and prevented from being transmitted further. The reflection of structure-borne sound is also referred to as damping of structure-borne sound.

The term structure-borne sound insulation is also used for other components, e.g. deflections, changes in cross-section, etc. [12]. However, with regard to constructive noise reduction, there are essentially two options that are of importance:

- Structure-borne sound insulation with the aid of barrier compounds,
- Structure-borne sound insulation through soft intermediate layers.

In this case, a very high impedance is generated with the barrier masses and a very low impedance with the soft intermediate layers. In both cases, structure-borne sound is reflected or damped at the interface.

This is best explained by the magnitude of the so-called reflection factor $|r|$, which is also called the ***law of adjustment of*** acoustics [10, 12]:

$$|r| = \left|\frac{Z_2 - Z_1}{Z_2 + Z_1}\right| = \left|\frac{1 - \frac{Z_1}{Z_2}}{1 + \frac{Z_1}{Z_2}}\right| \tag{5.18}$$

For a better understanding, Fig. 5.9 shows in simplified form the propagation of sound or vibration waves at the boundary of two different media.

Here are:

e	Incident energy, e.g. structure-borne sound energy of the source
τ	Transmitted or transmitted energy, e.g. transmitted structure-borne sound
Z_1, Z_2	Impedance of the medium 1 or 2
$0 \leq r \leq 1$	Reflection factor

For $Z_1 = Z_2$ is the reflection coefficient is equal to zero ($r = 0$), *perfect adaptation*, i.e. the sound or structure-borne sound is not reflected and is transmitted undiminished.

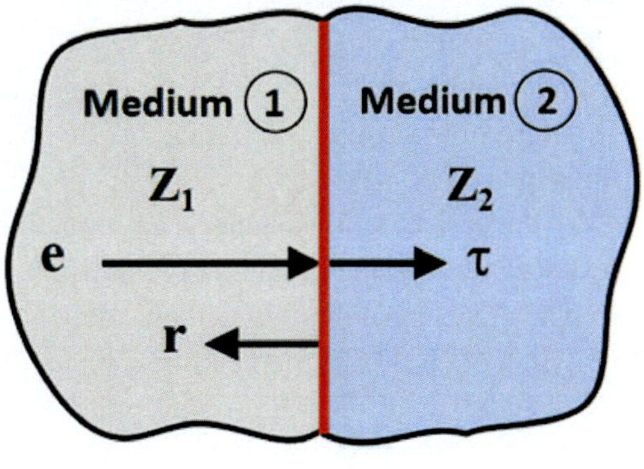

Fig. 5.9 Schematic representation of sound and vibration propagation at the boundary of two different media

A According to (5.18), it follows that regardless of whether $Z_1 \gg Z_2$ or $Z_2 \gg Z_1$, the reflection coefficient is greatest ($|r| \approx 1$), *perfect mismatch*. In the case of vibration transmission, this means that high structure-borne sound insulation is present when impedances on both sides of the interface are very different, e.g. steel (Z_1) and rubber (Z_2) with or $Z_1 \gg Z_2$ structure-borne sound transmission via a barrier mass or a high input impedance with $Z_2 \gg Z_1$.

In order to keep the noise generation of machines as low as possible, the transmission of structure-borne noise to neighbouring construction parts must be prevented. In principle, this can be achieved by creating a *perfect mismatch* at critical points by means of different impedances.

Soft intermediate layers and barrier masses considerably reduce the transmission of structure-borne sound in higher frequency ranges. Basically, the following rules must be observed when mismatching impedances:

- The mechanical input impedance of a soft interlayer (or an elastic bearing element) must be considerably smaller than the mechanical input impedance of the connected structural elements.
- The mechanical input impedance of a blocking mass must be significantly greater than the mechanical input impedance of the connected construction elements.

The effect of the insulation of structure-borne sound can very quickly be nullified by structure-borne sound bridges and secondary paths of structure-borne sound transmission.

As already described in Sect. 2.2.4, the mechanical input impedance Z_e is decisive for the structure-borne sound transmission. According to (2.26), the vibration velocity of the structure excited by the alternating force F at the initiation point "v_e" can be given as follows [13]:

$$v_e = \frac{F}{Z_e} \quad \text{m/s} \tag{5.19}$$

Z_e is a frequency-dependent quantity determined by the dynamic properties of the structure. If the input impedance is small at certain frequencies, the structure can be easily excited to vibrate at those frequencies. This is referred to as resonant frequencies, which usually also coincide with the natural frequencies of the structure. Depending on the excitation spectrum, the impedance dips of the mechanical input impedance can cause a high vibration or structure-borne sound excitation of the structure and consequently also generate a high noise level.

In the case of antiresonances, i.e. large input impedances, the structure can be excited to less structure-borne vibrations (velocity v).

The mechanical input impedance depends on the dynamic properties of the structure, the modulus of elasticity, the mass occupancy, the damping and the clamping conditions. It is not a function of operating conditions. In other words, low input impedances at certain frequencies are only critical if excitation frequencies that depend on the operating data are also present there.

In order to avoid resonance excitations, it is essential to ensure that the excitation frequencies originating from the operation of the machine do not lie in the range of the impedance dips, i.e. the natural structural frequencies. For the acoustic design of the force application points, it is therefore necessary to know the mechanical input impedance. This can best be determined by measurements of the dynamic mass, see (2.33), e.g. by impact tests. Using (5.19), the magnitude of the dynamic mass m_b can be determined as follows:

$$m_b = \frac{Z_e}{\omega} = \frac{F}{v_e \cdot \omega} = \frac{F}{a_e} \quad \text{kg} \tag{5.20}$$

Here are:

$\omega = 2 \cdot \pi \cdot f$ 1/s	Circular frequency
$a_e = v_e \cdot \omega$ m/s^2	Vibration acceleration at the point of discharge

The dynamic mass at the point of force application corresponds to the input impedance related to the angular frequency and indicates in kg the actual resistance of the structure to excitation forces. It is severely frequency-dependent and, with the exception of a compact mass, becomes smaller and smaller with increasing frequency for all other structures. This is one of the reasons why many structures can be relatively easily excited to structure-borne vibrations at high frequencies.

Figure 5.10 shows the dynamic mass of a flat steel plate (1000 × 500 × 6 mm) measured on a model, exposed on the short side. For comparison, the idealised curve according to (2.39) is also given as the mean value of the mechanical input impedance.

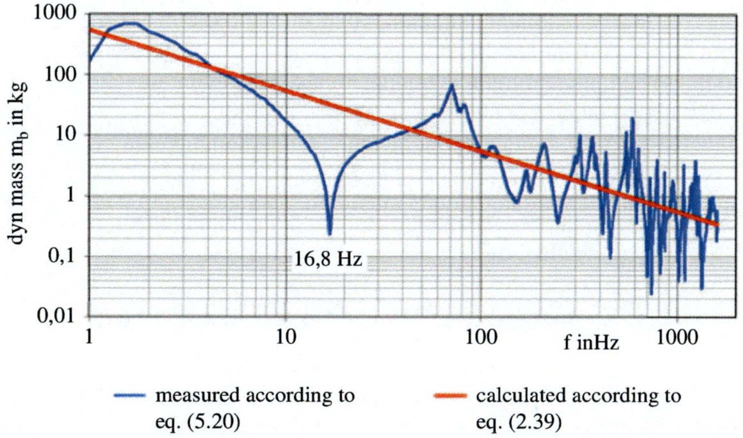

Fig. 5.10 Measured dynamic mass of a 6 mm steel plate

From this, on the one hand, one can see the position of the natural frequencies, and on the other hand, one obtains the actual impedance or dynamic mass, the resistance of the structure at the point of force application. At the first natural frequency, 16.8 Hz, is $m_b \approx 0.25$ kg. Without the natural frequency at this point, the actual resistance would be much higher, about 32 kg, the red curve.

As already described in Sect. 2.2.4.2, the first natural frequencies are particularly critical, since the probability of resonance excitation is high due to the low natural frequency density.

The reduction of vibration transmission can be achieved by various measures:

1. Avoidance of resonances caused by the excitation frequency coinciding with the natural frequencies by changing the natural frequencies. This can be achieved by detuning the structure at the force application point, e.g. by stiffening. Care must be taken to ensure that no excitation frequencies are present in the region of the altered natural frequencies. A distance of the relevant natural frequency of approx. ±10% is not critical with regard to resonance excitation. The 10% refers to the usual system damping that occurs in practice, i.e. a loss factor of $\eta \approx 0.03 \ldots 0.06$.
2. Change of the excitation frequencies by changing the operating parameters, e.g. speed, number of teeth, number of blades, etc. Here, too, care must be taken to ensure that the new excitation frequencies have sufficient distance to the natural frequencies.
3. Increase of the system damping compared to the existing damping. This primarily achieves an increase in the dynamic masses in the range of the natural frequencies and consequently also a reduction in the vibration velocity.
4. Increase of the dynamic mass at the point of introduction, e.g. by force-locking coupling of a compact mass at the point of introduction. The following requirements must be met:

(a)	The increase in dynamic mass shall be at least a factor of two higher than the existing dynamic mass at the coupling point.
(b)	The coupling points should be small (point-shaped coupling) in relation to the relevant wavelength. A point-like coupling is present if the dimensions of the coupling point are $<\lambda/6$. Usually, the governing wavelengths are the bending waves λ_B, see (2.36). Line and/or plane coupling is not suitable for this purpose.

The increase in dynamic mass m_b due to the additional mass m_z at the injection point also increases the input impedance Z_e:

$$Z_{e,z} = (m_b + m_z) \cdot \omega = (m_b + m_z) \cdot 2\pi \cdot f \quad \text{kg/s} \tag{5.21}$$

According to (5.19), the vibration velocity $v_{e,z}$ modified by the additional mass is obtained with the additional mass:

$$v_{e,z} = \frac{F}{(m_b + m_z) \cdot \omega} \quad \text{m/s} \tag{5.22}$$

The vibration velocity without additional mass is then obtained:

$$v_e = \frac{F}{m_b \cdot \omega} \quad \text{m/s} \tag{5.23}$$

The reduction in vibration velocity level achieved by the additional mass can then be determined as follows:

$$\Delta L_v = 20 \cdot \lg\left(\frac{v_e}{v_{e,z}}\right) = 20 \cdot \lg\left(\frac{m_b + m_z}{m_b}\right) \quad \text{dB} \tag{5.24}$$

It follows from this that increasing the dynamic mass at the point of introduction only leads to a reduction in the vibration velocity or its level if the additional mass is significantly higher, at least by a factor of 2, than the existing dynamic mass at the point of introduction. This presupposes, of course, that the dynamic mass at the point of force application is determined beforehand, e.g. by measurements. In Chap. 6 (Application examples), various possibilities for increasing the input impedance are also presented.

5.1.2.3 Influence of Damping on Structure-Borne Sound Transmission

Once structure-borne sound has been generated, the vibration energy must be extracted from the component, i.e. usually converted into thermal energy. This process is called damping. Damping is achieved by using lossy materials with a certain internal damping, by friction losses at contact surfaces or by coating weakly damping materials (metals) with strongly damping substances (high polymers, sand).

The mechanisms of damping are best explained by the oscillatory behavior of a damped single-mass oscillator (SMO). According to (2.27) the equation of motion is:

$$m \cdot \ddot{x} + x \cdot k_F + d \cdot \dot{x} = F \quad \text{N} \tag{5.25}$$

Here F is the periodic excitation force with the angular frequency $\omega = 2\pi f$:

$$F = \widehat{F} \cdot e^{j\omega t} \quad \text{N}$$

The solution of the differential equation (5.25) depends on the type of vibration – free or forced vibrations.

Free Vibrations

The characteristics for damping can best be explained by free vibrations. The solution of (5.25) for shock excitation, for $F = 0$, is [14–17]:

$$x = x_0 \cdot e^{-\delta \cdot t} \cdot \sin(\omega_d \cdot t) \quad \text{m} \tag{5.26}$$

With

$$\delta = \vartheta \cdot \omega_0 = \frac{d}{2 \cdot m} \quad \frac{1}{\text{s}} \qquad \text{Decay constant} \tag{5.27}$$

$$\vartheta = \frac{d}{2 \cdot m \cdot \omega_0} \qquad \text{Degree of damping} \tag{5.28}$$

$$d = 2 \cdot \vartheta \cdot \omega_0 \cdot m \quad \frac{\text{kg}}{\text{s}} \qquad \text{Damping coefficient} \tag{5.29}$$

$$\omega_0 = \sqrt{\frac{k_F}{m}} = 2 \cdot \pi \cdot f_0 \quad 1/\text{s} \qquad \text{Natural angular frequency of the undamped oscillation} \tag{5.30}$$

$$f_0 = \frac{1}{T_0} \quad \text{Hz} \qquad \text{Natural frequency of the undamped oscillation}$$

$$\omega_d = \sqrt{\omega_0^2 - \delta^2} = 2 \cdot \pi \cdot f_d \quad 1/\text{s} \qquad \text{Natural angular frequency of the damped oscillation} \tag{5.31}$$

$$f_d = \frac{1}{T_d} \quad \text{Hz} \qquad \text{Natural frequency of the damped oscillation}$$

T_0, T_d Period of oscillation in s

$$\eta = \frac{2 \cdot \delta}{\omega_0} = \frac{W_v}{2\pi \cdot W_r} \qquad \text{Loss factor} \tag{5.32}$$

W_v	Energy lost or converted into heat per vibration
W_r	Energy recovered per vibration

The loss factor "η" according to (5.32) is a suitable quantity for the quantitative recording of structure-borne sound attenuation.

With the help of the maxima of the equation of motion (5.26) it is possible to determine experimentally the decay constant "δ" and the loss factor "η". The times at which the positive maxima lie coincide with the maxima of the sine function, i.e.: $\sin(\omega_d \cdot t_{max}) = 1$.

$$t_{\max} = t_i = \frac{2n \cdot \pi + \pi/2}{\omega_d} \quad \text{s} \qquad (n = 0, 1, 2, \ldots) \tag{5.33}$$

The period T_d of the damped vibration according to (5.26) can be determined as follows:

$$T_d = \frac{2\pi}{\omega_d} \quad \text{s} \tag{5.34}$$

The positive maxima are then:

$$x_i = x_0 \cdot e^{-\delta \cdot t_i} \quad \text{m} \tag{5.35}$$

$$x_{i+n} = x_0 \cdot e^{-\delta \cdot t_{i+n}} \quad \text{m} \tag{5.36}$$

With

$$t_{i+n} = t_i + n \cdot T_d \quad \text{s} \tag{5.37}$$

For the determination of the decay constant, (5.35) and (5.36) are logarithmised:

$$\begin{aligned} \ln(x_i) &= \ln(x_0) - \delta \cdot t_i \\ \ln(x_{i+n}) &= \ln(x_0) - \delta \cdot t_{i+n} = \ln(x_0) - \delta \cdot (t_i + n \cdot T_d) \end{aligned}$$

respectively

$$\ln(x_i) - \ln(x_{i+n}) = -\delta \cdot t_i + \delta \cdot (t_i + n \cdot T_d) = n \cdot \delta \cdot T_d$$

The quantity is called $\delta \cdot T_d$ the logarithmic decrement "Λ". It can be determined from two amplitudes which are separated from each other over the time *n period* T_d:

$$\Lambda = \delta \cdot T_d = \frac{1}{n} \cdot \ln \frac{x_i}{x_{i+n}} = \frac{2\pi \cdot \delta}{\omega_d} = \frac{2\pi \cdot \delta}{\sqrt{\omega_0^2 - \delta^2}} \tag{5.38}$$

With the help of the logarithmic decrement, which can be determined relatively easily by measurement, the decay constant is obtained:

$$\delta = \frac{\Lambda \cdot \omega_0}{\sqrt{\Lambda^2 + 4 \cdot \pi^2}} \quad 1/\text{s} \tag{5.39}$$

The loss factor can then be determined using (5.32) as follows:

$$\eta = \frac{2 \cdot \Lambda}{\sqrt{\Lambda^2 + 4 \cdot \pi^2}} \tag{5.40}$$

Example 5.1
Figure 5.11 shows the amplitude ratio of a damped single-mass oscillator (SMO) for Hz $f_0 = 60$ and a loss factor $\eta = 0.06$ over time. Such vibration curves can be determined, for example, by stop tests or with the aid of a shaker. The amplitudes of the 2nd and 6th maxima are:

$$x_i = x_2 \Rightarrow \frac{x_{2,\max}}{x_0} = 0.79$$
$$x_{i+4} = x_6 \Rightarrow \frac{x_{6,\max}}{x_0} = 0.37; \quad n = 4$$

From this you get the logarithmic decrement:

$$\Lambda = \frac{1}{4} \cdot \ln \frac{0.79}{0.37} \approx 0.19$$

The loss factor and the decay constant are then obtained:

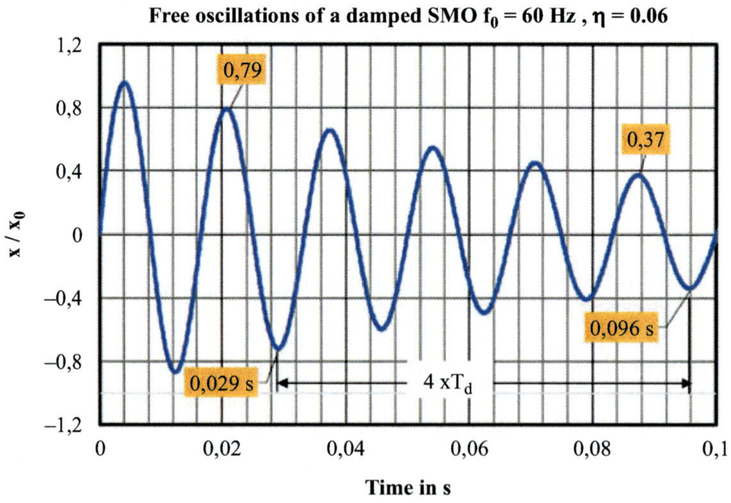

Fig. 5.11 Amplitude ratio of a damped single-mass oscillator

$$\eta = \frac{2 \cdot 0.19}{\sqrt{0.19^2 + 4 \cdot \pi^2}} \approx 0.06$$

$$\delta = \frac{\Lambda \cdot \omega_0}{\sqrt{\Lambda^2 + 4 \cdot \pi^2}} \quad 1/\mathrm{s}$$

$$4 \cdot T_\mathrm{d} = 0.096 - 0.029 = 0.067 \quad \mathrm{s}$$

$$T_\mathrm{d} = 0.01675 \quad \mathrm{s}$$

$$f_\mathrm{d} = \frac{1}{T_\mathrm{d}} = 59.7 \quad \mathrm{Hz}$$

$$\delta = \frac{\Lambda}{T_\mathrm{d}} = \frac{0.19}{0.01675} \approx 11.3 \quad \frac{1}{\mathrm{s}}$$

Using (5.31), the natural frequency of the undamped vibration can be calculated:

$$\omega_0 = \sqrt{\omega_\mathrm{d}^2 + \delta^2} = \sqrt{(2\pi \cdot 59.7)^2 + 11.3^2} \approx 375.3 \quad \frac{1}{\mathrm{s}}$$

$$f_0 = \frac{375.3}{2\pi} \approx 59.73 \quad \mathrm{Hz}$$

From this it follows that the essential damping values can be determined by decay tests. The natural frequencies of the damped vibrations are always slightly smaller than the natural frequencies of the undamped vibrations. For the usual damping in practice, the differences are negligibly small.

Forced Vibrations

The machine vibrations (structure-borne noise) occurring in practice are usually forced vibrations. Here, as already mentioned, the structure always vibrates at the frequency of the excitation. In the case of harmonic excitation, the solution of (5.25) is [10]:

$$x = \widehat{x} \cdot \mathrm{e}^{\mathrm{j}\omega t} \quad \mathrm{m} \tag{5.41}$$

Here is

$\omega = 2 \cdot \pi \cdot f$	Excitation circuit frequency in 1/s
f	Excitation frequency in Hz

Equation (5.41) states that the excited structure in forced vibrations always oscillates only with the excitation frequencies. Critical here are only the excitation frequencies which lie in the range of structure natural frequencies.

Depending on the frequency, a distinction is made between vibration and structure-borne sound transmission, which will not be discussed in detail here [18, 19].

The damping behavior of structures in forced vibrations differs significantly from free vibrations. Damping is basically effective only in the range of resonances, i.e. when the structure natural frequencies (the system constant) and the excitation frequencies match, depending on the operation of the machine. Figure 2.12 shows the principle effect of damping in the case of forced vibrations. From this it can be seen that damping is primarily effective only in the range of natural frequencies (impedance dips in frequency range II of Fig. 2.12). This means that the damping increases the impedance in the range of the natural frequencies and thus the amplitudes become smaller during resonance excitation.

In order for the damping measures to be effective, they must be significantly higher – at least by a factor of 5–10 – compared to the existing system damping. When using structure-borne sound damping, the following questions should therefore be clarified beforehand:

1. Do the structures to be damped perform resonant vibrations?
2. How high is the existing system attenuation?

System damping is the structural damping under real installation conditions. As a rule, the system damping is significantly higher than the material damping. This is caused, among other things, by additional friction surfaces that arise, for example, in screw connections.

In the case of broadband excitation (e.g. noise), the effect of attenuation is essentially perceived as a reduction in loudness. In the case of impulsive or abrupt excitation, the attenuation is perceived as a shortening of the noise duration.

If the structures in the resonance range are excited with a sine tone, the result is a level reduction:

$$\Delta L = 20 \cdot \lg \frac{\eta_{\mathrm{n}}}{\eta_{\mathrm{v}}} \quad \mathrm{dB} \tag{5.42}$$

η_{v}	Loss factor before measure
η_{n}	Loss factor after measure

If there is an excitation range that includes several resonances (normal case), the following applies:

$$\Delta L = 10 \cdot \lg \frac{\eta_{\mathrm{n}}}{\eta_{\mathrm{v}}} \quad \mathrm{dB} \tag{5.43}$$

A level reduction or noise reduction can only be achieved if the damping is noticeably increased (approx. factor 10) compared to the existing damping.

An increase of the loss factor and thus an increase of the damping can be achieved in practice as follows:

- Use of materials with high internal damping, e.g. plastic, rubber, ceramic, lead or copper have higher damping than steel.[1]
- Exploiting the friction between construction elements and dissipating the energy to the "outside", e.g. a bolted construction has a higher damping than a welded construction.
- Combination of undamped, force-carrying construction parts with strongly damped materials (e.g. coating with anti-drumming linings or use of composite sheet metal constructions)

5.1.2.4 Influence of Elastic Decoupling on Structure-Borne Sound Transmission

The task of elastic decoupling is to reduce the transmitted alternating forces, consequently also the vibrations, on load-bearing structures, e.g. foundations. Figure 5.12 shows the force excitation of a structure schematically [18].

Here are:

m	Effective mass of the machine on the spring element in kg
k	Spring stiffness of the insulation in N/m
$d = 2 \cdot m_1 \cdot \omega_0 \cdot \vartheta$	Damping coefficient of the insulation in kg/s, [14]
ϑ	Degree of damping of the insulation
$\omega_0 = 2 \cdot \pi \cdot f_0$	Natural angular frequency of the insulation in 1/s

The mode of action of elastic decouplings is essentially influenced by the transmitting frequency f. Two cases are to be distinguished here:

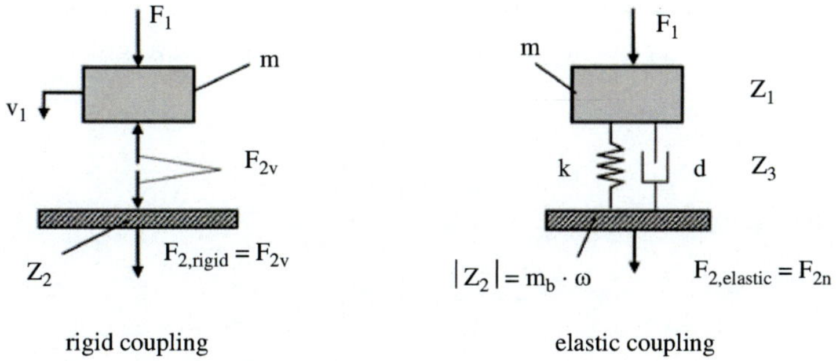

Fig. 5.12 Schematic representation of the force excitation of a structure

[1] As mentioned above, in practice, the damping of structures is usually determined largely by the system damping, structural damping under real installation conditions, and less by the internal damping or material damping.

(I)	Vibration isolation of low frequencies, $f <$ approx. 100 Hz
(II)	Structure-borne sound insulation of higher frequencies, $f >$ approx. 100 Hz

Vibration isolation, which is not discussed in detail here, is primarily intended to reduce the transmission of vibrations to the foundation or connected structures. With regard to vibration isolation, reference is only made to the relevant literature [10, 14–18].

With the help of structure-borne sound insulation, the noise generation of the connected structures is to be reduced in the first instance. In contrast to vibration isolation, where essentially the 1st and 2nd natural frequency of the structure or foundation is taken into account [14, 18], at higher frequencies the influence of numerous natural frequencies must be taken into account [20].

Due to the very small vibration displacement at higher frequencies (structure-borne sound), the dynamic behaviour and the effective mechanism of the spring elements also differ from the dynamic behaviour at lower frequencies ($f < 100$ Hz). In the case of structure-borne sound transmission, the isolating effect is not achieved by the phase difference between the upper and lower edges of the spring element, as in the case of vibration isolation, but primarily by elasticity and impedance differences at the coupling point [21].

The magnitude of the amplitude frequency response of the floor force under force excitation $|\alpha|$ can be represented in general terms for vibration and structure-borne sound transmission as follows [10]:

$$|\alpha| = \left|\frac{F_{2,\text{elastic}}}{F_{2,\text{rigid}}}\right| = \left|\frac{F_{2n}}{F_{2v}}\right| = \left|\frac{1 + \underline{Z_1}/\underline{Z_2}}{1 + \underline{Z_1}/\underline{Z_2} + \underline{Z_1}/\underline{Z_3}}\right| = \left|\frac{Z_3}{Z_3 + \frac{Z_1}{K_Z}}\right| \tag{5.44}$$

The quantities and impedances given in (5.44) are defined for structure-borne sound insulation as follows. For better comparison, the impedances are given as dynamic mass $m_\text{b} = |Z|/\omega$:

$$K_Z = 1 + \frac{Z_1}{Z_2} \tag{5.45}$$

$$Z_1 = \frac{F_1 - F_2}{v_1} = jm_{\text{b}1} \cdot \omega \quad \text{kg/s} \tag{5.46}$$

$$|Z_1| = |j \cdot m_{\text{b}1} \cdot \omega| = m_{\text{b}1} \cdot \omega \quad \text{kg/s}$$

$$m_{\text{b}1} = \frac{|Z_1|}{\omega} \approx m \quad \text{kg} \tag{5.47}$$

$$Z_2 = \frac{F_2}{v_2} \quad \text{kg/s}$$

$$m_{b2} = \frac{|Z_2|}{\omega} \quad \text{kg} \tag{5.48}$$

The impedance or dynamic mass of the structure is fundamentally a complex number / parameter. For structure-borne sound insulation, the amount of impedance or dynamic mass is of primary interest. The dynamic mass can best be determined by measurements under real clamping or installation conditions, e.g. by impact tests [7, 10]:

$$m_{b2} = \frac{|Z_2|}{\omega} = \left|\frac{F}{a_e}\right| \quad \text{kg} \tag{5.49}$$

F	Excitation force introduced into the structure by means of a shaker or impulse hammer
a_e	Vibration acceleration at the point of initiation caused by the excitation force F

In contrast to vibration isolation, where the mass of the spring elements is neglected, the influence of the spring mass cannot be neglected in the case of structure-borne sound isolation ($f > 100$ Hz). This is because the dynamic mass of the spring elements decreases with the square of the frequency. Physically, however, the dynamic mass of the spring element cannot become zero, at least not significantly smaller than the actual mass of the spring element. Considering the spring mass m_F, the impedance of the spring element can be given by the following approach [10, 22]:

$$\left.\begin{aligned} &Z_3 = \frac{F_2}{v_1 - v_2} = j m_F \cdot \omega + d - j\frac{k}{\omega} \\ &\text{bzw.} \\ &Z_3 = 2 \cdot m \cdot \omega_0 \cdot \vartheta_1 + j \cdot \left(m_F \cdot \omega - \frac{m \cdot \omega_0^2}{\omega}\right) \end{aligned}\right\} \quad \text{kg/s} \tag{5.50}$$

With

$$\eta = \frac{\omega}{\omega_0} = \frac{f}{f_0} \tag{5.51}$$

and

$$k = m \cdot \omega_0^2 \quad \text{N/m} \quad \text{bzw.} \quad \omega_0^2 = k/m \quad 1/\text{s}^2$$

the dynamic mass of the spring element can be determined as follows:

$$m_{b3} = \frac{|Z_3|}{\omega} = \frac{m}{\eta^2}\sqrt{4\vartheta^2\eta^2 + \left(1 + \frac{m_F}{m}\cdot\eta^2\right)^2} \quad \text{kg} \tag{5.52}$$

For the structure-borne sound insulation, one can replace the parameter K_Z by its magnitude $|K_Z|$ in (5.44) in a good approximation:

$$|\alpha| = \left|\frac{Z_3}{Z_3 + \frac{Z_1}{|K_Z|}}\right| \tag{5.53}$$

Using (5.46) and (5.50), after some transformations, the amplitude frequency response of the floor force under force excitation is obtained as a function of the frequency ratio η [10]:

$$|\alpha(f)| = \sqrt{\frac{\left(1 - \frac{m_F}{m}\cdot\eta^2\right)^2 + 4\vartheta^2\cdot\eta^2}{\left(1 - \frac{m_F}{m}\cdot\eta^2 - \frac{1}{|K_Z|}\cdot\eta^2\right)^2 + 4\vartheta^2\cdot\eta^2}} \tag{5.54}$$

Equation (5.54) represents an extended approach to structure-borne noise isolation. For $m_F = 0$ and $K_Z = 1$, as usually used in practice for classical vibration isolation, i.e. massless spring element and high foundation impedance ($Z_2 \approx \infty$), the relation for vibration isolation [14] is obtained from (5.54).

$$|\alpha(f)| = \sqrt{\frac{1 + 4\vartheta^2\cdot\eta^2}{(1-\eta^2)^2 + 4\vartheta^2\cdot\eta^2}} \tag{5.55}$$

The magnitude of $|K_Z|$ is best determined by measurements of the dynamic mass m_{b2} of the structure or foundation according to (5.49):

$$|K_Z| = \left|1 + \frac{Z_1}{Z_2}\right| \approx 1 + \frac{m}{m_{b2}} \tag{5.56}$$

For structural members such as slabs, beams, etc., m_{b2} can also be determined theoretically for idealised edge restraints. For plate-like structures, e.g. slabs, m_{b2} can be determined as follows [19]:

$$m_{b2} = \frac{4}{\pi}\cdot\sqrt{\frac{E\cdot\rho}{12\cdot(1-\mu^2)}}\cdot\frac{h^2}{f} \quad \text{kg} \tag{5.57}$$

with

m_{b2}	Dynamic mass of the structure or foundation
E	Modulus of elasticity of the plate in N/m^2
ρ	Density of the plate in kg/m^3
μ	Transverse contraction coefficient
h	Slab thickness in m
f	Frequency in Hz

With the help of (5.57) an analytical calculation of the amplitude frequency response according to (5.54) would be possible in principle. The disadvantage here is that the impedance dips at the natural frequencies cannot be taken into account.

If the foundation vibrations with rigid coupling of the machine (v_{before}) are known, the expected foundation vibration after isolation $|\alpha(f)|$ of the machine (v_{after}) can be determined with the aid of the magnitude of the amplitude frequency response:

$$v_{\text{after}}(f) = v_{\text{previously}}(f) \cdot \mid \alpha(f) \mid \quad \text{m/s} \tag{5.58}$$

For structure-borne sound transmission, the insertion loss $D(f)$ is usually given in dB instead of the magnitude of the amplitude frequency response $|\alpha(f)|$. If in (5.44) instead of the impedances their reciprocals, the so-called admittances or mobilities "h", are inserted, the insertion loss $D(f)$ can be determined as follows [19, 23]:

$$D(f) = 20 \cdot \lg \frac{1}{\mid \alpha(f) \mid} = 10 \cdot \lg \left| \frac{h_1 + h_2 + h_3}{h_1 + h_2} \right|^2 \quad \text{dB} \tag{5.59}$$

with

$$h_1 = \frac{1}{Z_1}; \qquad h_2 = \frac{1}{Z_2}; \qquad h_3 = \frac{1}{Z_3}$$

h_1, h_2, h_3	Admittances or mobilities in s/kg.

Example 5.2
To test the structure-borne sound insulation, a mass of kg $m = 10.36$ was elastically placed on a steel plate in a model test [18] as a substitute mass for the effective mass of the machine with the aid of a spring element.

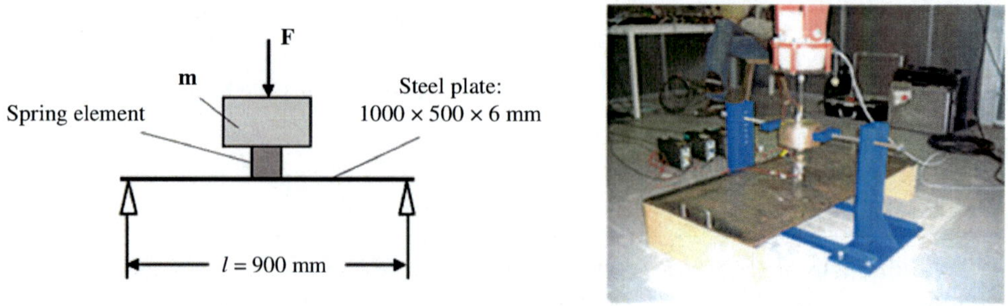

Fig. 5.13 Test setup for testing the structure-borne sound insulation

For the metrological determination of the amplitude frequency response of the floor force during force excitation, the floor forces (F_2), see Fig. 5.13, were measured with and without spring element or before and after insulation according to (5.60):

$$|\,\alpha\,| = \left|\frac{F_{2,\text{elastisch}}}{F_{2,\text{starr}}}\right| = \left|\frac{F_{2\text{n}}}{F_{2\text{v}}}\right| \tag{5.60}$$

The following spring elements were used:

Steel spring:

$$\begin{aligned} k_{\text{Steel spring}} &= 25000\,\text{N/m}; \qquad \vartheta \approx 0.03 \\ f_{0,\text{Steel spring}} &= \frac{1}{2\cdot\pi}\cdot\sqrt{\frac{k_{\text{Steel spring}}}{m}} \approx 7.5 \quad \text{Hz} \\ m_{\text{F}} &= 0.098\,\text{kg} \end{aligned}$$

Rubber element:

$$\begin{aligned} k_{\text{Rubber element}} &\approx 300,000\,\text{N/m}; \qquad \vartheta_1 \approx 0.06 \\ f_{0,\text{Rubber element}} &= \frac{1}{2\cdot\pi}\cdot\sqrt{\frac{k_{\text{Rubber element}}}{m}} \approx 27.1 \quad \text{Hz} \\ m_{\text{F}} &= 0.11\,\text{kg} \end{aligned}$$

Steel plate:
Plate dimensions: 1000 × 500 × 6 mm

$$E_{\text{steel}} = 2.1 \cdot 10^{11}\,\text{N/m}^2; \qquad \mu = 0.3$$
$$h = 6\,\text{mm}$$

Figures 5.14 and 5.15 show the amounts of the amplitude frequency responses $|\alpha(f)|$ of the floor force at low connection impedance (Fig. 5.13) measured and calculated according to (5.54) and (5.60) for the above-mentioned steel spring and the rubber element. For comparison, the corresponding values according to (5.55) are also plotted as classical vibrationisolation.

The dynamic mass m_{b2} required to calculate the amount given in $|K_Z|$ (5.56) was determined both by the measurement given in (5.49) and theoretically given in (5.57).

From this it follows that the calculated values according to (5.54) can be used as a good approximation for the structure-borne sound insulation, $f > 100$ Hz. Furthermore, it can be seen that the relations for the classical vibration isolation according to (5.55) cannot be applied at all for the structure-borne sound isolation. Compared to the classical vibrationon isolation, the approximate calculation according to (5.54) provides a significant improvement.

In Fig. 5.16, the results given in Fig. 5.14 for the steel spring are shown as the insertion isolation dimension $D(f)$ according to (5.59). As a comparison, the corresponding results of the classical vibration isolation according to (5.55) are also given with [10, 19].

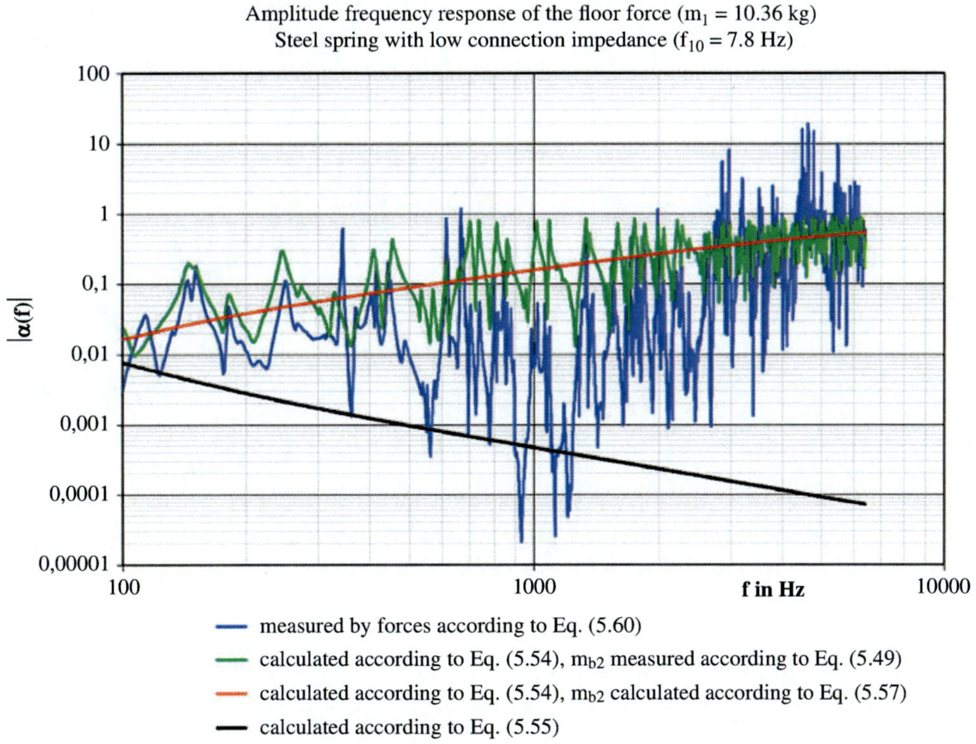

Fig. 5.14 Amplitude frequency responses $|\alpha(f)|$ of the floor force of the steel spring at low connection impedance

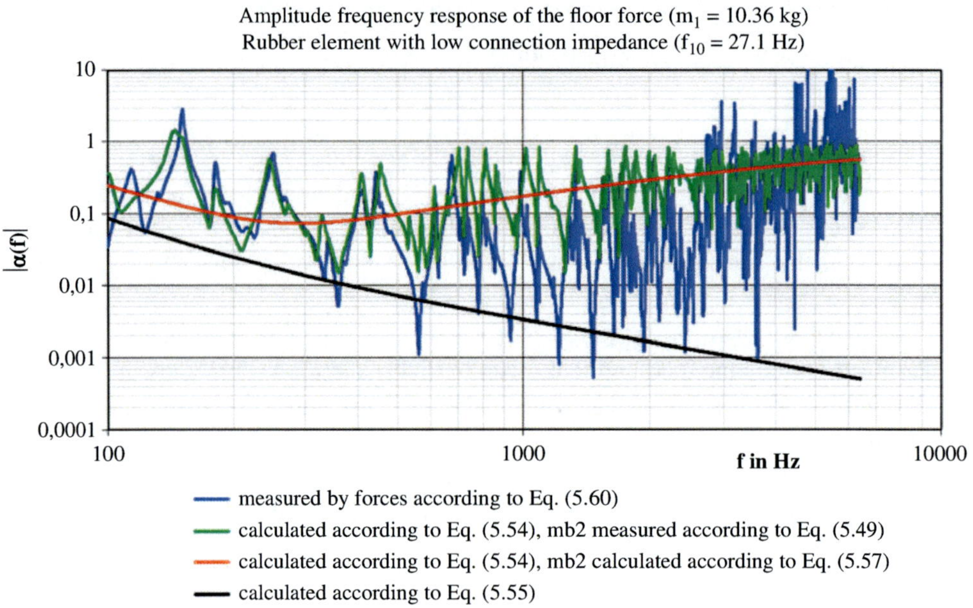

Fig. 5.15 Amplitude frequency responses $|\alpha(f)|$ of the floor force of a rubber element at low connection impedance

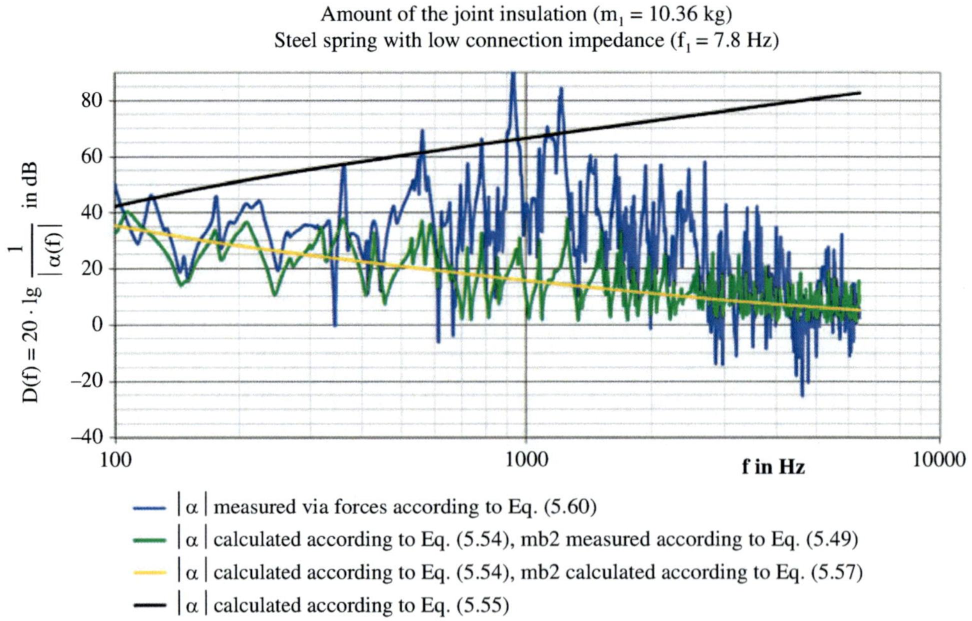

Fig. 5.16 Insertion loss according to (5.59) of a steel spring with low connection impedance

The differences in the high-frequency range are primarily caused by the natural vibration behaviour of the spring elements; further investigation results and explanations are compiled in [18, 19].

5.1.2.5 Influencing the Radiation of Structure-Borne Sound

As described in Sect. 2.2.6, the degree of radiation σ, which indicates the conversion of structure-borne sound into airborne sound, is decisive for structure-borne sound radiation.

The radiated structure-borne sound power is defined as follows, see (2.47):

$$P_K = \rho \cdot c \cdot \overline{\tilde{v}^2} \cdot S \cdot \sigma \quad \text{W} \tag{5.61}$$

With

$$P_0 = \rho_0 \cdot c_0 \cdot v_0^2 \cdot S_0 = 10^{-12} \quad \text{W}$$

the radiated structure-borne sound power level L_{WK} of subarea i is obtained, see also Sect. 3.2:

$$L_{WK,i} = 10 \cdot \lg \frac{P_{K,i}}{P_0} = 10 \cdot \lg \frac{\rho \cdot c}{\rho_0 \cdot c_0} + 10 \cdot \lg \frac{\overline{\tilde{v}_i^2}}{v_0^2} + 10 \cdot \lg \frac{S_i}{S_0} + 10 \cdot \lg \sigma_i \quad \text{dB} \tag{5.62}$$

With the assumption that for usual structure-borne sound radiation the difference between density and sound velocity of the environment and the corresponding reference values are negligible, i. e. $\rho \cdot c \approx \rho_0 \cdot c_0$, the structure-borne sound power level of the partial area S_i *is* obtained from (5.62):

$$\left.\begin{aligned} L_{WK,i} &\approx 10 \cdot \lg \frac{\overline{\tilde{v}_i^2}}{v_0^2} + 10 \cdot \lg \frac{S_i}{S_0} + 10 \cdot \lg \sigma_i \\ L_{WK,i} &\approx L_{v,i} + L_{S,i} + \sigma_i' \end{aligned}\right\} \quad \text{dB} \tag{5.63}$$

$\overline{\tilde{v}_i^2}$	RMS value of the mean square of velocity in m²/s²
S_i	Radiating surface in m²
$\rho \cdot c$	Sound characteristic impedance of the air in N s/m³
P_K	Radiated structure-borne sound power in W
$L_{WK,\,i}$	Radiated partial structure-borne sound power level in dB
$L_{v,i}$	Average velocity level in dB
$L_{S,i}$	Area measure in dB
σ_i'	Logarithm of the radiation factor of the surface S_i in dB

The total structure-borne sound power level of a machine is obtained by power level addition of all relevant partial sound power levels:

$$L_{W_K} = 10\lg \sum_{i=1}^{n} 10^{L_{W_{K,i}}/10} \quad \text{dB} \tag{5.64}$$

With the aid of (5.64) it is possible, among other things, to localise the partial surfaces and/or components which are decisive for the total structure-borne sound radiation.

The metrological determination of the partial structure-borne sound power levels is usually carried out by measuring the acceleration levels L_a:

$$L_{W_{K,i}} \approx L_{a_i} + L_{S_i} + \sigma'_i + 20\lg \frac{a_0}{v_0 \cdot \omega} \quad \text{dB} \tag{5.65}$$

With

$$L_{v_i} = 10\lg \frac{\overline{\tilde{v}_i^2}}{v_0^2} = 20\lg \frac{\overline{\tilde{v}_i}}{v_0} = 20\lg \frac{\overline{\tilde{a}_i} \cdot a_0}{v_0 \cdot a_0 \cdot \omega} = L_{a_i} + 20\lg \frac{a_0}{v_0 \cdot \omega} \quad \text{dB} \tag{5.66}$$

$$L_{a_i} = 20\lg \frac{\overline{\tilde{a}_i}}{a_0} \quad \text{dB} \quad \text{Acceleration level on the partial surface} A_i$$

$\overline{\tilde{a}_i}$	RMS value of the vibration acceleration as mean value for the partial area S_i in m/s^2
a_0	Reference acceleration (freely selectable), e.g.: $a_0 = 10^{-6}\,m/s^2$
$v_0 = 5 \cdot 10^{-8}\,m/s$	Quick reference

The radiation factor σ or logarithm of the radiation factor σ' can be calculated theoretically for simple radiator types [10, 12, 24, 25]. If the sound power level is known, it can also be determined by measurement using (5.65):

$$\sigma'_i \approx L_{W_{K,i}} - L_{a_i} - L_{S_i} - 20\lg \frac{a_0}{v_0 \cdot \omega} \quad \text{dB} \tag{5.67}$$

respectively

$$\sigma = 10^{\frac{\sigma'}{10}}$$

Due to the frequency dependence of the vibration acceleration, the radiation factor or the logarithm of the radiation factor is of course also strongly frequency-dependent.

Using the example of plate-like structures, which are usually responsible for the structure-borne sound radiation of machines (e.g. housings) or buildings (e.g. walls), the main influencing parameters are determined or described below.

Coincidence Effect, Radiation Factor

As already mentioned, the radiation coefficient indicates the proportion of structural vibrations that are radiated as airborne sound ("structure-borne sound"). Of particular importance here is the relationship between the airborne wavelength λ_L and the bending wavelength of the plate λ_B.

In plate-like structures, it is primarily the bending waves that are in question for structure-borne sound radiation. The bending wave velocity "c_B" in solid structures, e.g. machine housings, can be determined as follows [10, 12]:

$$c_B = \sqrt{2\pi} \cdot \sqrt[4]{\frac{B'}{m''}} \sqrt{f} \quad \mathrm{m/s} \tag{5.68}$$

With

B'	Stiffness of the structure in relation to the length in N m
m''	Mass occupancy in kg/m^2

For plane plates of thickness h is:

$$B' = \frac{E \cdot h^3}{12 \cdot (1 - \mu^2)} \quad \mathrm{N\,m} \tag{5.69}$$

$$m'' = \rho_P \cdot h \quad \mathrm{kg/m^2} \tag{5.70}$$

ρ_P	Density of the plate in kg/m^3
f	Frequency in Hz

The bending wavelength can then be determined as follows:

$$\lambda_B = \frac{c_B}{f} = \sqrt{2\pi} \cdot \sqrt[4]{\frac{B'}{m''}} \frac{1}{\sqrt{f}} \quad \mathrm{m} \tag{5.71}$$

In air, the speed of sound is essentially a function of temperature and is not frequency dependent:

$$c_L = \sqrt{\kappa \cdot R \cdot T} \quad \mathrm{m/s} \tag{5.72}$$

R	Specific gas constant, $R_{air} = 287$ N m/kg K
T	Absolute temperature in K
κ	Adiabatic exponent, $\kappa_{air} = 1.4$

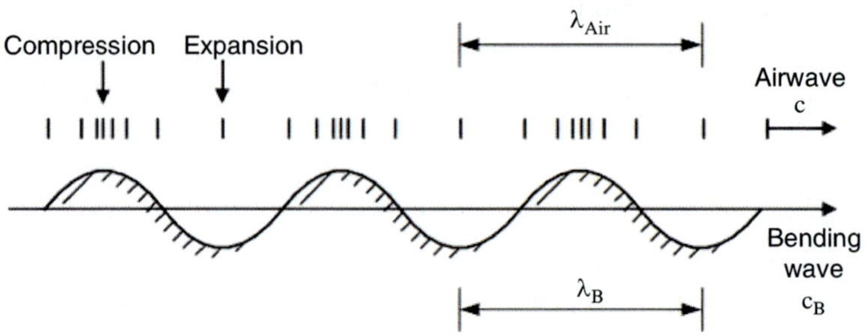

Fig. 5.17 Excitation of sound waves in air by bending waves of a plate ($\lambda_L = \lambda_B$)

The airborne sound wavelength λ_L is then obtained:

$$\lambda_L = \frac{c_L}{f} = \frac{\sqrt{\kappa \cdot R \cdot T}}{f} \quad \text{m} \tag{5.73}$$

As can be seen from (5.71) and (5.73), the bending and air wavelengths have different frequency dependence. The frequency at which the wavelengths and the propagation velocities are the same is called the coincidence frequency f_g. The coincidence with respect to the wavelengths as well as the propagation velocities is called coincidence between bending waves and sound waves in the surrounding air [10]. The transverse bending wave perpendicular to the plate surface produces exactly matching pressure fluctuations in the adjacent air and thus indexes the synchronously travelling sound wave in the air (Fig. 5.17).

The coincidence frequency f_g is obtained by equating the propagation velocities:

$$c_L = \sqrt{\kappa \cdot R \cdot T} = c_B = \sqrt{2\pi} \cdot \sqrt[4]{\frac{B'}{m''}} \sqrt{f_g} \quad \text{m/s}$$

$$f_g = \frac{1}{2\pi} c_L^2 \sqrt{\frac{m''}{B'}} \quad \text{Hz} \tag{5.74}$$

By (5.69) and (5.70), it follows from (5.74):

$$f_g = \frac{c_L^2}{2 \cdot \pi} \cdot \sqrt{\frac{m''}{B'}} = \frac{\sqrt{12 \cdot (1 - \mu^2)}}{2 \cdot \pi} \cdot c_L^2 \cdot \sqrt{\frac{\rho_P}{E}} \cdot \frac{1}{h} \approx 0,526 \frac{c_L^2}{c_{De} \cdot h} \quad \text{Hz} \tag{5.75}$$

At this:

$$c_{De} = \sqrt{\frac{E}{\rho_P}} \quad \text{m/s} \qquad \text{(Strain wave velocity)} \tag{5.76}$$

As can be seen from (5.75), the product $f_g \cdot h$ is a constant and is called the coincidence constant:

Table 5.1 Coincidence constant of some materials

$f_g - h$ in mm/s	Material
50,000	Lead
35,000	Plasterboard
30,000	Plasterboard, hardboard
22,000	Brick masonry
18,000	Concrete. Cement asbestos slabs
12,500	Steel, glass, aluminium

$$f_g \cdot h = \frac{\sqrt{12 \cdot (1-\mu^2)}}{2 \cdot \pi} \cdot c_L^2 \cdot \sqrt{\frac{\rho_P}{E}} \approx 0.526 \frac{c_L^2}{c_{De}} \quad \text{m/s} \tag{5.77}$$

The coincidence constant ($f_g \cdot h$) depends only on the sound velocities of the two media (usually air and metallic materials).

The **coincidence frequency** f_g is also significantly influenced by the material thickness *h. The smaller the wall thickness h, the greater the* **coincidence frequency**. The smaller the wall thickness *h,* the greater the coincidence frequency. Table 5.1 shows the average coincidence constant for different materials.

In Fig. 5.18, the coincidence frequency f_g is plotted as a function of plate thickness *h* and material.

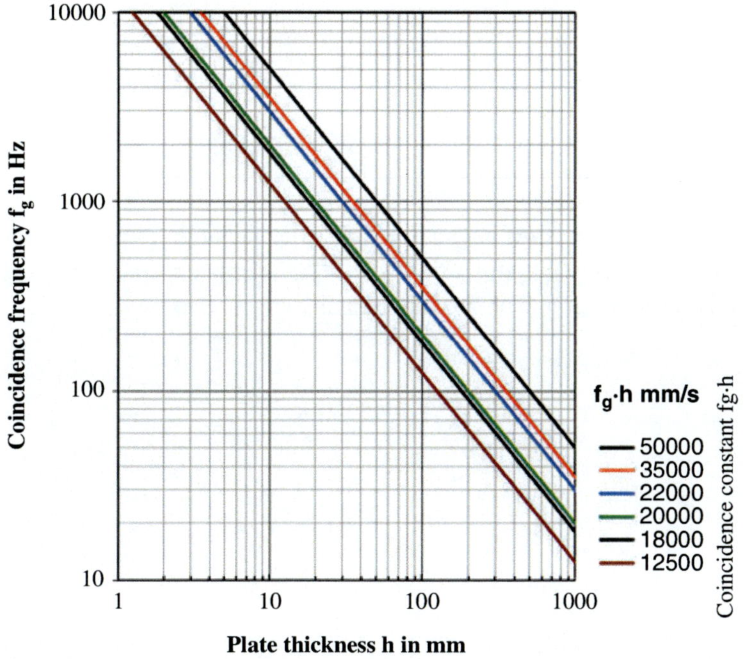

Fig. 5.18 Coincidence frequency f_g as a function of the plate thickness h in mm for different coincidence constants ($f_g \cdot h$)

The coincidence frequency for steel, aluminium and glass plates can be approximated using the following numerical equation:

$$f_{\mathrm{g}} \approx \frac{12000}{h} \quad \mathrm{Hz} \qquad h \quad \mathrm{in\,mm} \tag{5.78}$$

The degree of radiation is significantly dependent on the ratio of the bending wavelengths or on the position of the frequency f compared to the coincidence frequency f_{g} [12].

$$\frac{c_{\mathrm{B}}}{c_{\mathrm{L}}} = \frac{\lambda_{\mathrm{B}}}{\lambda_{\mathrm{L}}} = \sqrt{\frac{f}{f_{\mathrm{g}}}} \tag{5.79}$$

Here, a distinction is made between two cases:

(a) $\lambda_{\mathrm{L}} < \lambda_{\mathrm{B}}$

Here, the air wavelength λ_{L} is smaller than the wavelength of the advancing bending waves. In this case, a plane progressive sound wave field is built up in front of the plate excited by the bending waves, i.e. sound pressure and sound velocity are in phase. The sound wave in the air generated by bending waves can therefore be radiated almost completely, i. e.:

$$\text{für } f > f_g \quad \text{is} \quad \sigma \approx 1 \quad \text{or} \quad \sigma' \approx 0 \quad \mathrm{dB} \tag{5.80}$$

(b) $\lambda_{\mathrm{L}} > \lambda_{\mathrm{B}}$

In this case, coincidence is no longer possible. The radiation is clearly reduced. For the case, an acoustic short circuit now forms $\lambda_L \gg \lambda_B$ at the plate surface, which suppresses the radiation very strongly (Fig. 5.19). σ theoretically approaches zero.

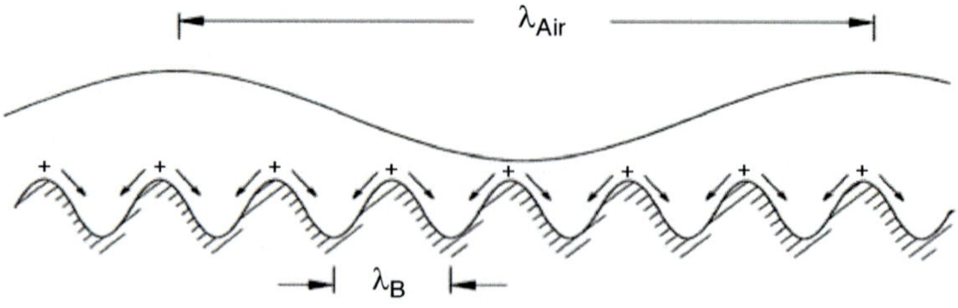

Fig. 5.19 Formation of the acoustic short circuit in front of a plate ($\lambda_L \ll \lambda_B$)

Fig. 5.20 Formation of the acoustic short circuit on a perforated plate excited by bending waves

Such a particularly effective acoustic short circuit can also be achieved specifically on a perforated plate excited to bending vibrations, since a direct pressure equalization between the front and rear sides can form at the perforation (Fig. 5.20).

Figure 5.21 shows the wavelengths of sound in air (20°C; $R = 287$ N m/kg K) and in a steel plate ($E = 2.1 \cdot 10^{11}$ N/m^2; $\rho_P = 7850$ kg/m^3; $h = 10$ mm) as a function of frequency.

In the frequency range ($f > f_g$), the radiation factor σ is assumed to be 1 for practical application ($\sigma \approx 1$). For the determination of the radiation factor below the cut-off frequency ($f < f_g$), the modes of the excited structure-borne sound vibrations are required. That is, in addition to the mean square of velocity on the plate, other data, such as phase position of the vibrations at various points on the plate, are necessary [10, 12, 25–27].

In the following, especially for plate-shaped structures with finite dimensions, some estimation formulas, which were determined empirically, are given for the radiation coefficient. The calculation formulas are transformed according to the results in [28, 29] and are valid for the entire frequency range.

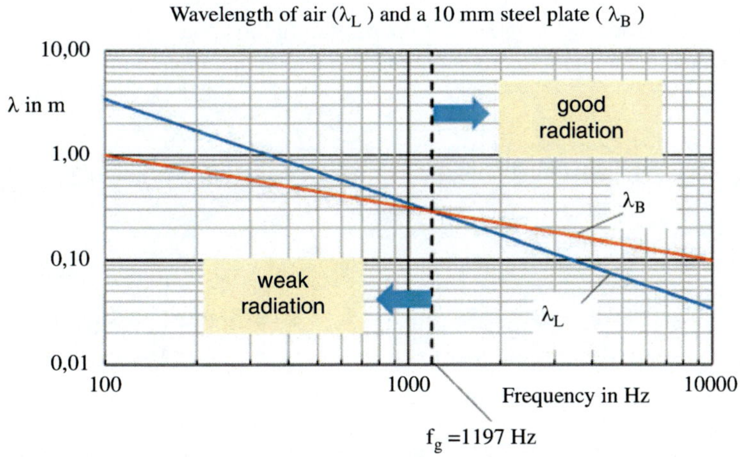

Fig. 5.21 Air and bending wavelength as a function of frequency

$$\left.\begin{aligned}
&\sigma' = 10 \cdot \lg(\sigma) \approx 0 && f \geq f_g\\
&\sigma' \approx 10 \cdot \lg\left(\frac{U \cdot c_L}{\pi^2 \cdot S \cdot f_g} \cdot \sqrt{\frac{f}{f_g}}\right) && f < \frac{f_g}{2}\\
&\sigma' \approx \sigma'^* \cdot \left(1 - \frac{\lg(2) + \lg\left(\frac{f}{f_g}\right)}{\lg(2)}\right) && \frac{f_g}{2} < f < f_g\\
&\text{with} \quad \sigma'^* \approx 10 \cdot \lg\left(\frac{U \cdot c_L}{\pi^2 \cdot S \cdot f_g \cdot \sqrt{2}}\right)
\end{aligned}\right\} \tag{5.81}$$

By this we mean:

σ	Radiation factor
$\sigma' = 10 \cdot lg(\sigma)$	Logarithm of the radiation factor in dB
U	Circumference of the plate supported moment-free on all sides in m
S	Panel area in m^2
c_L	Speed of sound of the surrounding air in m/s
f_g	Coincidence frequency ($\lambda_B = \lambda_L$) in Hz

Note

Equation (5.81) cannot be easily applied to complex machine structures. Their degree of radiation must usually be determined by measurements [30].

Example 5.3 The radiated A-structure-borne sound power level of two steel plates of different plate thicknesses with the same mean A-octave sound velocity level in the frequency range between 63 and 8000 Hz on the plate is sought.

Steel plates:

$$E = 2.1 \cdot 10^{11}\ \text{N/m}^2; \quad \rho = 7850\ \text{kg/m}^3; \quad \mu = 0.3$$

Plate dimensions:

$$L = 0.4\ \text{m}; \quad B = 0.35\ \text{m}; \qquad h_2 = 5\ \text{mm} \quad h_1 = 1\ \text{mm}$$
$$U = 2 \cdot 0.4 + 2 \cdot 0.35 = 1.50\ \text{m} \qquad S = 0.4 \cdot 0.35 = 0.14\ \text{m}^2$$

According to Eq. (5.74)

$$f_{g2} = 2394\ \text{Hz} \quad f_{g1} = 11971\ \text{Hz}$$

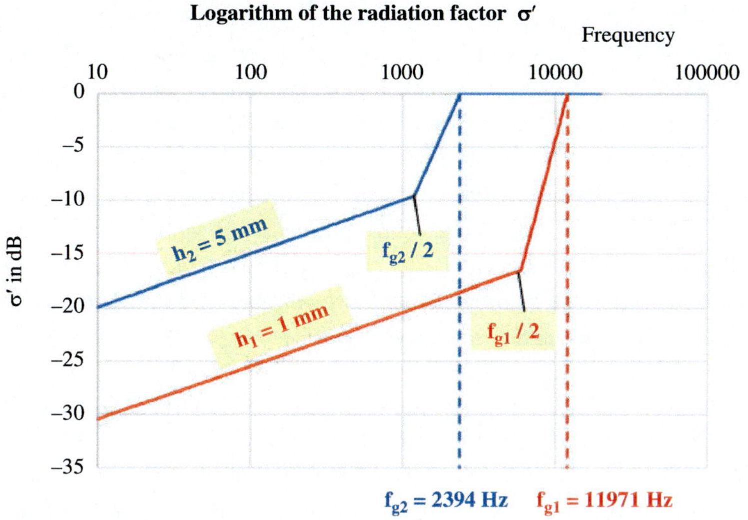

Fig. 5.22 Logarithm of the radiation factor σ' of two steel plates according to (5.81), plates supported moment-free on all sides

According to Eq. (5.81)

$$\sigma_2'^{*} = -9.6 \text{ dB} \quad \sigma_1'^{*} = -16.6 \text{ dB}$$

$$L_S = 10 \cdot \lg(0.14/1.0) = -8.5 \quad \text{dB}$$

Air:

$$t = 20\,°\text{C}; \quad R = 287 \ \text{Nm/kg K}; \quad \kappa = 1.4; \quad c_L = 343 \ \text{m/s}$$

Figure 5.22 shows the logarithm of the radiation factor of the steel plates. It follows from this that, all other conditions being equal (boundary stresses, excitations, etc.), thicker plates radiate more strongly than thinner ones. The reason for this is the lower coincidence frequency of the 5 mm steel plate compared to the 1 mm steel plate.

Table 5.2 shows the results of Example 5.3. From this it is easy to see that the radiated structure-borne sound power level of the 5 mm steel plate is significantly higher (13 dB(A)) than that of the 1 mm steel plate under otherwise identical conditions.

Table 5.2 Calculation results, Example 5.3

f_m	63	125	250	500	1000	2000	4000	8000	Total	Hz
$L_{v,Oct,A}$ (measurement)	94	87	95	86	84	82	86	78	98.7	dB(A)
$h_1 = 1$ mm										
σ'_1 (Eq. 5.81)	−26.5	−25.0	−23.5	−22.0	−20.5	−19.0	−17.5	−9.6	–	dB
$L_{W,Oct,A\text{-}1}$ (Eq. 5.62)	59.0	53.5	63.0	55.5	55.0	54.5	60.0	59.8	**67.7**	dB(A)
$h_2 = 5$ mm										
σ'_2 (Eq. 5.81)	−16.0	−14.5	−13.0	−11.5	−10.0	−2.5	0.0	0.0	–	dB
$L_{W,Oct,A\text{-}2}$ (Eq. 5.62)	69.5	64.0	73.5	66.0	65.5	71.0	77.5	69.5	**80.7**	dB(A)

Note

It should be noted that Example 5.3 is only intended to illustrate the influence of the radiation level due to the panel thickness. This means that an increase in wall thickness will generally lead to an increase in structure-borne sound radiation. In the example it was assumed that the structure-borne sound velocity level of the panels does not change.

An increase in the thickness of the plate not only leads to a deterioration in the radiation efficiency, it also changes the natural vibration behaviour and the mass occupancy of the plate, among other things. This inevitably changes the vibration behaviour of the plate. An unchanged structure-borne sound velocity level with increased plate thickness, as assumed in Example 5.3, therefore rarely occurs in practice.

As shown in Example 5.3, the radiation behaviour of structures is decisively determined by the level of the coincidence frequency. Since the structures can only fully radiate the structure-borne sound vibrations above a coincidence frequency of f_g ($\sigma \approx 1$), the highest possible coincidence frequency should be aimed at in the design of sound-radiating components with regard to the lower structure-borne sound radiation.

By (5.74), the coincidence frequency for plate-like structures is proportional to the mass occupancy m'' and inversely proportional to the length-related bending stiffness of the B' plate. With (5.69) and (5.70) it follows from (5.74):

$$f_g \sim \sqrt{\frac{m''}{B'}} = \sqrt{\frac{\rho_P \cdot h \cdot 12 \cdot (1-\mu^2)}{E \cdot h^3}} \sim \frac{1}{h} \tag{5.82}$$

Equation (5.82) shows that the coincidence frequency is inversely proportional to the plate thickness h. *This means that thinner plates have a higher coincidence frequency and therefore a worse radiation efficiency than thicker plates.* This means that thinner plates have a higher coincidence frequency and therefore a worse radiation efficiency than thicker plates, see also Example 5.3. In principle, it follows from this that the stiffened structures radiate structure-borne sound better than structures with lower stiffness.

5.1.3 Principles of Design Acoustics

Based on the findings and results in Sects. 5.1.1 and 5.1.2, the following principles for primary and constructive noise reduction can be stated:

Principle 1

Primary noise reduction can only be effectively implemented if the following questions have been answered beforehand, e.g. with the help of a sound engineering weak point analysis:

- **What** is to be noise reduced. This means that the sources responsible for the overall noise development must be localised.
- **How** should the noise be reduced? For this purpose, the sound generation mechanisms must be described and determined, i.e. whether airborne or structure-borne sound is decisive for the noise development. In addition, the frequency components determining the noise level are to be determined and the sound transmission described.
- **How much** noise reduction is necessary. In order to plan noise reduction in a targeted manner, the noise reduction measures for airborne and structure-borne noise, both as a total value and in level-determining frequency components, must be quantified.

2nd principle
The alternating forces that are decisive for the generation of sound should be stretched (in the course of time) as much as possible – especially in the case of pulse-like excitation. The force-time curves should be as smooth and continuous as possible. This means that sudden changes in the time-dependence of the forces, e.g. an increase and/or a decrease, must be avoided.

3rd principle
In design acoustics, it is not the mass or weight of a machine that is decisive for noise generation, but rather its dynamic mass at the points of force application.

4th principle
In the case of elastic decoupling, a distinction must be made between vibration and structure-borne sound insulation, as the mechanisms of action differ. Basically, an elastic element intended for vibration or structure-borne noise insulation is only effective if the dynamic mass of the coupling point (connection or foundation mass) is significantly greater, by a factor of approx. 5, than the mass of the machine to be decoupled.

5th principle
Structure-borne sound vibrations can only be fully radiated above a so-called coincidence frequency. In order to reduce noise generation, the sound-emitting structures should have as high a coincidence frequency as possible, i.e. they should have as high a mass occupancy as possible and low rigidity. In the case of plate-like structures whose structure-borne sound radiation determines the overall noise development of a machine, the principle applies:

As thin as possible and as thick as necessary.

Finally, it is pointed out that in addition to the objective noise reduction of a machine by primary or constructive measures, in practice the influencing of the subjective perceptions of the noise development of the machine is also considered for noise reduction. Here, the objective is essentially not the reduction of the A-total sound power level, but the modification or adaptation of the frequency spectrum of the machine to the subjective perceptions of humans.

The subjective modification of the noise is referred to as "noise design" or "sound quality". The consideration of individual subjective sensations, which will not be discussed in more detail here, is not the subject of primary noise reduction.

5.2 Secondary Measures

Primary noise reduction measures are primarily intended to reduce the emission level (sound power level). Secondary noise reduction measures aim to reduce the immission level (sound pressure level) while the emission level remains unchanged. The mechanisms of action of various secondary measures that are most frequently used in practice are described below.

5.2.1 Enclosure, Acoustic Enclosure

The propagation of airborne sound between the noise source and the place of immission is to be prevented or at least reduced by means of a sound enclosure (sound-insulating enclosure of noise sources).

The effectiveness of an enclosure is indicated by the so-called insertion loss. It can be determined in a simple way by measuring the sound power level of a machine both without and with an enclosure. Since the attenuation effect of an enclosure is strongly frequency-dependent, the frequency-related insertion loss is determined according to [31].

The noise reduction that can be achieved by encapsulating the machine is limited, among other things, by the existing external noise level in the machine's installation room. Therefore, the level reduction to be aimed for in the enclosure should be matched to this. In the following sections, some requirements for individual enclosure components are described and explained by examples.

5.2.1.1 Basic Construction of an Acoustic Enclosure

The resistance of an enclosure to airborne sound depends mainly on the mass per unit area of the enclosure wall and is referred to as sound attenuation. As already mentioned, the sound attenuation is frequency-dependent. It increases with frequency and mass per unit area. In addition, the bending stiffness of the enclosure wall also has an influence on the reduction effect [7].

The sound insulation of a enclosure is primarily achieved by sound reflection at the sound-hard enclosure wall. This causes an increase in the sound pressure level inside the enclosure. The level increase due to sound reflection considerably reduces the reduction effect of a sound enclosure. Therefore, for an optimal design of a sound enclosure, the inside of the enclosure must be sound-absorbing. Sound-absorbing materials are open-

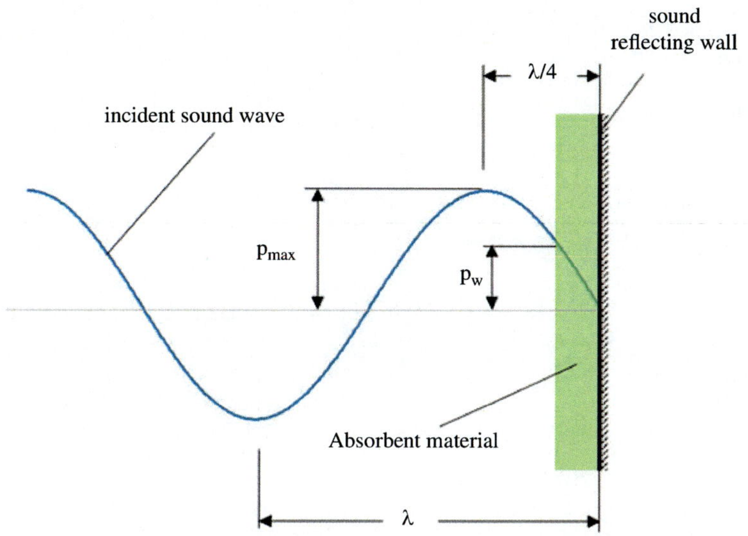

Fig. 5.23 Schematic representation of the mode of action of an absorption material

pored fibrous structures or foams that convert the incoming sound energy into heat (sound damping). The effectiveness of the sound-absorbing lining depends on its absorption coefficient, which is determined by the thickness of the absorption layer and the frequency of the sound to be attenuated [27, 32].

Figure 5.23 schematically shows the mode of action of the absorption material inside the enclosure. The effective amplitude of the sound "p_w", which is still in the absorption material, is decisive for the absorption. The highest effect is achieved when the largest amplitude "p_{max}" is also located in the absorption material.

The optimum thickness of the absorption material results:

$$d_{\text{optimal}} = \frac{\lambda}{4} = \frac{c}{4 \cdot f} \quad \text{m} \tag{5.83}$$

With $c = 340$ m/s for air follows from (5.83):

$$f = 100 \text{ Hz}; \qquad d_{\text{optimal}} = 0.85 \text{ m}$$
$$f = 1000 \text{ Hz}; \qquad d_{\text{optimal}} = 0.085 \text{ m}$$

It follows that the lower the level-determining frequencies, the thicker the absorption material must be in order to reduce the proportion of sound reflection within the enclosure. The usual absorption thickness in practice is between 50 and 100 mm. Only in special cases, e.g. enclosures for low-frequency noise sources, thicker layers than 100 mm are required. Closed-cell rigid foams with good thermal insulation properties such as Styropor® are not suitable for sound absorption linings.

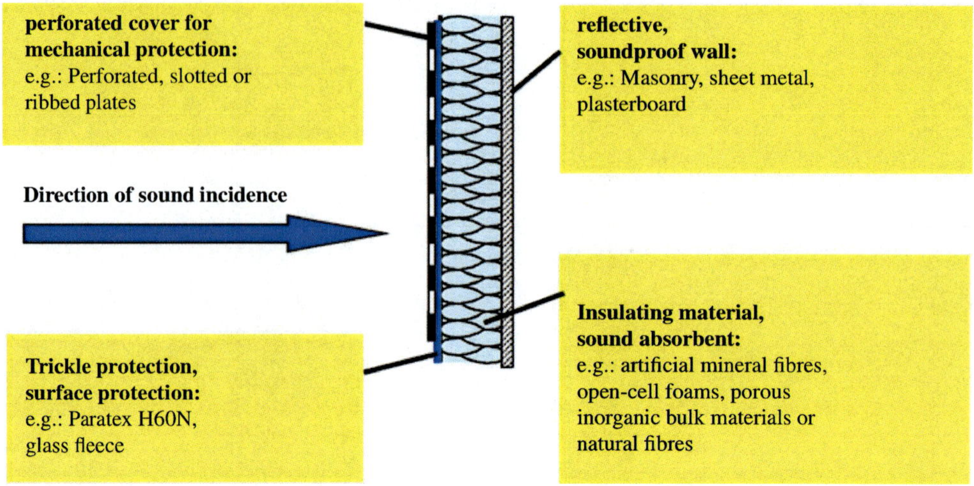

Fig. 5.24 Basic structure of a sound-absorbing enclosure wall

In addition to the thickness, the absorption material must be constructed in such a way that the incident sound waves or sound energy can penetrate the material and be converted as completely as possible into heat. For this purpose, the absorption material must have a certain input impedance, for which, among other things, the so-called flow resistance is decisive[7, 32].

Figure 5.24 shows the basic structure of a enclosure wall.

To protect the absorption material from contamination, oiling or moisture, it can be covered with plastic film, about 20–0 μm thick, if the surface of the material is not sufficiently sealed.

The wrapping of the mineral fibre mats by foil or fleece also serves as trickle protection, i.e. to prevent fibres from being released. Perforated sheeting, 1–1.5 mm thick, e.g. made of galvanised steel or aluminium with at least 30% perforated area, is usually used as mechanical protection for the absorption material. If perforated sheets are to be painted, this must be done before they are applied so that the effectiveness of the absorbent materials is not impaired by the application of paint.

5.2.1.2 Openings, Leaks

In the case of enclosures with numerous openings that cannot be sufficiently sealed acoustically by silencers, the overall insulation is limited and the greater the proportion of free openings in the total surface area, the lower the insulation.

Slots can also significantly impair the damping effect of the enclosure. This has a greater effect for high frequencies than for low frequencies.

The theoretically maximum achievable average insertion loss can be estimated from the ratio of the acoustically unsealable **O**pening area S_O to the **E**nclosure surface area S_E as follows:

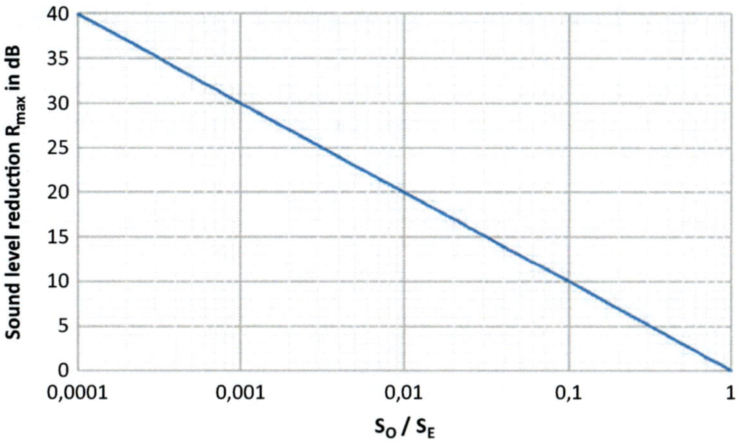

Fig. 5.25 Theoretically maximum achievable average sound level reduction as a function of the ratio S_O/S_E (**O**pening area to **E**nclosure surface)

$$R_{max} = 10 \cdot lg \frac{1}{S_O/S_E} \text{ dB} \tag{5.84}$$

Figure 5.25 shows (5.84) graphically.

In practical application, the maximum achievable sound level reduction is smaller than the attenuation values according to (5.84).

It is easy to see from Fig. 5.25 that the higher the attenuation values of a sound enclosure, the more sensitive it is to opening areas and leaks. In a sound enclosure for 10 dB level reduction, opening areas up to 10% of the enclosure surface are not critical. On the other hand, in the case of a enclosure for 30 dB level reduction, the opening areas may amount to a maximum of 1% of the enclosure surface.

Windows, doors and flaps – the sources of leaks – should therefore be reduced to the absolutely necessary number or be well sealed. To avoid or mitigate openings, remote operation of the machine from the outside should be made possible where feasible.

Simple window panes made of 4 mm thick glass have a mass per unit area roughly equivalent to that of 1.5 mm thick sheet steel, and thus also have a correspondingly high level of insulation. If safety glass is to be used where there is a risk of damage, multi-layer glass or similar is preferable to the lighter acrylic glass. Double windows are only required if the level reduction of the enclosure is to exceed 20 dB(A).

Slots or openings should be provided in the enclosure wall for the material feed, which should be acoustically sealed, e.g. with silencers, which should have approximately the same insulation values as the insertion insulation dimension of the acoustic enclosure. The same applies to any supply and exhaust air openings that may be present. Sealing the gap between the enclosure and the floor is also important, as this is where the largest gap usually occurs. For this reason, a seal, e.g. made of cellular rubber, 10 mm thick and at least

40 mm wide, Shore hardness 20–30°, is advisable when placing the enclosures on the hall floor.

In general, using (5.85), the resulting sound reduction index R_{res}. of a wall whose partial surfaces or components – e.g. doors or windows – have different insulation values R_i can be determined as follows:

$$R_{res} = -10 \cdot \lg\left[\frac{1}{S_{total}} \cdot \sum_i S_i \cdot 10^{\frac{-R_i}{10}}\right] \quad \text{dB} \tag{5.85}$$

R_i	Sound reduction index of the components or partial areas in dB
S_i	Partial areas in m^2
S_{total}	Total area of the acoustic enclosure in m^2

Example 5.4

We are looking for the resulting sound reduction index R_{res} of a sound enclosure with the following data:

Hood dimensions:	$L \times B \times H = 4.5\,\text{m} \times 3.8\,\text{m} \times 3.0\,\text{m}$
Total opening area:	$S_O = 0.4\ \text{m}^2$

$$R_{\text{Enclosure element}} = 45\,\text{dB}$$
$$S_{total} = 2 \cdot 4.5 \cdot 3.0 + 2 \cdot 3.8 \cdot 3.0 + 4.5 \cdot 3.8 = 66.9\,\text{m}^2$$

The resulting sound reduction index of the hood can be determined according to (5.85) as follows:

$$R_{res} = -10 \cdot \log\left\{\frac{1}{66.9} \cdot \left[(66.9 - 0.4) \cdot 10^{-\frac{45}{10}} + 0.4 \cdot 10^{-0}\right]\right\} = 22.21 \quad \text{dB}$$

The maximum achievable average sound level reduction of this sound enclosure is obtained according to (5.84) or Fig. 5.25:

$$S_O/S_{total} = 0.4/66.9 = 0.006$$
$$R_{max} = 10 \cdot \lg\frac{1}{S_O/S_E} = 10 \cdot \lg(0.006) = 22.23 \quad \text{dB}$$

5.2.1.3 Structure-Borne Sound

As already mentioned, the primary task of an enclosure is to reduce the propagation of airborne sound between the noise source and the place of immission. For this purpose, it is essential to ensure that the enclosure is not excited by the structure-borne sound of the noise source, e.g. via so-called bridges for structure-borne sound, to cause structure-borne sound vibrations.

The overall attenuation effect of an enclosure can be significantly reduced by structure-borne sound transmission. In extreme cases, even a enclosure that is intended to reduce the sound level can cause an increase in level.

All connections between the enclosure and the machine must not be rigid, but elastic. The same applies to the installation of the enclosure, especially if the foundation is excited by structure-borne sound vibrations from the noise source. In such cases, the enclosure must be installed in such a way that it is insulated from vibrations and/or structure-borne noise.

The principle of the acoustic enclosure can also be used for all machine protective cladding that is not designed for load-bearing forces. In other words, the protective enclosure of a machine can be designed as a sound enclosure. For this purpose, the machine guard must not have any connection to the machine frame, if possible. If the protective enclosure has to be connected to the machine frame at certain points, it must be designed to be insulated against structure-borne noise, otherwise it cannot act as a sound enclosure.

Figure 5.26 shows the basic structure and mode of action of a enclosure [7, 27].

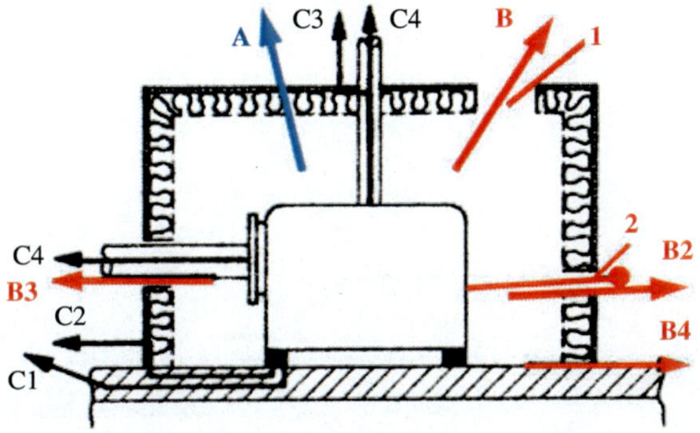

1, 2 Ventilation opening, operating lever

A Airborne sound transmission via the capsule wall

B Airborne sound transmission through the leaks and unavoidable openings

C Structure-borne sound transmission and radiation as airborne sound

Fig. 5.26 Mode of operation of a sound enclosure

A sound enclosure is always intended for transmission path A. All other transmission paths (1; 2; B and C) reduce the insertion loss of the enclosure. Particularly critical are the structure-borne sound transmission paths C, which can even lead to an increase in level.

If openings are necessary for ventilation or heat dissipation, they must be closed acoustically with the aid of suitable silencers so that the insulating effect of the enclosure is not reduced. For this purpose, the silencer must have at least the same dimension of insertion loss as the enclosure wall elements. This also applies to any shaft and pipe penetrations. These must be sealed and insulated against structure-borne noise.

5.2.2 Silencer

Silencers are devices that reduce the propagation of sound in ducts, pipes and openings without preventing or hindering the transport of the medium [33].

Figure 5.27 shows the basic design of a sound silencer and the possible sound propagation paths.

The effect of a sound silencer is limited by sound bypasses. With the absorption silencer usually used in practice, insertion loss values of up to approx. 30 dB(A) can be achieved. With special silencers, such as those used in vehicles, insertion loss values of up to approx. 80 dB(A) are possible. In this case, the high insertion loss (in addition to sound attenuation through sound absorption) is primarily achieved by installing resonators that are tuned to specific frequencies. The basic design of some silencers is explained below.

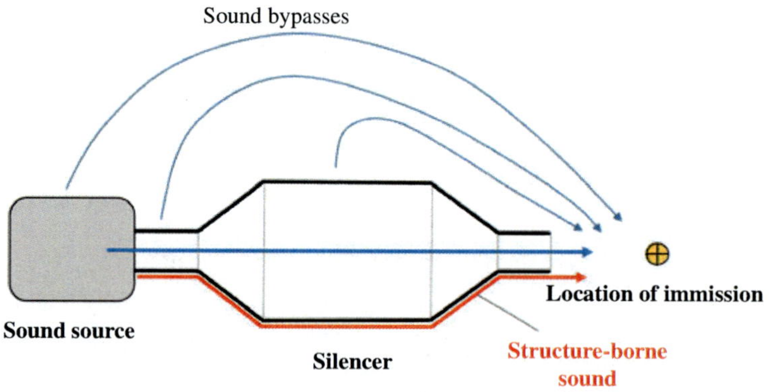

Fig. 5.27 Basic design of a sound attenuator and different sound propagation paths

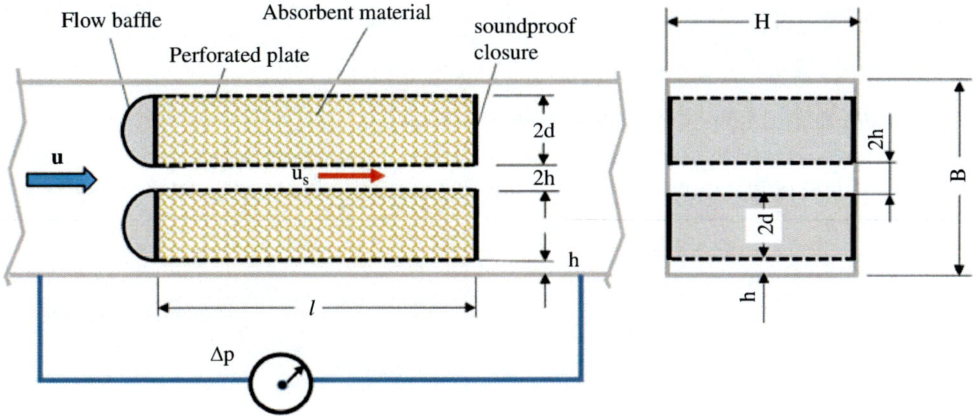

Fig. 5.28 Basic design of a splitter attenuator

5.2.2.1 Absorption Silencer

The level reduction of an absorption silencer is achieved by converting part of the sound energy into heat by installing sound-absorbing materials in the flow channel. The most commonly used absorption silencers are the so-called splitter silencers. Figure 5.28 shows the basic design of a splitter attenuator [33–35].

Here are:

u	Flow velocity in the duct or pipeline in m/s
u_s	Gap speed, speed between the baffles in m/s
2d	Baffle thickness in m
2 h	Gap width in m
l	Silencer length in m
B	Silencer width in m
H	Silencer height in m
Δp	Pressure loss of the silencer in N/m^2

The damping is achieved by the sound waves penetrating the baffles perpendicular to the direction of flow along the baffles, where the sound energy is converted into heat. For this purpose, the surface of the damping material must be permeable to sound. To prevent the penetrating sound from passing through the absorbing material, the ends of the baffles must be sealed soundproof.

In addition, the damping material must have an optimal length-related flow resistance r_s [10, 27, 36]. The flow resistance R_S is defined as the ratio of the pressure difference Δp_S between the front and back sides of the material and the volume flow q_V passing through the material cross-sectional area A_S:

$$R_S = \frac{\Delta p_S \cdot A_S}{q_V} \quad \mathrm{N\,s/m^3} \tag{5.86}$$

The length-related flow resistance r_s is then obtained:

$$r = \frac{R_S}{d} \quad \mathrm{N\,s/m^4} \tag{5.87}$$

d	Insulation thickness in m

The optimal length-related flow resistance [27, 36] can be approximately restricted to the following range:

$$5000 \ \mathrm{N\,s/m^4} < r_s < 50{,}000 \ \mathrm{N\,s/m^4}$$

Components and Characteristics of a Splitter Attenuator

Splitter Thickness 2d

The baffle thickness 2d depends primarily on the frequency or wavelength of the sound to be attenuated. The optimum thickness would be, as already explained in Sect. 5.2.1.1, λ/4. The baffle thickness usually used in practice is between 50 and 250 mm. At frequencies below approx. 200 Hz, other types of silencers, e.g. resonators, are often used, mainly because of the very large silencer volume (see Sect. 5.2.2.2), especially when tonal frequencies are present.

Gap Speed u_{gs}

A silencer can only reduce sound events which are higher than the pure flow noise in the duct (see Sect. 3.1.7). Therefore, by installing silencer baffles, the flow velocity in the gap u_{gs} must not be greater than the flow velocity in the duct, i.e.:$u_{gs} \leq u$. To achieve this, the cross-section is usually widened in the area of the splitters see Fig. 5.29.

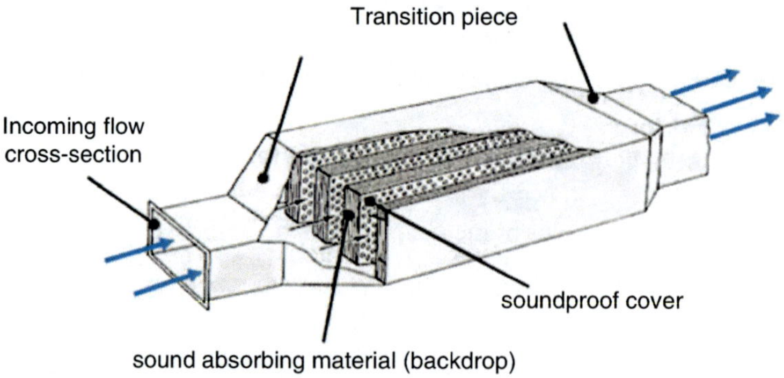

Fig. 5.29 Construction of a common splitter attenuator

Gap Width 2 h

The gap width determines the effectiveness of a sound attenuator at high frequencies. If the wavelength of the sound to be attenuated is less than or equal to the gap width, then the main part of the sound energy is transmitted or not attenuated. The corresponding frequency is therefore referred to as the passband frequency f_d:

$$f_{\mathrm{d}} \geq \frac{c}{2 \cdot h} \quad \mathrm{Hz} \tag{5.88}$$

respectively

$$\lambda_{\mathrm{d}} = \frac{c}{f_{\mathrm{d}}} \leq 2 \cdot h \quad \mathrm{m} \tag{5.89}$$

The frequency f_{d} indicates up to which frequency a splitter attenuator can be optimally used.

Perforated Plate

The perforated plate with a perforation proportion of approx. 30% is sound-permeable, it serves as mechanical protection and is intended, among other things, to prevent the flowing medium from taking the absorption material with it through abrasion. To prevent fibres from being released, the surface of the absorption material is additionally covered with trickle protection.

Splitter Length *l*

The baffle length is decisive for the total attenuation of the sound attenuator. Due to the always existing sound bypasses, see Fig. 5.27, the attenuation is limited. In practice, a total level reduction of max. 30 dB(A) can be achieved with the usual splitter attenuators.

Pressure Loss Δ*p*

The pressure loss of a silencer, which is decisive for the energy consumption, is primarily determined by the flow velocity in the gap. Other essential parameters are the baffle length *l*, the density of the medium and the number of baffles.

Depending on the volume flow, it is possible to define the gap velocity by selecting a suitable the gap width 2 h as well as the number and height of the splitters and thus to set the desired pressure drop. The lower the pressure drop of a silencer, the larger the silencer volume, so that an optimum must be found between pressure drop, the available volume / size and the costs.

Flow Deflector

The flow baffle has the task, among other things, of reducing turbulence during inflow and thus reducing the pressure losses of the silencer.

Estimating the Attenuation of an Absorption Silencer

The attenuation of an absorption silencer can be calculated most simply with the aid of Piening's formula [37, 38]:

$$D = 1.5 \cdot \alpha \cdot \frac{U}{S} \cdot l \quad \text{dB} \tag{5.90}$$

Here are:

α	Frequency-dependent absorption coefficient of the absorption material for perpendicular sound incidence
U	Absorbent lined perimeter in m
S	Free cross-sectional area of the duct in m^2
l	Length of the silencer in m

With the help of (5.90), it is possible to calculate the damping relatively easily if the absorption coefficient is known. However, the influence of the sound radiation, i.e. the frequencies above the passband frequency f_d according to (5.88), is not taken into account. With the help of the so-called trapezoidal diagrams [27] one can determine the damping over the entire frequency range, whereby the reading from the diagram is complicated.

In the following, an empirical estimation procedure for the calculation of the damping is reproduced, which was determined based on theoretical considerations and oriented to Piening's formula, (5.90) [27, 38]. According to this, the damping can be calculated as follows:

$$D = D_h \cdot \frac{l}{h} \quad \text{dB} \tag{5.91}$$

With

$$D_h = 1.5 \cdot \alpha \cdot \frac{U}{S} \cdot h \quad \text{dB} \tag{5.92}$$

D_h is the so-called normalised attenuation. For an average flow resistance of, $R_S = 1200$ N s/m^3 the normalised damping D_h is limited to a maximum of 1.5 dB. The theoretically achievable attenuations, also for other flow resistances, are somewhat higher, especially in the medium frequency range, but are not quite achieved in practice.

Considering that dB $D_{h,\,max} = 1.5$ is, the normalised damping D_h can be approximated by (5.92) for different frequency ranges as follows [38]:

$$\left.\begin{aligned} f<f_u: &\quad D_h = 1.5\cdot\left(\tfrac{f}{f_u}\right)^n\cdot\frac{U\cdot h}{S}\quad \text{dB}\\ f_u<f<f_o: &\quad D_h = 1.5\cdot\frac{U\cdot h}{S}\quad \text{dB}\\ f>f_o: &\quad D_h = 1.5\cdot\left(\tfrac{f_o}{f}\right)^2\cdot\frac{U\cdot h}{S}\quad \text{dB}\end{aligned}\right\} \tag{5.93}$$

With

$$\left.\begin{aligned} f_u &= \eta_u\cdot\frac{c}{2\cdot h}\quad \text{Hz}\\ f_o &= 1.5\frac{c}{2\cdot h}\quad \text{Hz}\\ \left.\begin{aligned}\eta_u &= 0.19\cdot\Lambda^{-0.72}\\ n &= 1.75\cdot\Lambda^{-0.12}\end{aligned}\right\} &\Lambda = 0,1\ldots 2\end{aligned}\right\} \tag{5.94}$$

Here Λ is the so-called normalised lining depth:

$$\Lambda = d/h \tag{5.95}$$

In Fig. 5.30, D_h is shown according to (5.93) and (5.94) for different parameters Λ, as a function of the normalised frequency η.

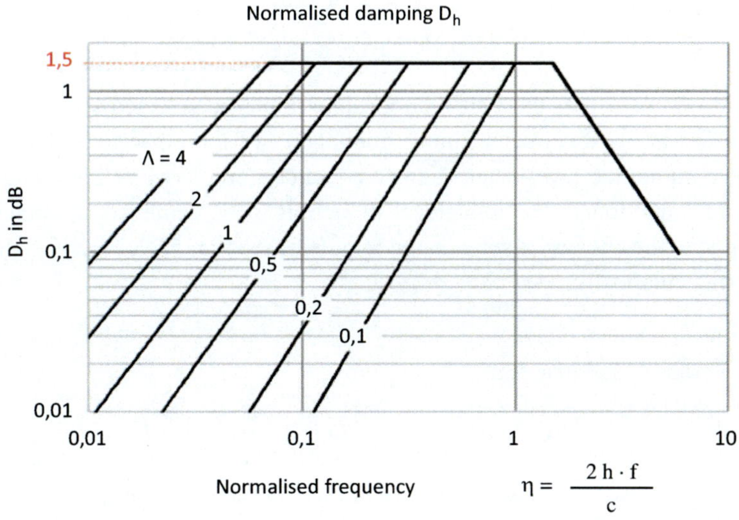

Fig. 5.30 Normalised attenuation D_h for absorption silencers as a function of the normalised frequency η for different Λ-parameters

$$\eta = \frac{f}{f_\mathrm{d}} = \frac{2h \cdot f}{c} \tag{5.96}$$

Note

The attenuation of a splitter silencer according to Fig. 5.28 is primarily determined by the splitter thickness 2d, the gap width 2 h and the attenuator length l *in* addition to the absorption material. The baffle height and the number of baffles are primarily responsible for setting the gap velocity u_s and for achieving the desired pressure drop of the silencer, respectively. For baffle silencers commonly used in practice (see Fig. 5.29), in (5.93) the factor: $\frac{U \cdot h}{S} = 1$.

Example 5.5

The noise of a centrifugal fan, see numerical example in Sect. 3.1.8.1, is to be reduced by $\Delta L = 15$ dB(A) with the aid of a splitter silencer. The dimensions of the silencer are to be selected so that the gap velocity in the silencer is not greater than 15 m/s.

The internal A-octave sound power level of the fan is:

Frequency in Hz	63	125	250	500	1000	2000	4000	8000	Total	
$L_{WA,Oct}$	75.3	88.1	96.2	108.0	99.4	94.7	86.3	73.9	**109.0**	dB(A)

and the volume flow $\dot{V} = 5750\ \mathrm{m^3/h}$.

The calculation is performed for a splitter thickness of $2d = 0.15$ m, a splitting width of m $2h = 0.2$ and a length of $l = 1.2$ m.

With the help of (5.91), (5.93) as well as $\Lambda = 0.15/0.2 = 0.75$ the attenuation of the silencer and the expected level reduction can be calculated, see Table 5.3.

It should be noted that the baffle dimensions must be selected so that the maximum attenuation lies at the level-determining frequencies of the noise to be attenuated. In the numerical example, the level-determining frequencies are between 250 and 2000 Hz.

As already mentioned, the total level reduction to be achieved is limited to approx. 30 dB(A) due to the always existing sound bypasses, see Fig. 5.27, even if higher values are achieved mathematically.

Table 5.3 Results of Example 5.5

Frequency in Hz	63	125	250	500	1000	2000	4000	8000	Total	
D_h	0.07	0.24	0.83	1.50	1.50	1.50	0.63	0.16	–	dB
D	0.8	2.8	10.0	18.0	18.0	18.0	7.6	1.9	–	dB
$L_{WA,Oct.} - D$	74.5	85.3	86.2	90.0	81.4	76.7	78.7	72.0	**93.1**	dB(A)
								ΔL	**15.9**	dB(A)

By selecting the appropriate baffle height and the number of baffles, it is then possible to determine the desired gap velocity and thus also the pressure loss of the silencer. With a baffle height $H = 0.3$ m, see Fig. 5.31, and two baffles, one then obtains:

Absorption silencers are used in various forms. Figure 5.32 shows some common basic forms of absorption silencers. The sound attenuation can be estimated with the dimensions given in Fig. 5.32 using (5.90), (5.91), (5.92), (5.93), (5.94), (5.95) and (5.96).

Pipe silencers, which can be manufactured relatively inexpensively, are most frequently used in practice along with splitter silencers. Figure 5.33 shows some circular silencers [27, 35].

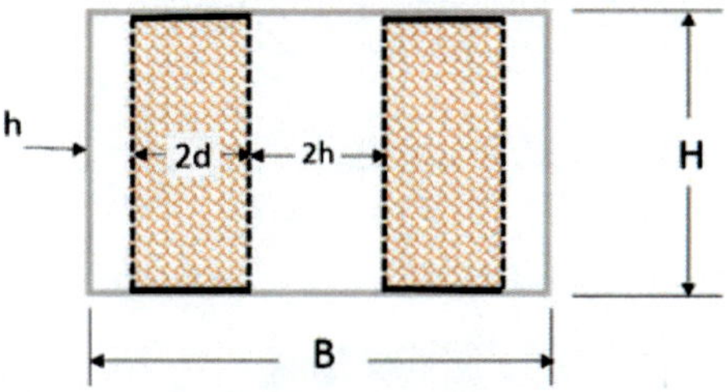

$$S = 4 \cdot h \cdot H = 0{,}12 \ \ m^2$$

$$u_s = \frac{5750}{3600 \cdot 0{,}12} = 13{,}3 \ \ m/s$$

$$B = 4 \cdot h \cdot d = 0{,}7 \ \ m$$

$$2h = 0{,}2 \ \ m \ ; \ 2d = 0{,}15 \ \ m$$

$$H = 0{,}3 \ \ m$$

Fig. 5.31 Splitter silencer, Example 5.5

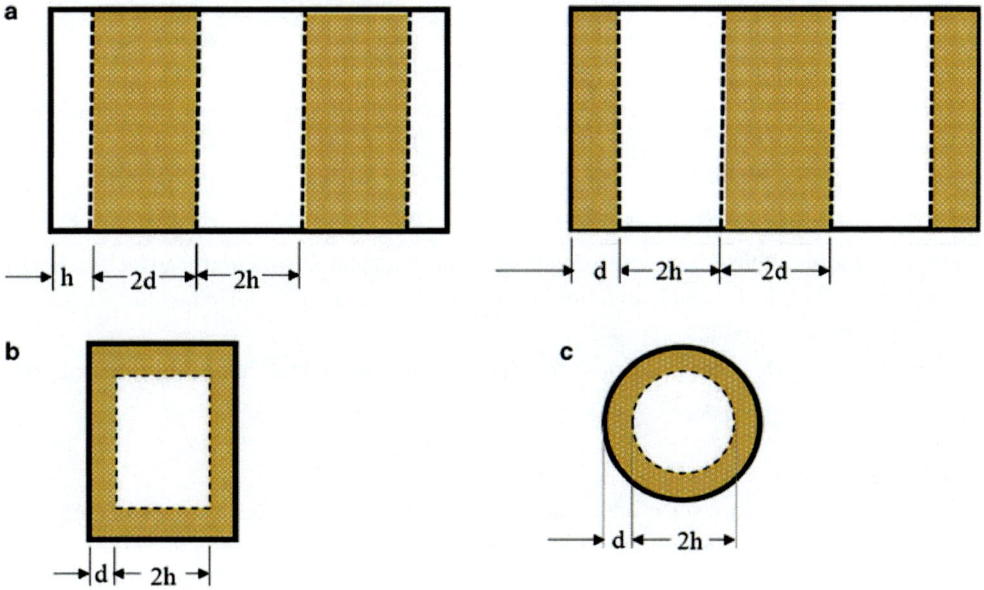

Fig. 5.32 Various basic shapes of absorption silencers; (**a**) splitter silencer, (**b**) rectangular duct lined on all sides, (**c**) **circular** silencer

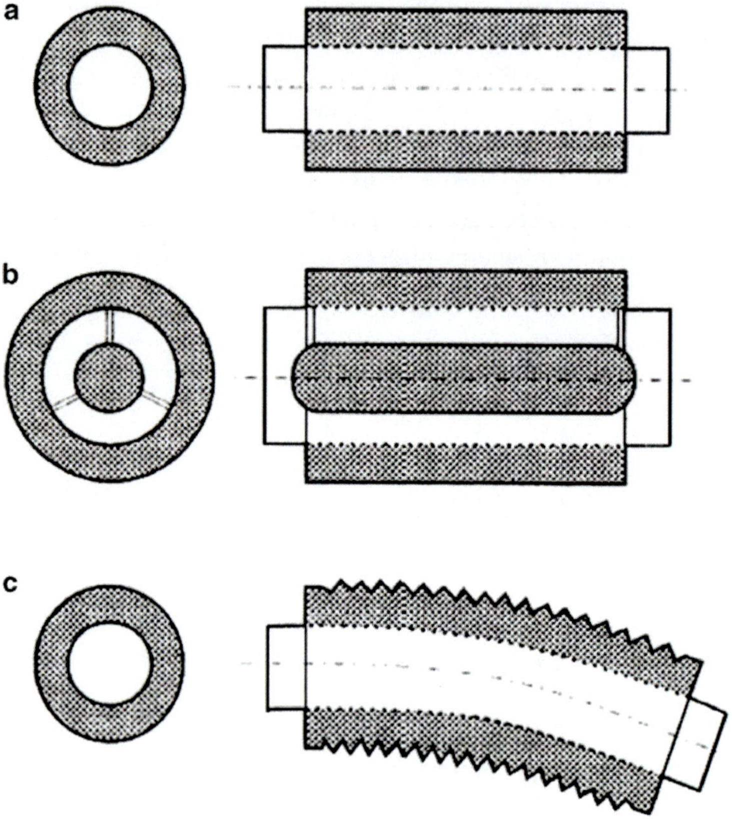

Fig. 5.33 Designs of circular silencers; (**a**) fixed circular silencer, (**b**) fixed circular silencer in the core, (**c**) flexible circular silencer

5.2.2.2 Resonance Silencer

The absorption silencers have a broadband effect and are particularly effective at high frequencies. For low frequencies, especially if they have a tonal character, absorption silencers are not suitable.

For this purpose, special resonance silencers are used which are designed for defined frequencies.

Plate Resonators

The plate resonators are selective spring-mass systems that are excited to resonance. In the resonance range they have the optimum effect [10].

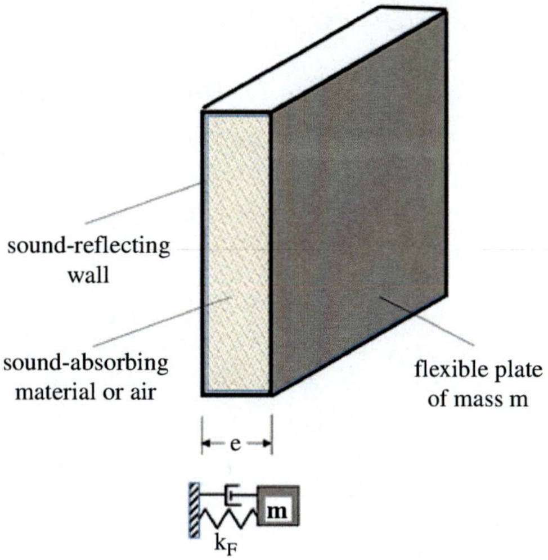

Fig. 5.34 Schematic structure of a plate resonator

In principle, all freely vibrating, thin and flexurally elastic panels of a room boundary – such as window panes, thin wooden panels and metal panels – have a sound-absorbing effect. They are excited to resonate by the impinging sound waves and can convert part of the vibration energy into heat. Figure 5.34 shows the basic structure of a plate resonator.

The plate resonator can be modeled as a damped mass oscillator for the determination of the 1st natural frequency. The natural frequency $f_{0,\mathrm{P}}$ of the plate with area S can be determined as follows, neglecting its bending stiffness and damping [10]:

$$f_{0,\mathrm{P}} = \frac{1}{2\pi} \cdot \sqrt{\frac{k_\mathrm{F}}{S \cdot m''_\mathrm{P}}} \quad \mathrm{Hz} \tag{5.97}$$

With

$$m''_\mathrm{P} = \rho_\mathrm{P} \cdot h_\mathrm{P} \quad \mathrm{kg/m^2}$$

$$k_\mathrm{F} = \frac{E_\mathrm{dyn} \cdot S}{e} \quad \mathrm{N/m} \tag{5.98}$$

k_F	Spring stiffness or spring constant of the damping material
ρ_P; h_P.	Density in and kg/m³ thickness of the plate in m
$E_\mathrm{dyn} = c_\mathrm{D}^2 \cdot \rho_\mathrm{D}$	Modulus of elasticity of the damping material in N/m²
m''_P	Mass assignment of the plate

Here is

c_D; ρ_D	The speed of sound in m/s and density of the damping material in kg/m^3

With an air cushion as damping material, the 1st natural frequency of the plate can be determined as follows:

$$f_{0,\mathrm{P}} = \frac{1}{2\pi} \cdot \sqrt{\frac{Z_{\mathrm{air}} \cdot c_{\mathrm{air}}}{e \cdot m''_{\mathrm{P}}}} \quad \mathrm{Hz} \tag{5.99}$$

With

$Z_{\mathrm{air}} = c_{\mathrm{air}} \cdot \rho_{\mathrm{air}}$	Impedance of the air in N s/m^3
c_{air}	Speed of sound in air in m/s

Figure 5.35 shows the basic frequency response of a plate resonator. Without sound absorbing material, the resonator has a narrow band effect only in the range of the natural frequency. With damping material, the damping is somewhat more broadband with reduced effect in the range of the natural frequency.

Fig. 5.35 Schematic frequency response of a plate absorber

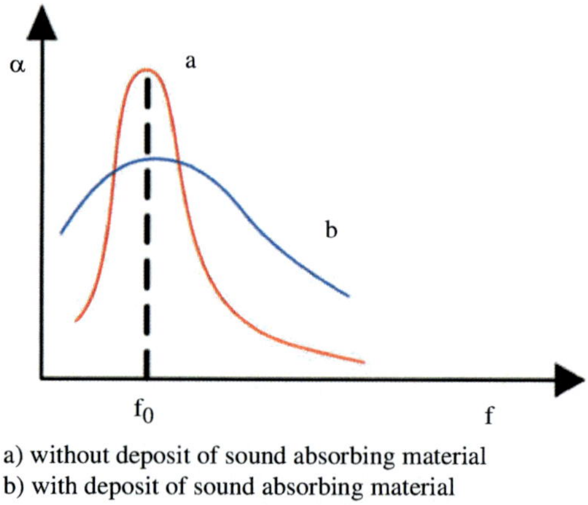

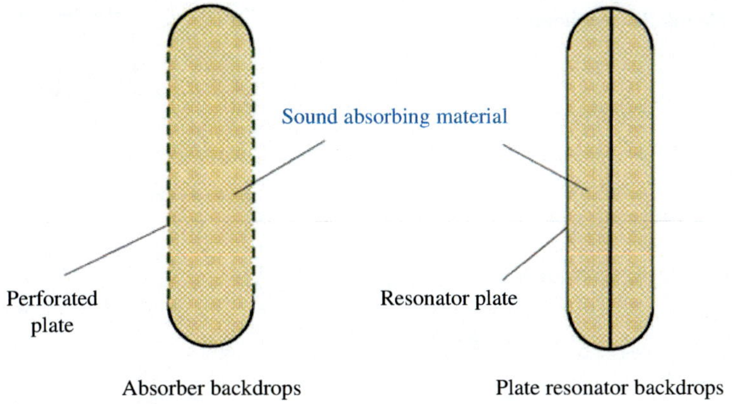

Fig. 5.36 Various scenery elements

In order to enable sound attenuation in the entire frequency range, splitter attenuators are used in practice together with both absorber splitters and resonator splitters. Figure 5.36 shows such baffle elements with the same dimensions.

Helmholtz Resonator

Sound energy can also be converted into heat at lower frequencies by means of a Helmholtz resonator. Here, damping is achieved with the aid of an air-spring system consisting of an air volume V in conjunction with a neck-like constriction of cross-sectional area S and length l (Fig. 5.37).

The air in the throat forms the mass, the air volume V the spring of the resonator. The relevant natural frequency of the Helmholtz resonator is then obtained [10]:

$$f_{0,H} = \frac{c_{\text{air}}}{2\pi}\sqrt{\frac{S}{V \cdot l}} \quad \text{Hz} \tag{5.100}$$

The result must still be corrected, since practically in the resonator neck a larger air mass than $m = S \cdot l \cdot \rho$ resonates. The real, equivalent air mass is $S \cdot l^* \cdot \rho$. For circular holes $l^* \approx l + 0.8 \cdot d$. Here d is the hole diameter. For non-circular holes is given in l^* [39, 40].

Fig. 5.37 Schematic diagram of a Helmholtz resonator

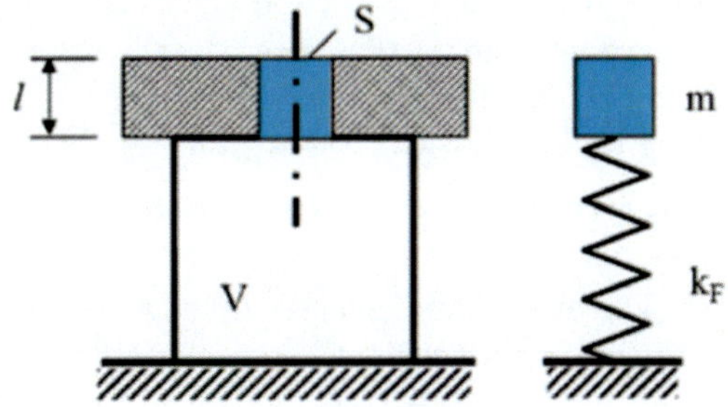

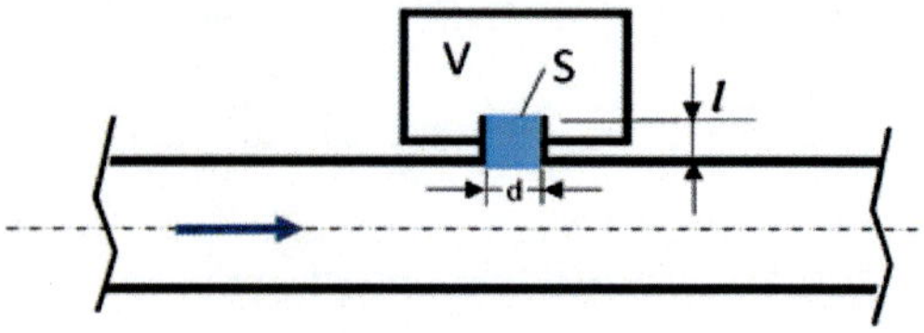

Fig. 5.38 Helmholtz resonator on a pipeline or duct

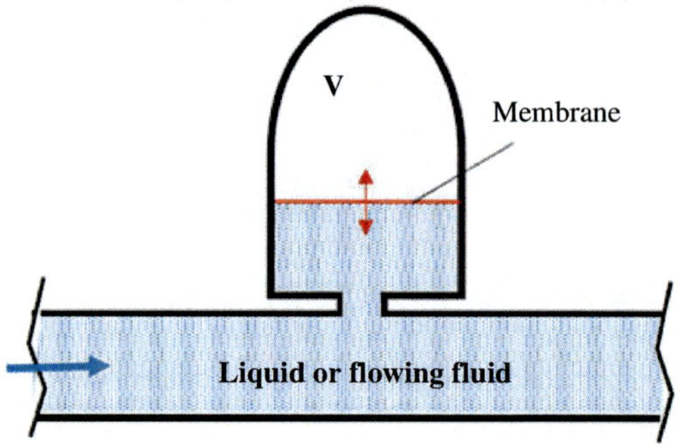

Fig. 5.39 Schematic structure of a pulsation damper for liquids

Figure 5.38 shows the basic structure of a Helmholtz resonator on a pipeline or channel for gaseous media.

The natural frequency of the Helmholtz resonator is then obtained:

$$f_{0,\mathrm{H}} \approx \frac{c_\mathrm{F}}{2\pi}\sqrt{\frac{S}{Vl^*}} \quad \mathrm{Hz} \tag{5.101}$$

c_F	Speed of sound of the fluid in the pipeline in m/s

The principle of the Helmholtz resonator is also used as a pulsation damper for pipes through which liquids flow. Figure 5.39 shows the schematic structure of a pulsation damper for liquids. Here, the liquid is separated from a certain gas volume V (usually nitrogen) by means of a membrane. Due to the compressibility or spring stiffness and the volume of the gas, the natural frequency of the pulsation damper is tuned to the pressure pulsation to be damped in the pipe. By oscillating the gas volume at the frequency of the pressure pulsation, vibration energy is extracted from the system or the amplitude of the pulsation is reduced.

The λ/4 Resonator

The λ/4 resonator corresponds to a tube of length l, one end of which is sonically closed, see Fig. 5.40 [10].

The natural frequencies of the λ/4 resonator can be determined as follows:

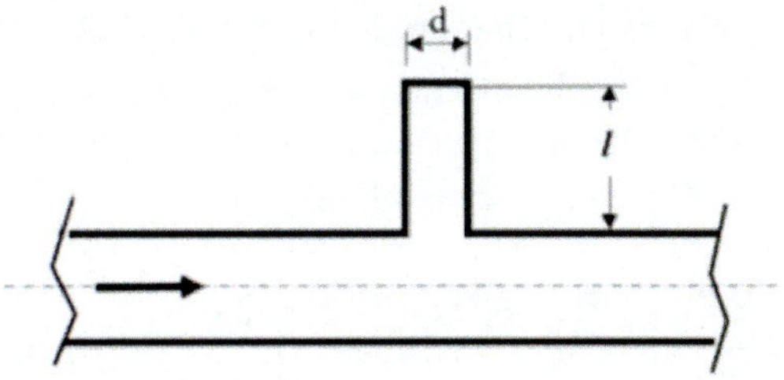

Fig. 5.40 λ/4 resonator connected to a tube

$$f_{n,\lambda/4} = \frac{c_F}{4 \cdot l + \Delta l} \cdot (2n+1) \quad \text{Hz} \qquad n = 0, 1, 2, 3, \ldots \tag{5.102}$$

$$\Delta l = 0.4 \cdot d \quad \text{m}$$

c_F	Speed of sound of the fluid in the pipeline in m/s

From (5.102) it can be seen that the damping of a λ/4-resonator can be effective not only at λ/4, but also at multiples of λ/4.

Figure 5.41 shows the sectional view of a special resonator baffle consisting of various λ/4- and Helmholtz resonators with sound-absorbing material behind. Such resonators are

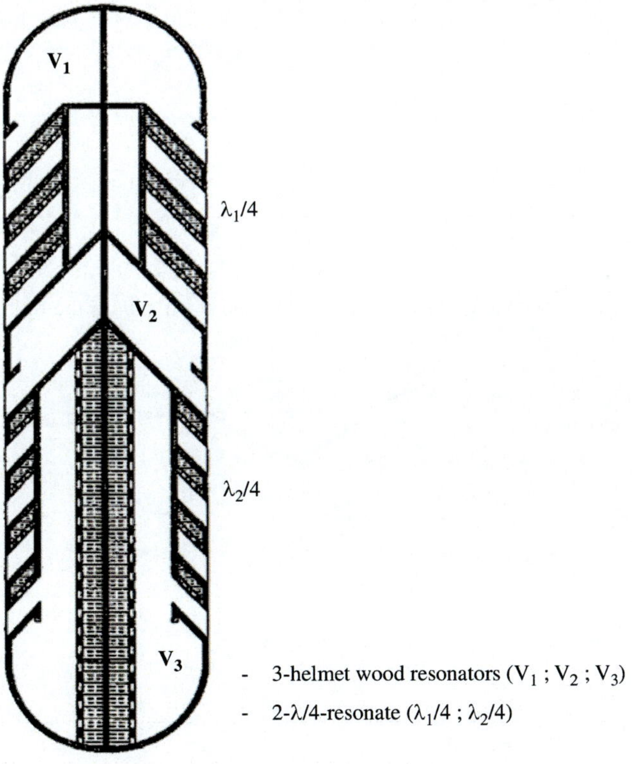

Fig. 5.41 Schematic illustration of a resonator backdrop

of course much more cost-intensive than simple baffle silencers, they are primarily used for the attenuation of gaseous media, e.g. air or flue gas, when low frequencies <250 Hz with tonal character prevail. The design of such resonators requires special experience and manufacturing capabilities.

5.2.2.3 Reflective Silencers

The sound waves are reflected, especially at low frequencies, from changes in cross-section or from the volume behind them. This happens when the diameter and length of the volume are at least a factor of 2 greater than the dimensions of the pipe [10].

Figure 5.42 shows a pipe connected to a sound source whose length is $\frac{5 \cdot \lambda}{4}$ even. λ is the wavelength of the sound wave generated by the sound source. This causes a resonance to form in this pipe.

If a sound attenuator volume is placed in the pressure antinode in front of the muzzle (Fig. 5.43), the sound field changes drastically. Among other things, this is related to the characteristics of the sound source. A position with a pressure antinode in the pipe becomes a position for a pressure node. The entire sound field in the pipe changes. In this context, less sound is emitted at the mouth. It can be seen in Fig. 5.43 that the wave has a node at the left edge of the muffler. However, the right edge represents the maximum – pressure antinode – for the transmitted sound wave. Accordingly, only very little is transmitted.

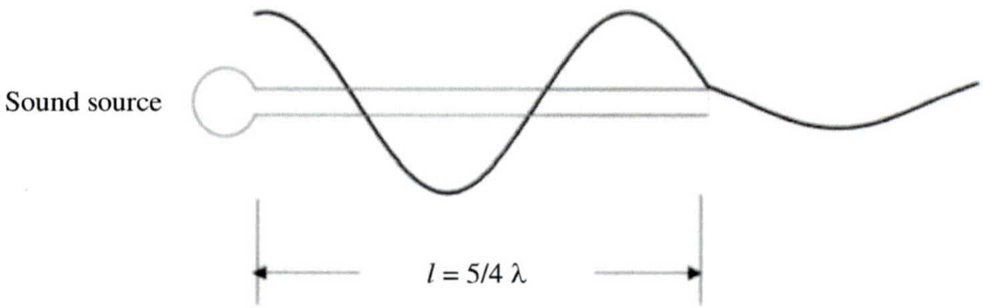

Fig. 5.42 Duct system with sound source on the left and open end on the right

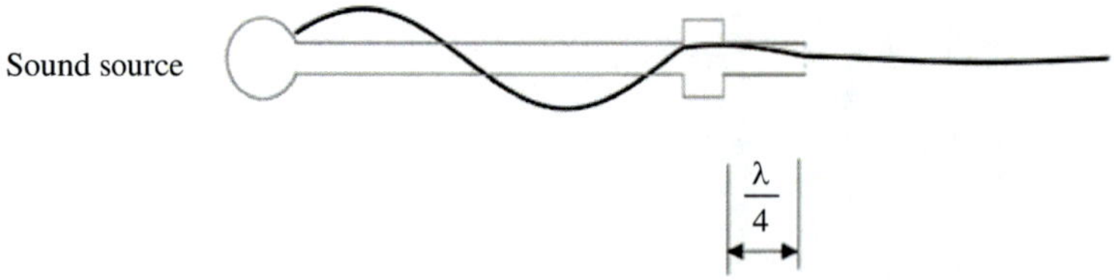

Fig. 5.43 Duct system with sound source on the left and open end on the right as well as a sound attenuator volume at a distance λ/4 from the duct end

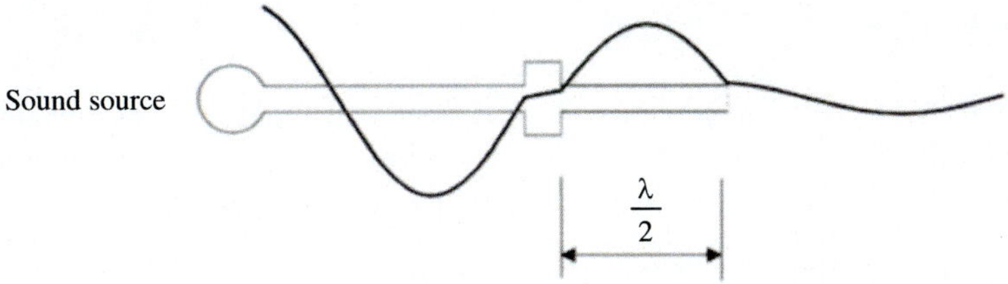

Fig. 5.44 Duct system with sound source on the left and open end on the right as well as a sound attenuator volume at a distance $\lambda/2$ from the duct end

If a silencer is placed in a pressure node, the sound field hardly changes and the silencer has almost no sound-reducing effect (Fig. 5.44). The sound pressure is small on both sides of the silencer, but increases again after the silencer.

From this it follows that a sound attenuator volume in front of a duct mouth has the best effect when it is placed in a pressure bulge of the standing wave. A silencer in a pressure node has only a small effect for this frequency.

The situation is different with an absorption silencer. On the one hand, such a silencer is no longer small in relation to the wavelength. Furthermore, a part of the incoming sound wave is absorbed. This absorbed part is then missing from the later reflected sound wave. Since the reflection is weakened, no distinct standing wave can form. Accordingly, as soon as the absorption sound attenuator absorbs sound in an appreciably way (attenuation >10 dB), it is insensitive to its installation position, in contrast to the reflection sound attenuator.

Figure 5.45 shows the schematic structure of a reflection silencer.

The design of a reflection silencer is much more complicated than the silencers discussed so far. This is because the sound field in the duct changes due to sound reflections both before and after the reflection point. Therefore, when designing a reflective silencer, the input and output spectra must always be matched to each other. For this purpose, a prototype is built and measured for each design variation. Based on the measurement results, the reflective silencers can then be optimised step by step. Depending on the task and the desired level reduction, 2 to 15 prototypes may be required [10, 40–42].

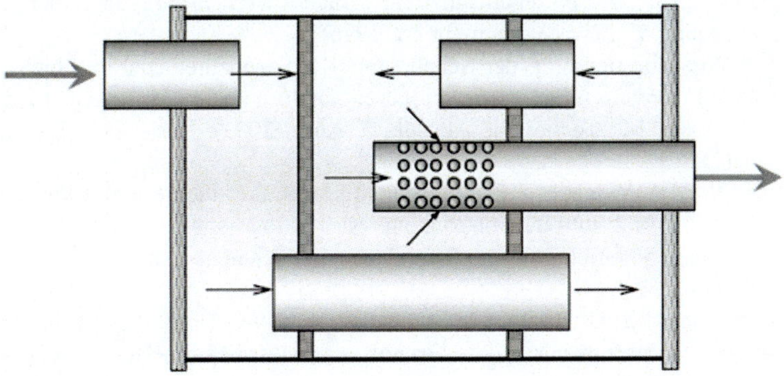

Fig. 5.45 Schematic structure of a reflection silencer

References

1. VDI-3720, B 1, Konstruktion lärmarmer Maschinen und Anlagen, Konstruktionsaufgaben und -methodik, 2014
2. Sinambari, Gh.R., Fallen, M., u. a.: Akustische und schwingungstechnische Probleme im Anlagenbau unter besonderer Berücksichtigung von Rohrleitungen und strömungsführenden Komponenten, Nr. E-35-727-134-9, HDT Essen, 1999.
3. Krüger, J.; Leistner, P.: Aktive Kulissenschalldämpfer in Strömungskanälen. Tagungsbeitrag, 23. Jahrestagung Akustik DAGA, 1997
4. Sinambari, Gh.R., Kunz, F., Thorn, U.: Active noise cancellation in large section air ducts with subdivisions. 7. International Congress on Sound and Vibration, Germany, 2000
5. Sinambari, Gh.R., Kunz, F., Siegel L, Thorn U.: Aktive Schalldämpfer für Kanäle mit großen Durchmessern. Konstruktion **4** (2000)
6. Sinambari, Gh.R., Kunz, F.: Primäre Lärmminderung durch akustische Schwachstellenanalyse. VDI-Bericht Nr. 1491 (1999)
7. Sinambari, Gh.R., Thorn, U., Tschöp, E.: "Konstruktionsakustik", Seminarveranstaltung. IBS-Seminar, St. Martin/Pfalz, 2016
8. Thorn, U., Wachsmuth, J.: Konstruktionsakustische Schwachstellenanalyse eines Straddle Carriers Typ Konecranes 54 DE und Erarbeiten von prinzipiellen Lärmminderungsmaßnahmen. VDI-Berichte 2118 (2010)
9. Randall, R.B.: Frequency Analysis. Application of B & K Equipment (1979)
10. Sinambari, Gh.R.; Sentpali, S.: Ingenieurakustik, 6. Aufl. Springer Vieweg, Wiesbaden (2020)
11. VDI 3720-7: Noise abatement by design; assessment of alternating powers at noise (1989) (withdrawn)
12. Cremer, L., Heckl, M.: Körperschall, 2. Aufl. Springer, Berlin (1996)
13. Sinambari, Gh.R.: Einflussparameter bei körperschallbedingter Geräuschentwicklung einer Maschine. 37. Jahrestagung, DAGA, 2011
14. VDI-Richtlinie 2062, Bl. 1: Vibration insulation – Terms and methods (2011)
15. Hering, E., Martin, R., Stohrer, M.: Physik für Ingenieure, 7. Aufl. (1999)
16. Magnus, K., Popp, K.: Schwingungen, 7. Aufl. (2005)
17. DIN 1311-2: Schwingungen und schwingungsfähige Systeme, Teil 2: Lineare, zeitinvariante schwingungsfähige Systeme mit einem Freiheitsgrad (2002)
18. Sinambari, Gh.R.: Ein erweiterter Ansatz zur Schwingungs- und Körperschallisolierung: Teil 1: Z. Lärmbekämpfung, Bd. 6, 2011, Nr. 2; Teil 2: Z. Lärmbekämpfung, Bd. 6, 2011, Nr. 3
19. Sinambari, Gh.R.: Geräuschreduzierung durch Körperschallisolierung. 38. Jahrestagung, DAGA, 2012
20. Sinambari, Gh.R.: Körperschallisolierung unter Berücksichtigung der mechanischen Eingangsimpedanz. Z. Lärmbekämpfung 43 (1996)
21. Thorn, U.: Vorausbestimmung der Schall- und Körperschallreduzierung durch Einfügen von mechanischen Impedanzen bei realen Strukturen. Diplomarbeit, FH Bingen (1995)
22. Cremer, L., Möser, M.: Technische Akustik, 5. Aufl. (2003), Möser, M.: Technische Akustik, 9. Aufl. (2012)
23. ISO/TR 11688-2: A Acoustics – Recommended practice for the design of low-noise machinery and equipment – Part 2: Introduction to the physics of low-noise design (1998)
24. Cremer, L.: Die wissenschaftlichen Grundlagen der Raumakustik, 1. Teil, 2. Aufl., Hirzel, Stuttgart (1978)
25. Morse, P.M., Ingard, K.U.: Theoretical Acoustics. Princeton Univ. Press, Princeton (1986)
26. Kollmann, F.G.: Maschinenakustik – Grundlagen, Messtechnik, Berechnung, Beeinflussung, 2. Aufl. Springer (2000)

27. Schirmer, W. [Hrsg.]: Technischer Lärmschutz. VDI-Verlag (1996)
28. Gebhard, B.: Bestimmung der Schallleistung, Schallabstrahlung einer krafterregten plattenförmigen Struktur nach verschiedenen Messverfahren und deren Bewertung. Diplomarbeit, FH Bingen, FB Umweltschutz (1996)
29. Gerbig, C.: Ermittlung der Schallabstrahlung einer krafterregten kastenförmigen Struktur nach verschiedenen Messverfahren und deren Bewertung. Diplomarbeit, FH Bingen, FB Umweltschutz (1996)
30. Mäurer, A: Ermittlung des Abstrahlgrades von komplexen Maschinenstrukturen am Beispiel einer Drehkolben-Vakuumpumpe. Diplomarbeit, FH Bingen, FB Umweltschutz (2002)
31. ISO 11546-2: Acoustics – Determination of sound insulation performances of enclosures – Part 2: Measurements in situ (for acceptance and verification purposes) (1995)
32. VDI 2711: Noise reduction by enclosures (1978) (withdrawn)
33. VDI 2081-1: Air-conditioning – Noise generation and noise reduction (2022-04)
34. VDI 2081-2: Air-conditioning – Noise generation and noise reduction – Examples (2022-10)
35. VDI 2567: Noise control by silencers (10/1994) (withdrawn)
36. Fuchs, H.V.: Schallabsorber und Schalldämpfer, 3. Aufl. Springer, Heidelberg (2010)
37. Piening, W.: Schalldämpfung der Ansauge und Auspuffgeräusche von Dieselanlagen auf Schiffen. VDI-Zeitschrift 81 (1937)
38. Brandstatt, P.; Fuchs, H.: Erweiterung der Pieningschen Formel für Schalldämpfer. Z. Lärmbekämpfung 44 (1997)
39. Fasold, W., Sonntag, E., Winkler, H.: Bau- und Raumakustik. VEB Verlag für Bauwesen, Berlin (1987)
40. Heckl, M., Müller, H.A.: Taschenbuch der Technischen Akustik, 2. Aufl. Springer, Berlin (1994)
41. Bechert, D.W.: Sound Absorption caused by vorticity shedding, demonstrated with a jet flow. J. Sound Vib. **70** (1980)
42. Cargill, A.M.: Low frequency acoustic radiation from a jet pipe – A second order theory. J. Sound Vib. **83**(3) (1982)

Practical Examples

In the following, the essential results of some practical examples are given, in which a targeted noise and vibration reduction was planned or carried out. For these application examples, an acoustic weak point analysis was carried out in advance, as dealt with in this book. Here, the systematic approach to noise reduction, especially in the development of primary noise reduction measures, is primarily to be clarified.

This is a small selection from the examples which were developed under the direction or with the cooperation of the author during his activity as managing director of the company IBS[1] and professor at the FH Bingen[2] and which have already been published [1, 4, 8, 11, 12, 17–19, 24, 25]. Other examples, which would certainly be helpful in understanding the relationships as given in this book, can unfortunately not be reproduced for reasons of confidentiality. However, it should be mentioned that without the many personal experiences from numerous practical applications, even if they are not published, the writing of this book would not have been possible.

Subsequently, several field-tested examples that are important for the understanding of low-noise design or for general noise reduction are reproduced. These examples, in which the author was not directly involved, were mainly developed in the period between 1970 and 1990, [3, 28–37].

[1] 1988–2014: IBS Ingenieurbüro für Schall- und Schwingungstechnik GmbH, Frankenthal, Germany from 1993 part-time, www.ibs-akustik.de.

[2] 1993–2013: FH Bingen, FB 1 u. 2, Germany Sound and vibration protection, emission technology, construction acoustics.

G. R. Sinambari, *Design Acoustics*,
https://doi.org/10.1007/978-3-658-40183-2_6

6.1 Drive Unit

The sound radiation of a drive unit of a coal conveyor belt, Fig. 6.1, is to be evaluated and optimised by means of design acoustic investigations [1]. In this context, the total noise development is to be reduced by approx. 10 dB(A). For the description of the actual condition, the total and partial sound power levels of the drive unit and its components were determined both in the form of the total A-value and the spectral distribution (third-octave spectrum). This was done mainly by sound intensity measurements, see Chap. 5, [2].

Figure 6.2 shows the A-total sound power levels of the drive unit and its components measured. It follows from this that the drive foundation, which only has a load-bearing function, contributes almost as much to the total sound radiation of the drive unit as the gear unit.

Figure 6.3 shows the A-weighted third-octave sound power spectra of the drive unit and its components.

From this it can be seen that the A total sound power level of the drive unit is determined by the frequency band from approx. 160–1250 Hz. It is also clear that the level component of the frequency bands up to approx. 800 Hz is primarily emitted by the gearbox and the drive foundation and from approx. 1250 Hz primarily by the engine.

The excitation frequencies of the gear meshing are responsible for the sound generation. Due to different conveyor belt loads, the engine speed varies from approximately to $f_n = 16.2$ 16.6 Hz ($n = 972$–996 1/min). The tooth meshing frequencies are:

- First stage $f_1 \approx 373$–381 Hz
- Second stage $f_2 \approx 174$–179 Hz

Fig. 6.1 Drive unit type 2000 kW with the examined components gearbox, engine, brake and drive foundation

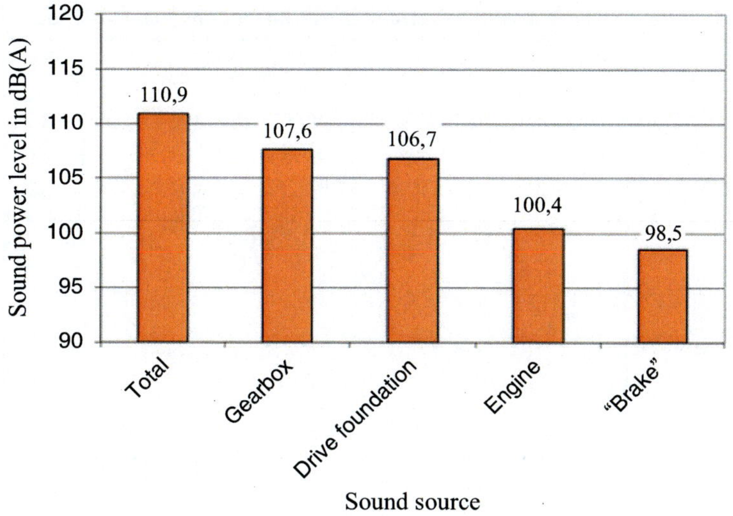

Fig. 6.2 A-total sound power level of the drive unit

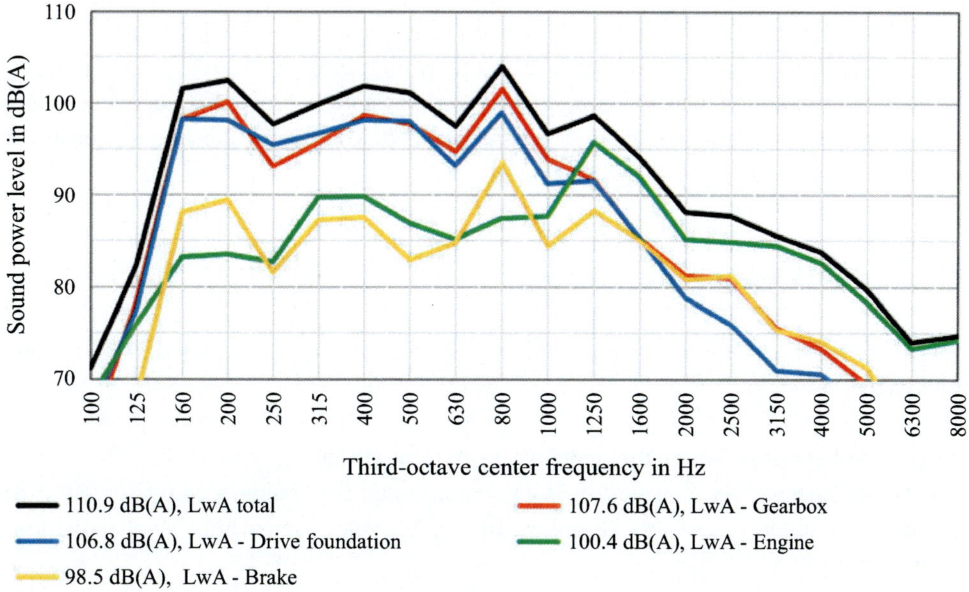

Fig. 6.3 Third-octave sound power spectra of the drive unit and its components

These frequencies and their harmonics (integer multiples of the tooth natural frequencies) serve as excitation frequencies. As a result, plate fields of the gearbox housing as well as all structures that are rigidly connected to the gearbox, e.g. drive foundation, are forcibly excited to vibrations. With regard to noise generation, plate fields or components whose natural frequencies lie in the range of the excitation frequencies are particularly critical due to resonance excitation.

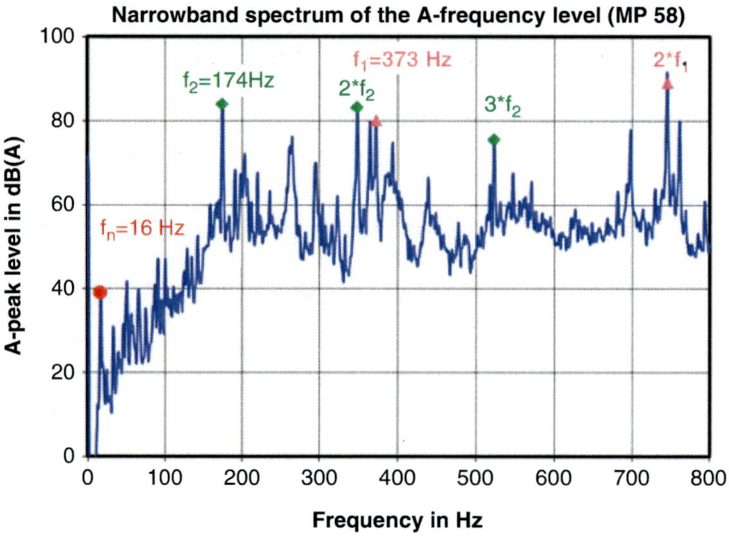

Fig. 6.4 Narrowband spectrum of A-peak levels, MP 58

Figure 6.4 shows an example of the narrowband spectrum of the A-velocity levels of a measurement point on the foundation (MP 58, see Fig. 6.6).

By determining the natural frequencies, it was possible to prove that many components of the drive foundation are excited to resonance vibrations, which are also responsible for its high noise development.

6.1.1 Noise Reduction Planning

In Table 6.1, based on the measured emission characteristics, the planned noise reduction plan is given.

Step 1: Noise reduction of the gearbox by $\Delta L_F = 15\ dB(A)$
The noise generated by the gearbox as a purchased part is to be reduced by 15 dB(A) with the aid of a sound enclosure. This measure could achieve an overall level reduction of approx. 2.6 dB(A).

Step 2: Noise reduction of the drive foundation by $\Delta L_F = 12\ dB(A)$
The noise development of the drive foundation is to be reduced by 12 dB(A) through constructive measures, based on an acoustic weak point analysis. The overall level reduction achieved would then be approx. 7.3 dB(A). After the noise reduction of the gearbox and the drive foundation, the engine is the loudest source with $L_{WA,\ \mathrm{Motor}} = 100.4$ dB(A). A further reduction of the A-total sound power level therefore only makes sense if the noise development of the engine is reduced.

Table 6.1 Noise reduction design of the drive unit

Noise reduction plan					
State	ACTUAL state	Gearbox $\Delta L_G = 15$ dB(A)	A-Foundation $\Delta L_F = 12$ dB(A)	Engine $\Delta L_E = 5$ dB(A)	Brake $\Delta L_B = 3$ dB(A)
L_w A, Gearbox	107.6	**92.6**	92.6	92.6	92.6
L_W A, $drive_f$ undament	106.7	106.7	**94.7**	94.7	94.7
L_w $A,_g$ es engine	100.4	100.4	100.4	**95.4**	95.4
L_w A,Brake	98.5	98.5	98.5	**98.5**	**95.5**
L_w A, Total	**110.9**	**108.2**	**103.6**	**101.9**	**100.7**
Noise reduction in dB (A)	–	2.6	7.3	9.0	10.2

Step 3: Noise reduction of the engine by $\Delta L_E = 5$ $dB(A)$ and of the brake by $\Delta L_B = 3$ dB(A)

In order to achieve the desired noise reduction of 10 dB(A), the noise development of the engine must be reduced by 5 dB(A) and the noise development of the brake by 3 dB(A). Since an overall encapsulation of the drive unit is out of the question due to maintenance work, the noise reduction of these components, which are purchased parts, can only be realized in cooperation with the manufacturers.

Figure 6.5 shows the results of the noise reduction planning, which were created with the help of the acoustic weak point analysis. This made it possible to determine quantitatively at which component how much noise reduction is necessary to achieve the objective.

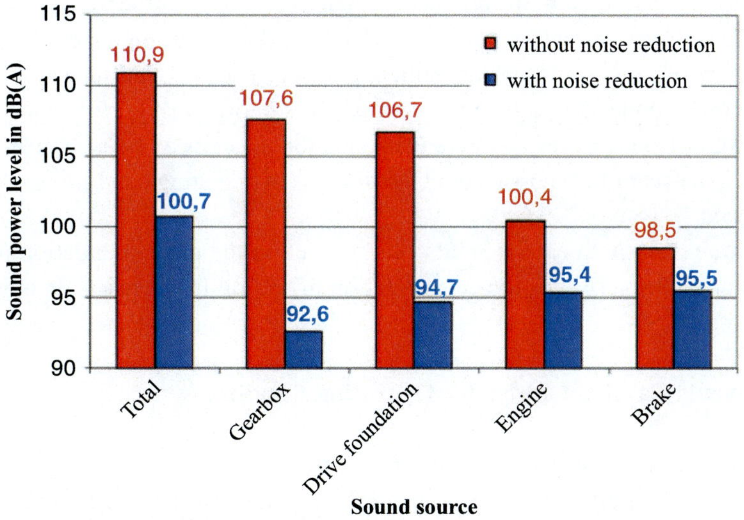

Fig. 6.5 Noise reduction plan for the drive unit

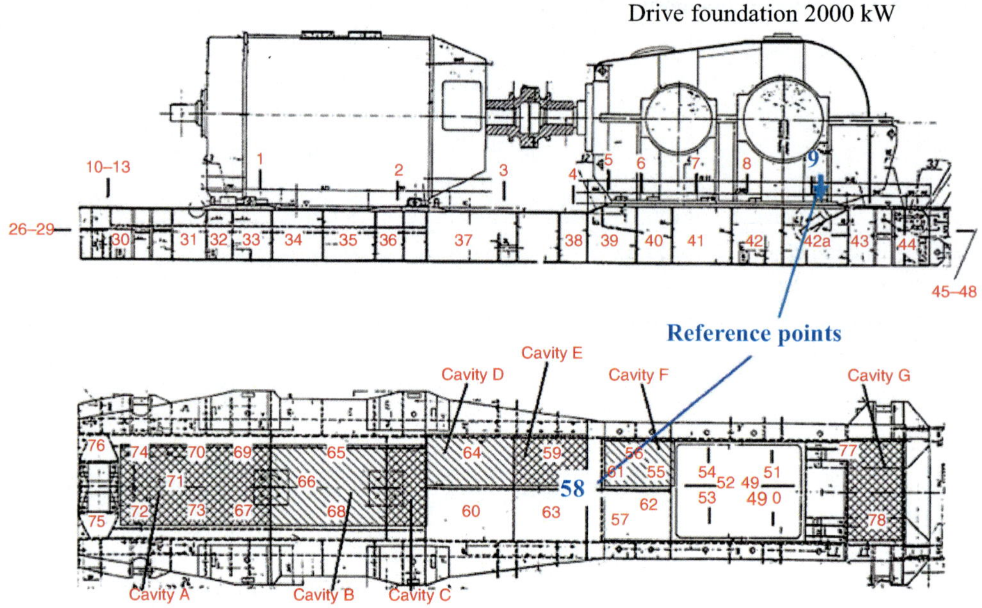

Fig. 6.6 Location of measuring points for sound intensity and structure-borne sound measurements

6.1.2 Acoustic Weak Point Analysis of the Drive Foundation

As already determined by the noise reduction planning, the noise development of the drive foundation is to be reduced by approx. 12 dB(A). For the identification of acoustic weak points and the elaboration of primary design measures, further sound and vibration investigations were carried out on the drive foundation during operation of the coal conveyor. The location of the measurement points at which sound intensity, airborne and structure-borne sound measurements were taken is shown in Fig. 6.6. Measuring points 9 and 58 were also measured as reference points for all series of measurements. This made it possible to detect or take into account fluctuations in operating conditions, e.g. changes in load or speed.

In addition, measurements were also carried out on the drive foundation under laboratory conditions, primarily for the determination of natural frequencies, damping and input impedances.

Measurements Carried Out Under Operating Conditions

- Sound intensity measurements to determine the total and partial sound power levels
- Structure-borne sound measurements on the surface of the drive foundations to determine the structure-borne sound power levels as well as at coupling points of the aggregates (engine, brake, gearbox) to determine the structure-borne sound transmission

- Structure-borne sound measurements at two reference points of the drive foundation for all measurement series to check the operating conditions. This ensures the comparability of the measurements, which were carried out at different times. In the present case, this check is particularly important, as it is possible to eliminate or take into account the fluctuations in the measured values caused by operation, e.g. by loading fluctuations of the conveyor belt, during the evaluation.
- Impact tests at the coupling points to determine the structure-borne noise transmission and at various plate fields of the drive foundation. The measurements were carried out at standstill.

For the localization of the acoustic weak points of the drive foundation, their partial sound power levels were determined by sound intensity and structure-borne sound measurements. Figure 6.7 shows the corresponding emission values as A-total sound power levels and Fig. 6.8 shows the third-octave spectra of some components with the highest A-sound power levels.

The A partial sound power levels were mainly determined by sound intensity measurements. For the partial areas of the drive foundation for which sound intensity measurements were not possible due to high sound reflection or lack of accessibility, the partial sound power level was determined by structure-borne sound measurements.

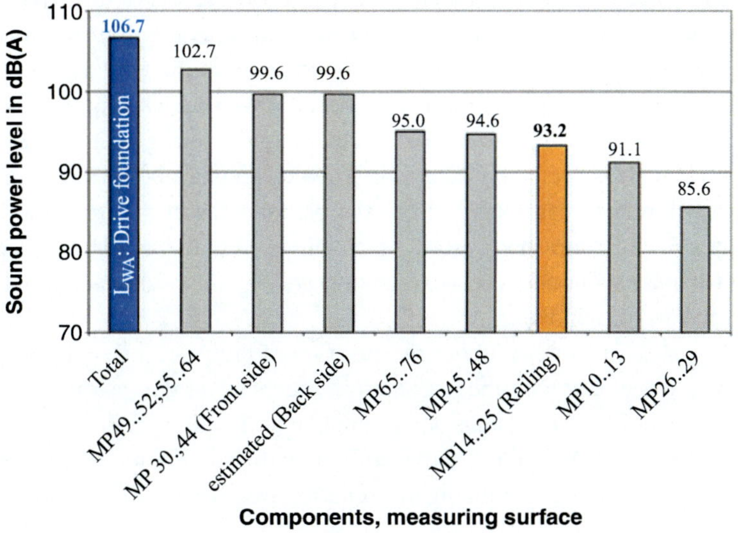

Fig. 6.7 A-sound power balance of the drive foundation

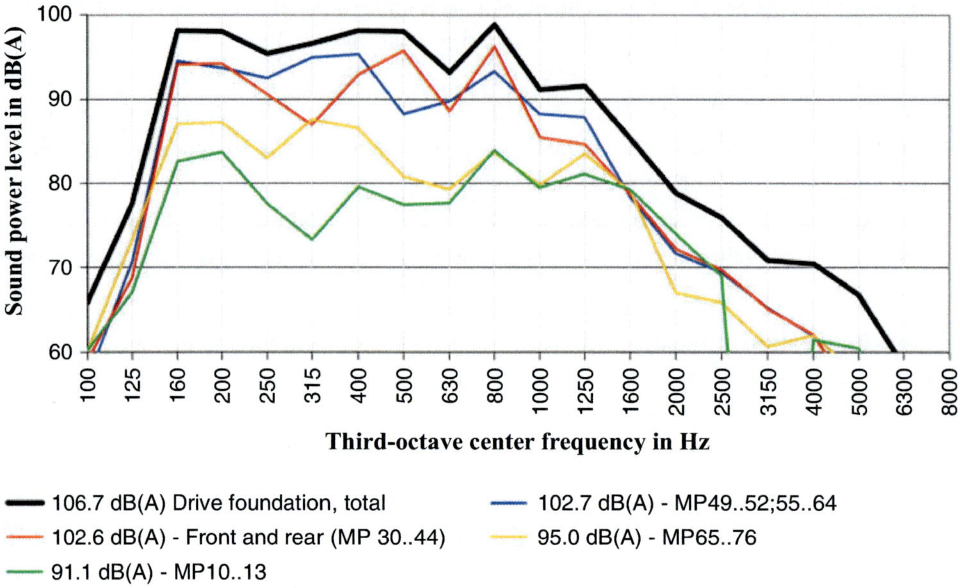

Fig. 6.8 Third-octave spectrum of the A-total and partial sound power levels of the drive foundation

The measurements show that the partial sound power levels of the drive foundation in the area below the gearbox (MP 49–52, 55–64) are considerably higher than in the area of the engine (MP 65–76) and thus contribute significantly to the total sound radiation.

The spectral representation of the partial sound power levels (Fig. 6.8) shows that the noise development of the drive foundation in the frequency range between approx. 160 Hz and approx. 1250 Hz is decisively determined by the structure-borne sound radiation of the partial areas below the gearbox as well as the partial areas in the front and rear. Furthermore, it can be seen from this, among other things, that the increase in the total sound power level at 500 Hz and 800 Hz is caused primarily by the sound radiation from the front and rear of the drive foundation.

Figure 6.9 shows the one-third octave spectrum of the A total sound power level of the drive foundation together with the gear mesh frequencies of the gearbox.

From this it can be seen, among other things, that the gear meshing frequencies of the first (f_1) and second stages (f_2) of the gearbox and their harmonic frequencies ($n \cdot f_1$ and $n \cdot f_2$, with $n = 1, 2, 3, \ldots$) are the main excitation frequencies of the drive foundation.

Measurements Carried Out Under Laboratory Conditions

For this purpose, the drive foundation was set up in a hall and excited by means of a shaker or an impulse hammer as well as a small loudspeaker, see Fig. 6.10.

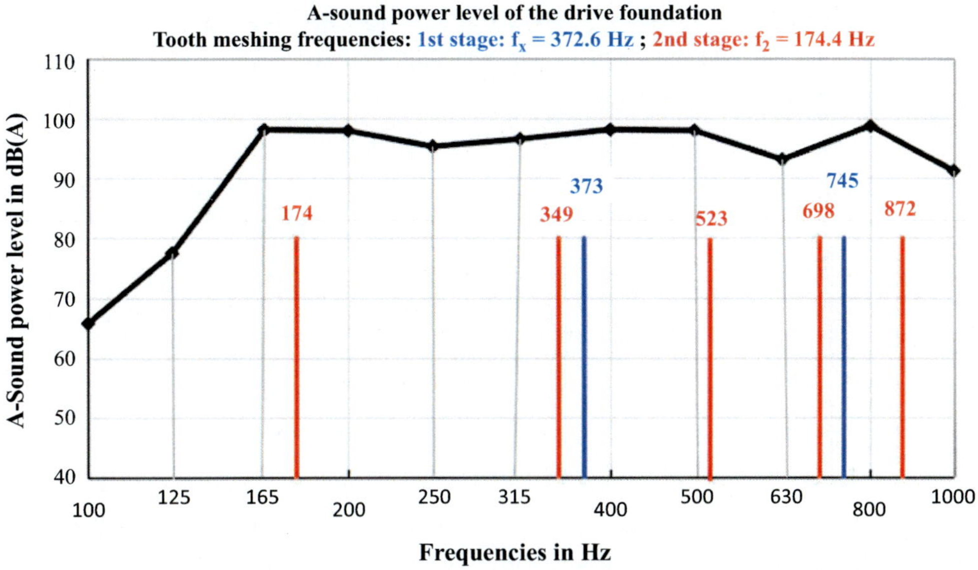

Fig. 6.9 Third-octave spectrum of the A-sound power level of the drive foundation and the position of the gear meshing frequencies of the gearbox

Fig. 6.10 Test setup for laboratory tests of the drive foundation

In detail, the following measurements were made:

- Force application by means of shaker or impulse hammer for the determination of input impedances at the coupling points of the gear as well as natural frequencies and damping of plate fields
- Excitation of cavities by means of loudspeakers for the determination of cavity resonances

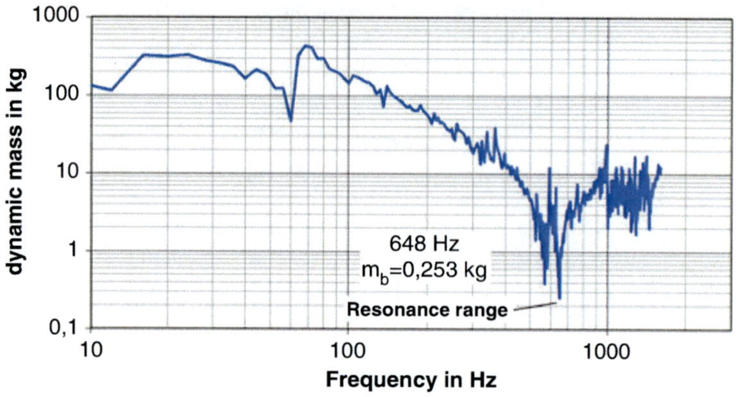

Fig. 6.11 Measured dynamic mass on the flange surface of an IPE 450 beam, gearbox coupling point

Figure 6.11 shows an example of the measured dynamic mass at a coupling point of the gearbox (flange surface of an IPE 450 carrier) as a function of frequency.

From this it can be seen that a relatively massive carrier, which is statically perfectly adequately dimensioned, has a dynamic mass of $m_\mathrm{b} \approx 0.25$ kg at the coupling point of the gearbox at 648 Hz as a vibration resistance. The same carrier has a dynamic mass (vibration resistance) of approx. 300 kg at frequencies below 100 Hz. It is therefore easy to understand that, depending on the exciting frequency, the drive foundation can be excited to considerable structure-borne sound vibrations or noise generation with low alternating forces [2].

In summary, it can be stated that the noise development of the drive foundation is mainly caused by structure-borne sound transmission from the gearbox to the drive foundation. Due to the high number of natural frequencies of the plate fields and cavities as well as the relatively low fluctuating rotational frequency of the engine, several plate fields and cavities of the drive foundation are excited to resonance vibrations. The total sound radiation is primarily determined by the structure-borne sound radiation or the secondary airborne sound and is decisively influenced by resonance vibrations which occur when the excitation frequencies of the gear unit coincide with the natural frequencies of the components of the drive foundation, the plate fields and the cavities.

6.1.3 Noise Reduction Measures

Based on the results of the acoustic weak point analysis, the following noise reduction measures were proposed to reduce the overall noise development of the drive unit. Principle design noise reduction measures were developed for the drive foundation, which should be taken into account in the new design of the drive foundation.

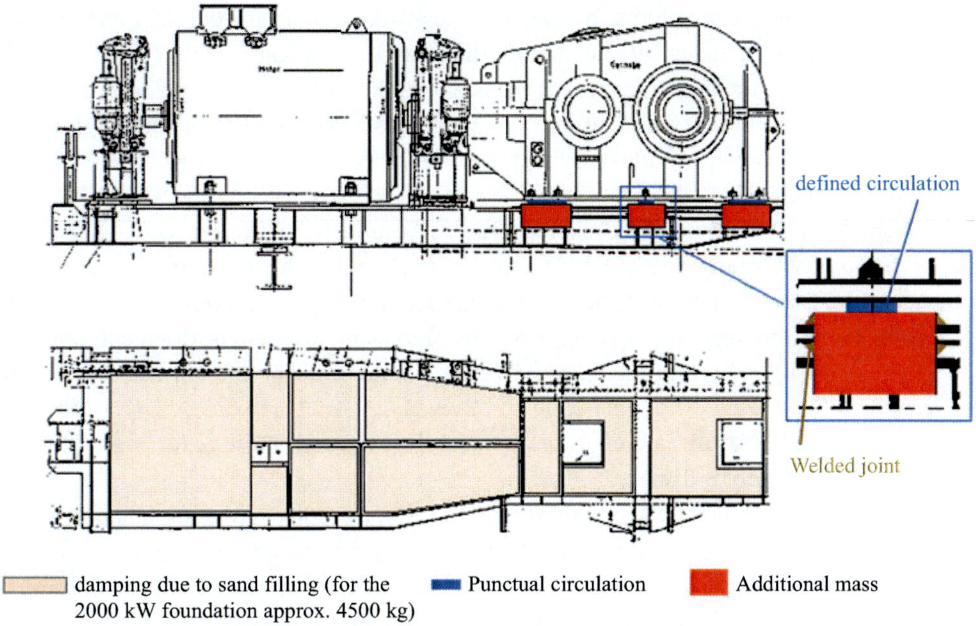

Fig. 6.12 Schematic representation of the measures for the drive foundation

1. For effective reduction of the total sound radiation of the drive unit, the gear unit should be encapsulated. The insertion loss of the acoustic enclosure should be at least 15 dB (A).
2. Increase the dynamic masses at the coupling points of the gearbox to reduce the introduction of structure-borne noise.
3. In order to clearly define the force application points, the coupling of the gear unit to the drive foundation should be designed in a point-like manner.
4. Reduction of impedance dips, especially in components that are excited to resonance, by increasing damping, e.g. by applying strip damping or by filling the cavities with dry quartz sand.
5. Close the opening below the gearbox. The resulting cavity should be damped by sand filling (quartz sand).

Figure 6.12 shows a schematic diagram of the above measures for a planned drive foundation, including the new gearbox with the same gearing and speeds [1, 3].

The effectiveness of primary design measures, e.g. by increasing the input impedance or by damping resonance-excited plate fields, depends essentially on whether it is possible to achieve a noticeable increase of at least a factor of 2 in relation to the existing values, e.g. dynamic mass, damping, etc.

In the present case, an overall noise reduction of the drive foundation of approx. 12 dB (A) was determined on the basis of the results of the weak point analysis in the noise reduction planning. Quantitative proposals were also drawn up for the principal measures so that they could be taken into account in the new construction.

The increase of the dynamic mass at the gear coupling points was realized by a compact mass block ($m = 10$ kg), which was to be force-fitted below the flange surface of the IPE 450 carrier. This was to significantly increase the impedance dips, above approx. 500 Hz, see Fig. 6.11 [2].

In order for the increase in dynamic mass to be effective, the previously linear force application of the gearbox to the drive foundation should be made point-shaped. This can be achieved, for example, with the aid of a washer at all coupling points.

Quartz sand with a grain size of approx. 1 to 2 mm, which should be dried beforehand by heating, was suggested for damping. Drying should prevent the formation of lumps over time.

The proposed measures are expected to result in an overall noise reduction of the drive unit of approximately 8 dB(A).

A further noise reduction of the drive unit listed in the noise reduction planning is only possible or reasonable if the noise development of the engine and the brake is reduced in cooperation with the manufacturers. As already mentioned, a total encapsulation of the drive unit is out of the question because of the maintenance work.

6.2 Packaging Machine

The noise emission of a cartoner type C 150 of the company Uhlmann Pac-Systeme GmbH & Co. KG (Fig. 6.13) shall be reduced to such an extent that the emission sound pressure level at the workplace of the packaging machine does not exceed a value of $L_{pA} = 75$ dB (A) [4].

Table 6.2 shows the emission sound pressure levels at the cartoner's workplaces determined in accordance with DIN EN ISO 11202 and the required noise reduction.

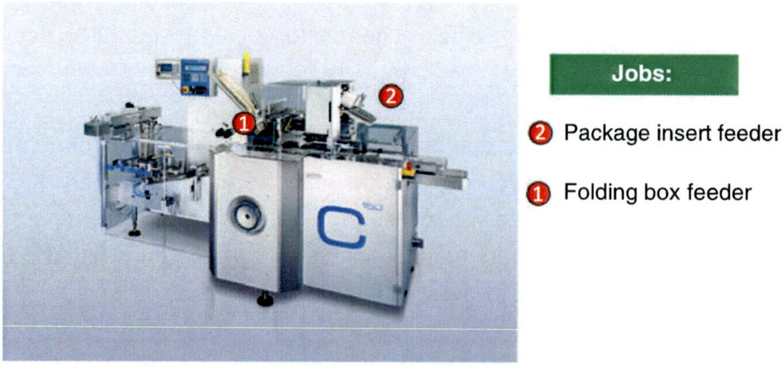

Fig. 6.13 Cartoner type C 150 (Uhlmann Pac-Systeme GmbH & Co. KG)

Table 6.2 Emission sound pressure level at the workplaces for the cartoner type C 150

Jobs	Emission sound pressure level L_{pA} Full load operation (150 cycles/min)	Required noise reduction
Folding box feeder	81 dB(A)	6 dB(A)
Package insert feeder	83 dB(A)	8 dB(A)

The noise reduction of 6 or approx. 8 dB(A) is oriented towards the required target value of 75 dB(A). Figure 6.13 shows the examined cartoner type C 150 including the mentioned workplaces [4].

6.2.1 Design Acoustic Weak Point Analysis

In order to reduce the noise emission of the packaging machine, a sound and vibration weak point analysis was carried out on the cartoner [5]. Based on this, primary and secondary noise reduction measures were developed to achieve the objective.

The relevant partial sound sources of the cartoner are:

- Opening blister knocker
- Folding box feeder
- "Chimney opening" in roof area
- Vacuum pump
- Leaflet feeder (leaflet feeder)
- Folding box ejector
- Machine guard

Figure 6.14 shows the relevant partial sound sources of the cartoner.

The total and partial sound power levels of the cartoner were determined using the sound intensity method according to DIN EN ISO 9614-2.

The partial sound power levels of the floor slots and the other opening areas were determined according to the total suppression method, in which the partial sound power of individual sound sources is calculated indirectly by level subtraction. For this purpose, the floor slots and other opening areas were provisionally sealed. The partial sound power level of the machine protective cladding was determined by structure-borne sound measurements according to DIN 45635-8.

Figure 6.15 shows the A-total and partial sound power levels of the cartoner C 150. It can be seen that the total sound energy radiated by the cartoner is distributed over many partial sound sources with approximately the same partial sound power level.

The energetic summation of the partial sound power levels results in an A-total sound power level of $L_{WA,\ \text{total},\ R} = 89.9$ dB(A). The metrologically determined A-total sound

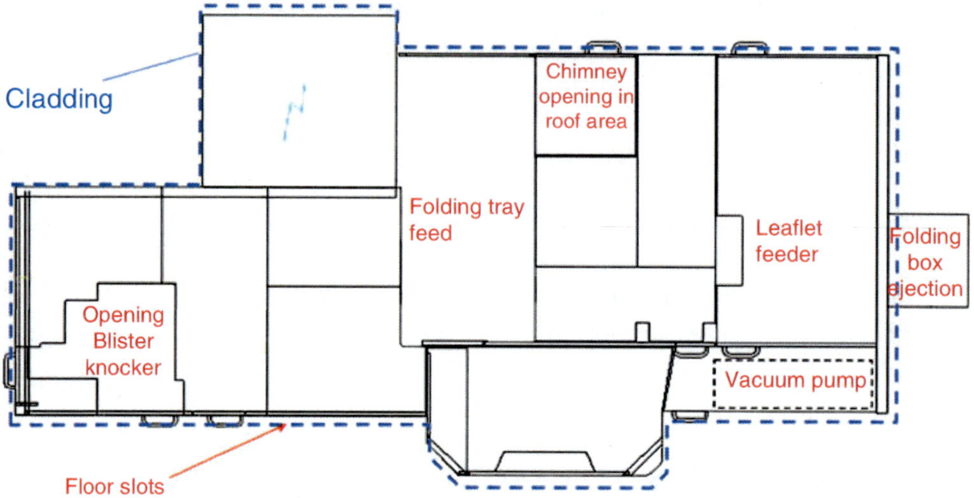

Fig. 6.14 Relevant partial sound sources (cartoner type C 150)

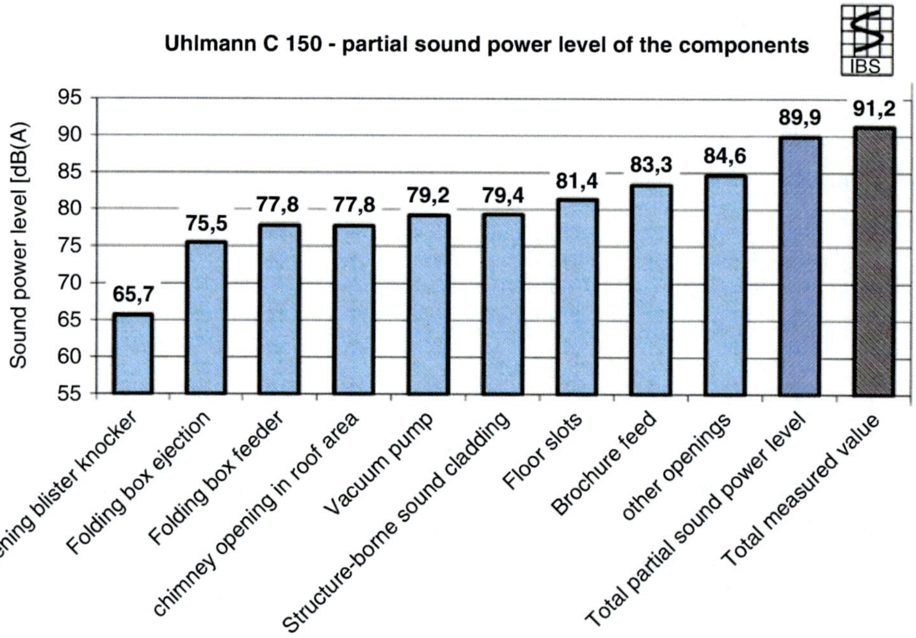

Fig. 6.15 Sound power level balance of the cartoner type C 150

power level of $L_{WA,\ \text{total},\ M} = 91.2$ dB(A) is 1.3 dB(A) higher. This deviation is due to unavoidable measurement inaccuracies in the determination of the sound power level of the individual partial sound sources and possible systematic deviations in the balancing of

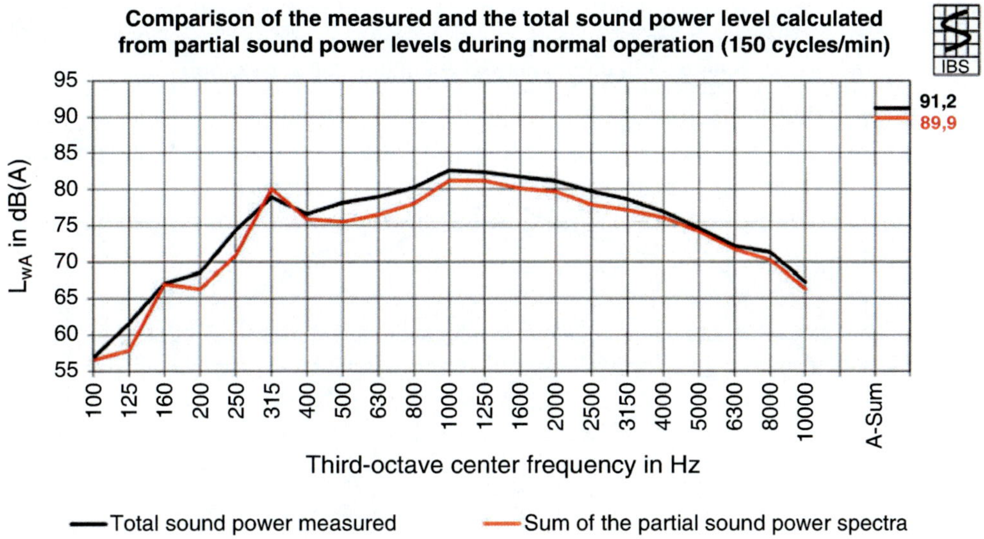

Fig. 6.16 Comparison of the frequency spectra of the measured and the total sound power level calculated from the partial sound power levels of the cartoner type C 150 at 150 cycles/min

individual partial sound power levels, but is still within the usual range for such measurements.

Figure 6.16 compares the one-third octave spectrum of the measured total sound power level ($L_{WA,\ \mathrm{fm},\ \mathrm{M}}$) with the one-third octave spectrum of the total sound power level ($L_{WA,\ \mathrm{fm},\ \mathrm{R}}$) energetically added over the individual partial sound power levels. As can be seen, the accuracy of the weak point analysis in the individual one-third octave bands is within the range of 2 to 5 dB to be aimed at, see Chap. 5.

Figure 6.17 compares the third-octave spectra of relevant partial sound sources with the third-octave spectrum of the total sound power level of the cartoner.

For reasons of clarity, the partial sound power levels of the partial sound sources opening blister knocker, folding box ejector, folding box feed, chimney opening in roof area, floor slots, brochure feed as well as the other openings in the protective cladding were added energetically to the partial sound power "all openings".

In summary, it can be stated that the sound emission of the cartoner during normal operation in the low-frequency range is significantly influenced by the vacuum pump.

The level increases in the third-octave band 160 Hz and 315 Hz are due to the pulsation frequency of the vacuum pump (f_0 = 170 Hz) and the 1st harmonic of the pulsation frequency (f_1 = 340 Hz). In the high-frequency range, the sound emissions are mainly due to the opening surfaces in the protective cladding, with the folding process of the instruction leaflet and the other opening surfaces in particular making a prominent contribution. The structure-borne noise emission is approx. 10 dB(A) lower than the airborne noise emission of the opening surfaces of the machine protective cladding. The airborne sound radiation is therefore decisive for the noise development of the cartoner.

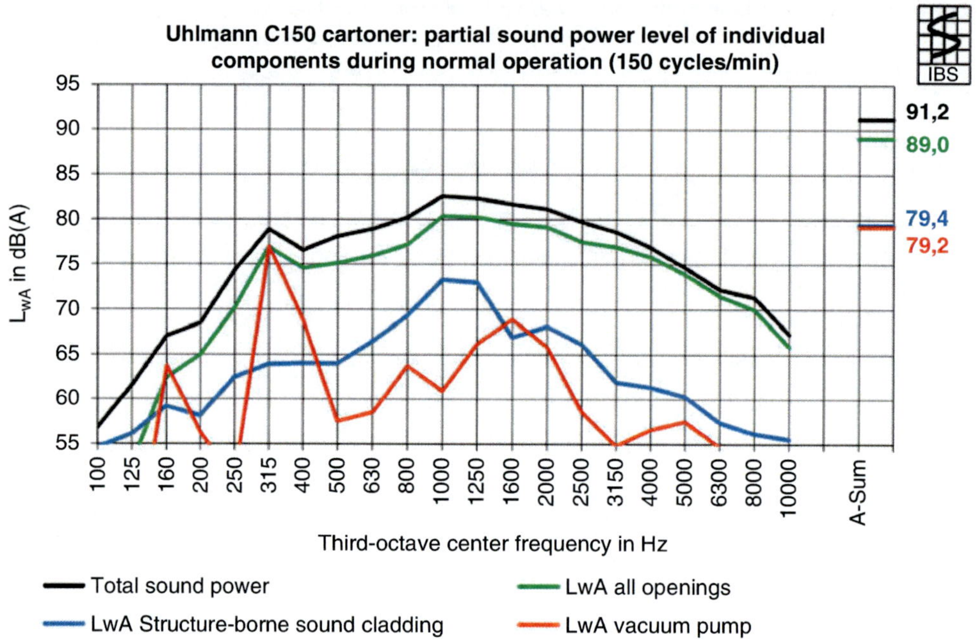

Fig. 6.17 Third-octave spectrum of the relevant partial sound sources compared to the total sound power level of the cartoner type C 150

6.2.2 Noise Reduction Plans

When developing noise reduction measures and drawing up priority plans, various scenarios are often considered in order to achieve the targeted level reduction. If, for example, some partial noise sources are "no longer to be tackled" for functional or economic reasons, the energy balance can be used to check whether the necessary noise reduction can be achieved by measures at other partial noise sources.

With the help of the sound power balance it is possible to estimate and evaluate the effectiveness of "in-petto solutions", i.e. reduction measures which the machine manufacturer already has in the drawer or which can be implemented with little effort [5].

A noise reduction plan coordinated with the manufacturer for an overall level reduction of approx. 7 to 8 dB(A) at the workplaces is reproduced below. The following noise reductions are planned:

- Dispensing with the vacuum pump. In principle, it is possible here to generate or provide the necessary process vacuum using a different technique.
- The partial emission of the brochure feed is to be reduced by 10 dB(A). This can be achieved, for example, by using an acoustically optimised folder with leaflet cover and improved sealing measures.

- Acoustic sealing of the chimney opening in the roof area and the floor slots. The required insertion attenuation is 15 dB(A).
- The folding box supply, where residual opening areas remain for functional reasons, shall be reduced by 7 dB(A).
- A level reduction of 10 dB(A) is envisaged for the prospect feed, where the residual opening areas can be kept smaller.
- The other opening surfaces in the area of the workplaces must be designed with optimised seals in such a way that a sound power level reduction of 5 dB(A) is achieved in total.
- At the folding box ejector, an effective sound power level reduction of 3 dB(A) is to be achieved by shielding.
- Reduction of the structure-borne sound power level of the cladding surfaces by approx. 7 dB(A).

Figure 6.18 shows the above-mentioned noise reduction plan and the planned noise reduction measures [4].

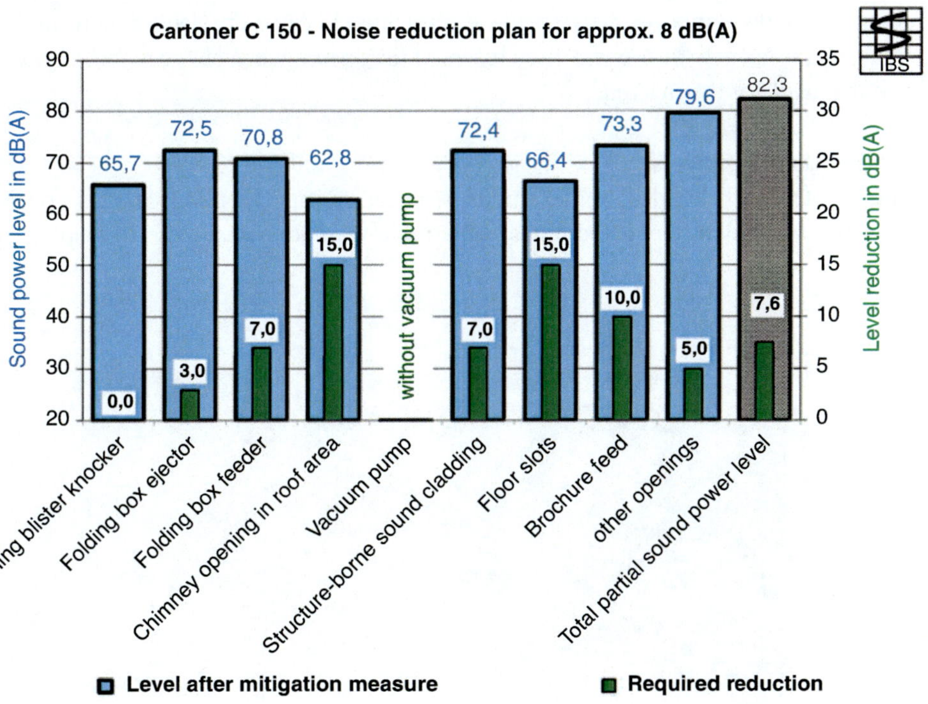

Fig. 6.18 Noise reduction plan for noise reduction of the cartoner type C 150 by approx. 8 dB(A)

6.2.3 Noise Reduction Measures in Principle

Some of the principal noise reduction measures are described below:

Noise Optimisation of the Brochure Table

The elongation of the force-time curve of the noise-intensive blowing process, which is intended to loosen the stack of leaflets, can be achieved, for example, by rounding off the edge of the leaflet table on the side of the transport roller, see Fig. 6.19. In addition to the rounding off, a further level reduction can be achieved by fitting grooves in the area of the leaflet table edge.

The opening area below the brochure table should be closed, for example, by a hinged door fitted with seals.

In addition to the measures mentioned, the opening area between the stack of instruction leaflets and the inside of the machine should be sealed. A corresponding sealing plate, which should be tightly connected to the surrounding housing if possible, is outlined in hatched lines in Fig. 6.19.

The standard configuration of the folder provided by the manufacturer is designed to be acoustically favorable in this respect because the entire insert stack is surrounded by a frame that is open at the top and also closes off the opening area between the brochure stack and the inside of the machine.

Sealing the Floor Slots

Floor slots can be sealed, for example, with an apron. Common designs consist of a mounting plate with elongated holes, which allow an adjustment e.g. to sloping ground,

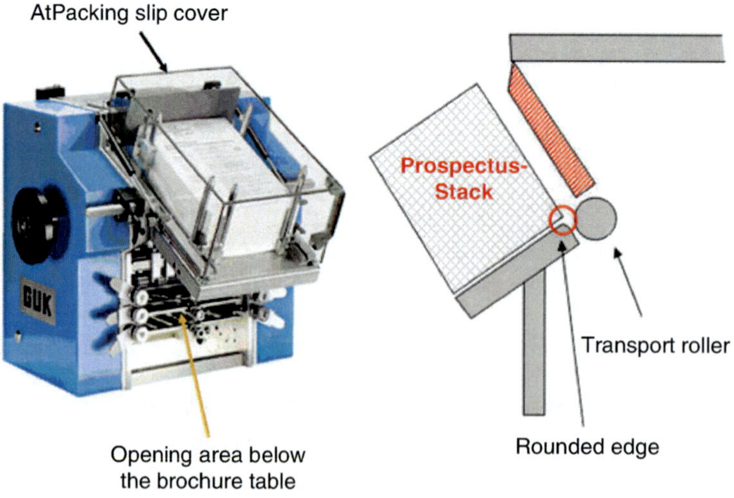

Fig. 6.19 Principle sketch of reduction measures brochure supply

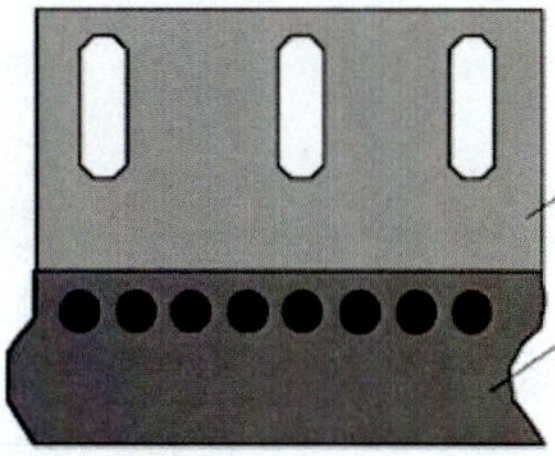

Fig. 6.20 Principle sketch sealing bottom slot

and a silicone lip mounted to it (see Fig. 6.20). Alternatively or additionally, the cartoner should be designed with a closed base plate.

Structure-Borne Sound Radiation of the Machine Protective Cladding

In order to reduce the structure-borne sound power level of the cladding surfaces accordingly, it is recommended to provide a load-bearing frame for the machine protective cladding which is consistently elastically decoupled from the base frame of the machine and thus has no bridges for structure-borne sound.

6.3 Straddle Carrier (Mobile Crane)

In order to meet future requirements, according to customer specifications the characteristic emission sound pressure level should be max. L_{pA20m} 65 dB(A) amount to. L_{pA20m} is measured at a distance of 20 m from the centre of the straddle carrier, type 54 DE (Fig. 6.21), according to the specifications of the in-house measuring regulations of EUROGATE Container Terminal Bremerhaven GmbH [6, 7].

6.3.1 AcousticWeak Point Analysis

For the determination of the acoustic weak point analysis, the A-partial sound power levels were determined for all acoustically relevant components of the straddle carrier by sound intensimetry (total and third-octave band levels). The partial sound power levels were determined for the following operating parameters [8, 9]:

- Lifting (16 m/min) and lowering (18 m/min) of the hoist unit
- Driving at maximum speed (approx. 24 km/h)

The following components of the Straddle Carrier, type 54 DE are acoustically relevant:

- Hydraulics
- Generator
- Engine
- Air inlet grille (inlet)

Fig. 6.21 Straddle carrier type 54 EN

- Air outlet (outlet)
- Exhaust
- Winch and winch drive
- Floor pan
- Landing gear complete (left and right landing gear carrier, travel drives, wheels 1 to 8)

Figure 6.22 shows the measured A-sound power levels of the components in the operating condition "Lifting & lowering the hoist".

As can be seen from the diagram, the noise development in this operating condition is primarily determined by the sound radiation of the air inlet grille and from the exhaust. Figure 6.23 shows the third-octave spectra of the A-sound power levels of the components in the operating state "Lifting & lowering the hoist".

It can be seen that the frequency spectrum of the air inlet grille is very broadband and its sound radiation is responsible for the noise generation of the straddle carrier at higher frequencies ($f > 160$ Hz). The noise development in the one-third octave band at 100 is almost exclusively determined by the tonal sound radiation of the exhaust.

Figure 6.24 shows the A-sound power levels of the relevant components in the operating mode "**driving with $\mathbf{v_{max} = 24}$** km/h".

As can be seen from the diagram, the noise development in the operating mode "driving with $v_{max} = 24$ km/h" is primarily determined by the sound radiation from the air intake grille and from the chassis.

Figure 6.25 shows the third-octave spectra of the A-part sound power levels of the components in the operating condition "driving with $v_{max} = 24$ km/h".

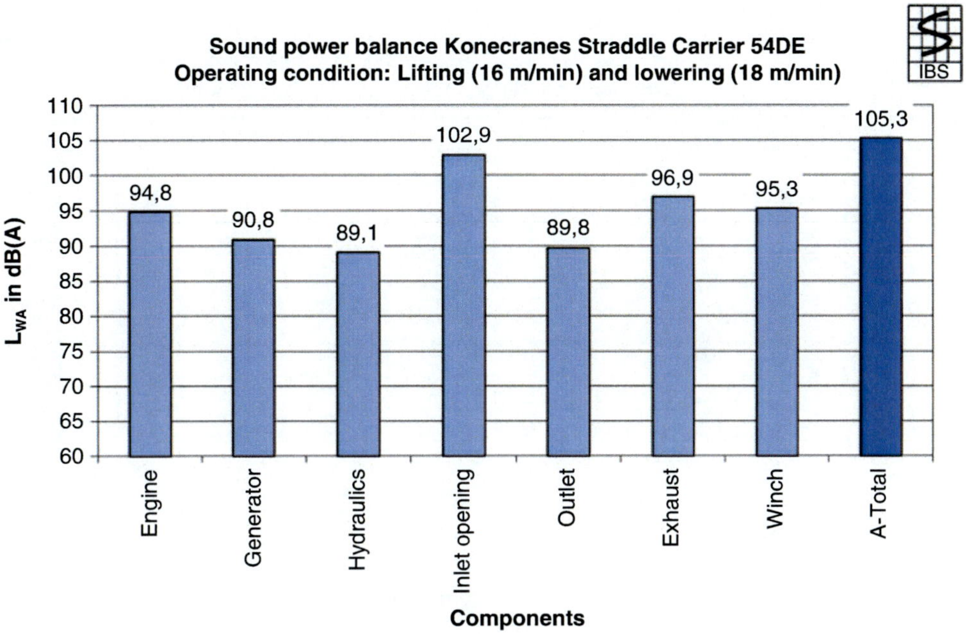

Fig. 6.22 A-sound power level for the operating state "Lifting and lowering of the hoist"

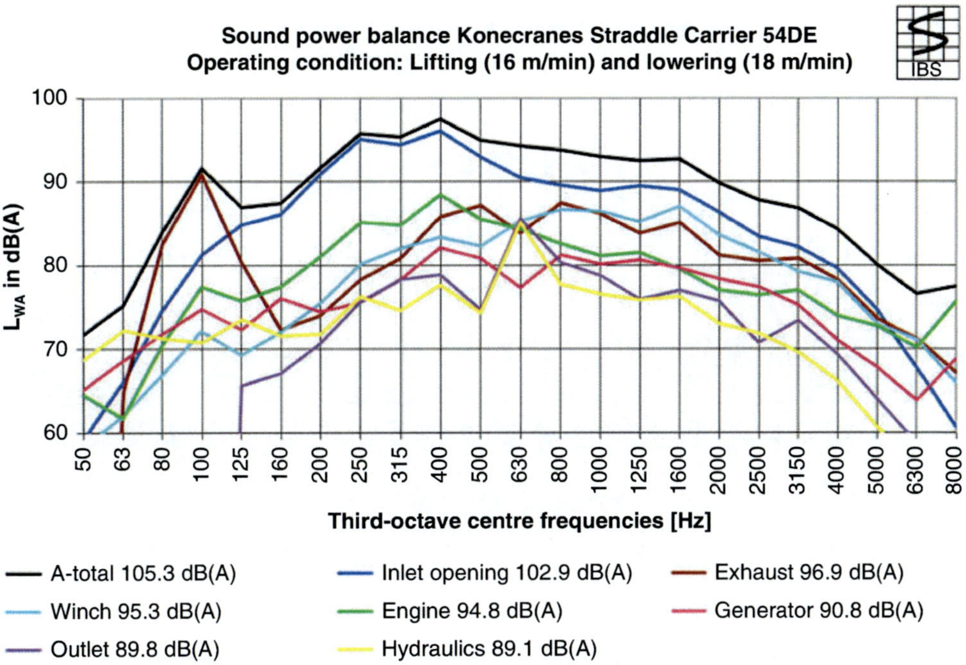

Fig. 6.23 Third-octave spectrum of the A-sound power levels of the components of the straddle carrier in the operating condition "Lifting and lowering of the hoist"

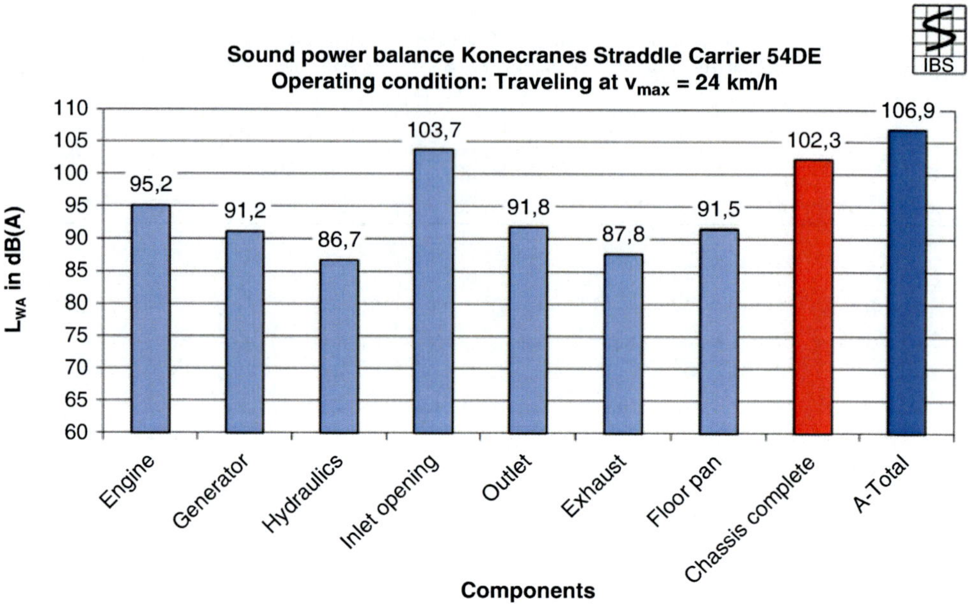

Fig. 6.24 A-sound power level for the operating condition "Driving with $v_{\text{max}} = 24$ km/h"

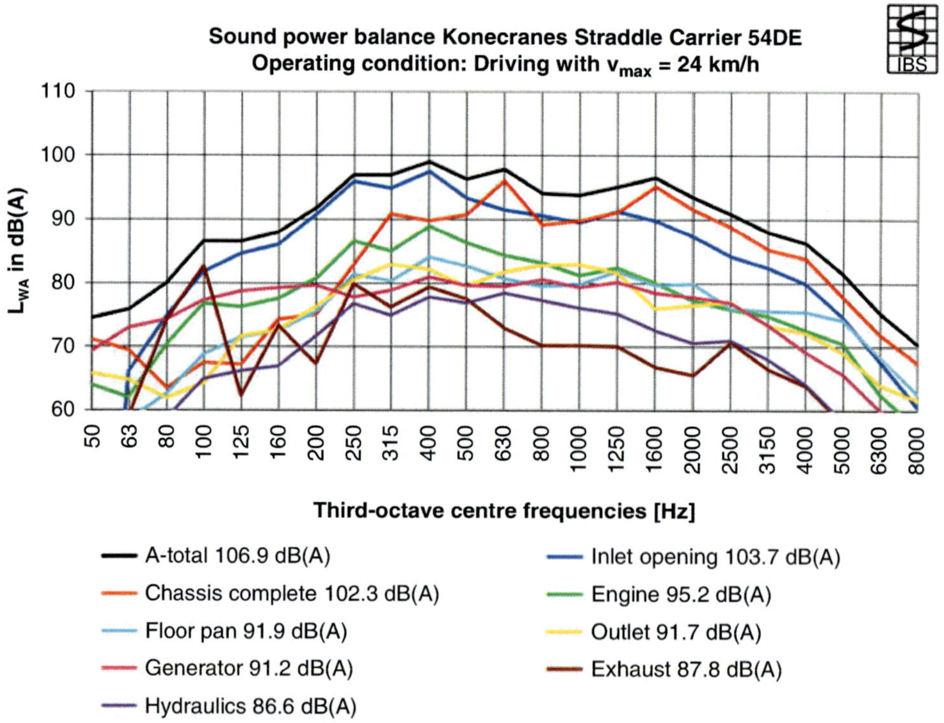

Fig. 6.25 Third-octave spectrum of the A-sound power levels of the components of the straddle carrier in the operating condition "Driving with $v_{\text{max}} = 24$ km/h"

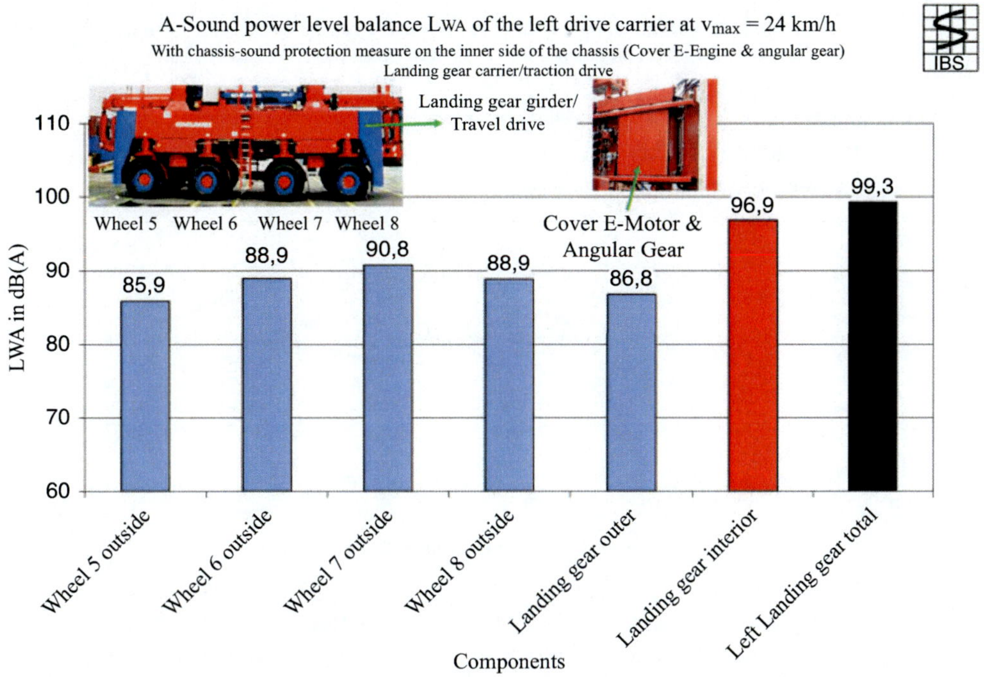

Fig. 6.26 Partial sound power level of the landing gear, "Driving with $v_{max} = 24$ km/h"

It can be seen that the frequency spectrum of the air inlet grille is again very broadband and its sound radiation is responsible for the noise generation of the straddle carrier in the 125 Hz $< f <$ 500 Hz frequency range. The noise emission of the exhaust is again responsible for the tonal components in the one-third octave band 100 Hz. In the middle and upper frequency range ($f >$ 500 Hz), the sound radiation of the straddle carrier in the operating state "driving with $v_{max} = 24$ km/h" is determined by the chassis, but also proportionally by the air intake grille.

Since the sound radiation of the undercarriage has a decisive influence on the overall noise development, the partial and total sound power levels of the undercarriage and the causes of the sound generation were determined by the acoustic weak point analysis.

Figure 6.26 shows the measured A-part sound power levels of the chassis components of the left travel support in the operating condition "**travelling with v_{max}**".

It can be seen that the noise emission of the landing gear is largely determined by the sound radiation from the inside of the landing gear.

Since the left and right travel units are of the same design, it can be assumed that the sound power balance determined for the left travel unit can be transferred to the right travel unit. The left and right girders together emit an A-part sound power level of $L_{WA,\ chassis\ complete} = 99.3 + 3 = 102.3$ dB(A) in the operating mode "driving with $v_{max} = 24$ km/h", see Figs. 6.24 and 6.25.

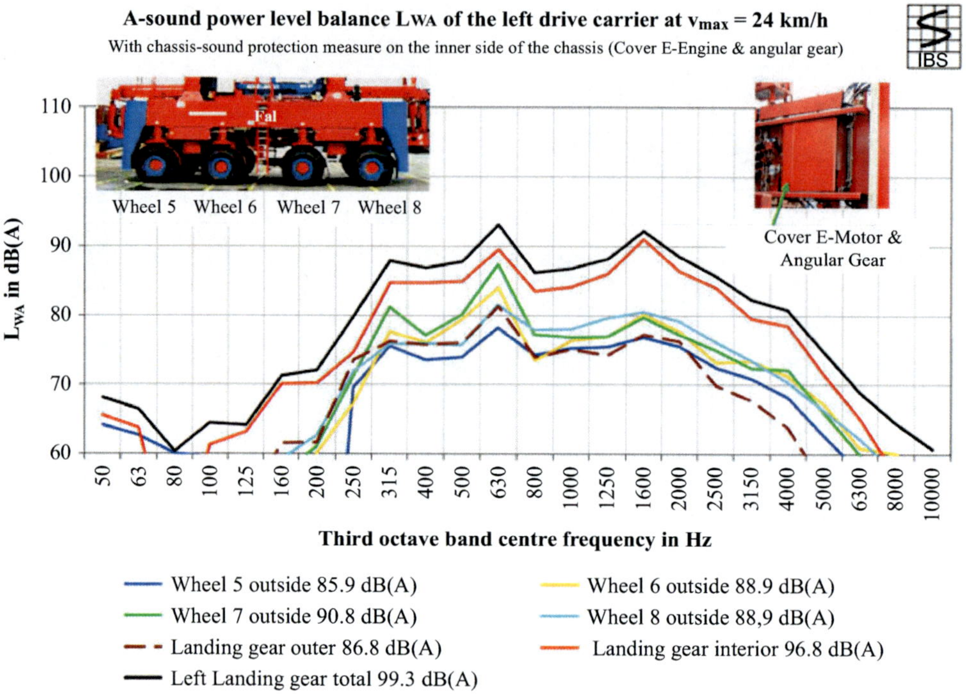

Fig. 6.27 Third-octave spectra of the A-sound power levels of the driving beam on the left in the operating condition "**Driving with** $v_{\text{max}} = 24\,\text{km/h}$"

Figure 6.27 shows the third-octave spectra of the A-part sound power levels of the carrier in the operating mode "driving with $v_{\text{max}} = 24\,\text{km/h}$".

Figure 6.28 shows the narrowband spectrum of the A sound pressure level at a distance of 20 m, maximum level L_{AFmax} [8].

Level increases in the narrowband spectrum can be assigned to the ignition frequency of the engine (f_{E}), the meshing frequency of the bevel gear (f_{K}) and the meshing frequency of the angular gear (f_{W}) and their respective harmonics. The level-determining component during the pass-by is the gear meshing frequency of the bevel gear f_{K}, with which each wheel drive is equipped (2 wheel drives on each inner side of the chassis). Figure 6.29 shows the wheel drive including the respective shaft speeds for the operating state "travel with $v_{\text{max}} = 24\,\text{km/h}$".

From the measured partial sound power levels, the average partial sound immission of the respective partial sound source at a distance of 20 m ($L_{p\text{A}20\,\text{m},\,j}$) can be determined by calculation $\overline{L_{p\text{A},j}}$ as the measurement area sound pressure level [8], see Fig. 6.30.

$$\overline{L_{p\text{A},j}} = L_{W\text{A},j} - L_S \tag{6.1}$$

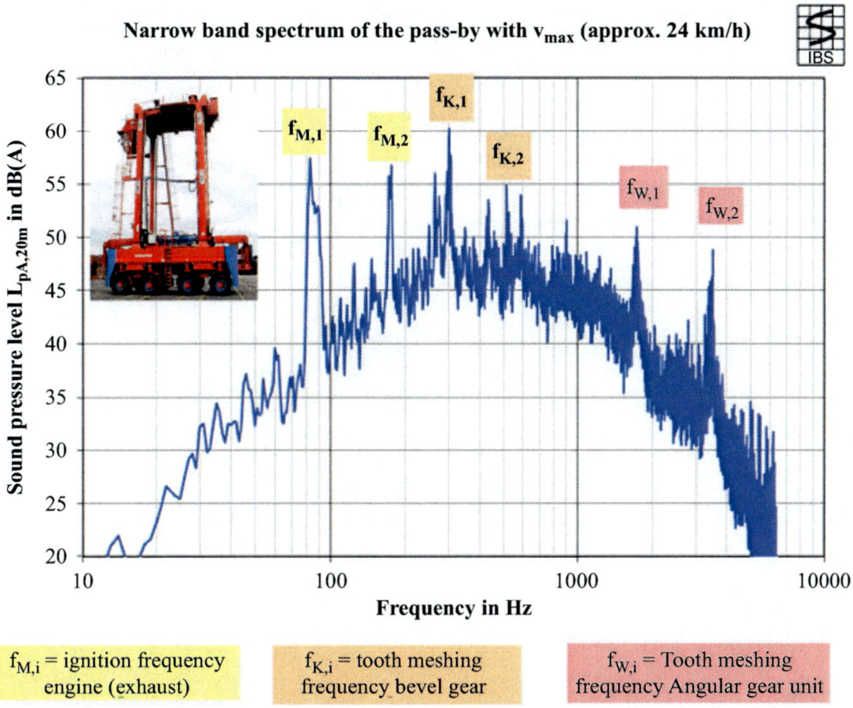

Fig. 6.28 Narrowband spectrum of the A-sound pressure level at 20 m from the centre of the straddle carrier as maximum level (L_{AFmax})

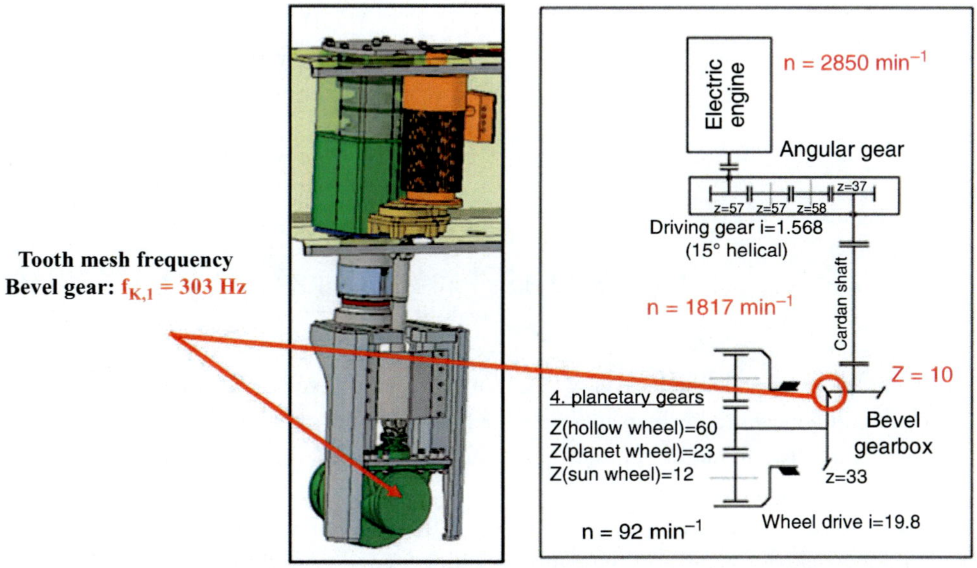

Fig. 6.29 Schematic diagram of the wheel drive and shaft speeds for $v_{max} = 24$ km/h

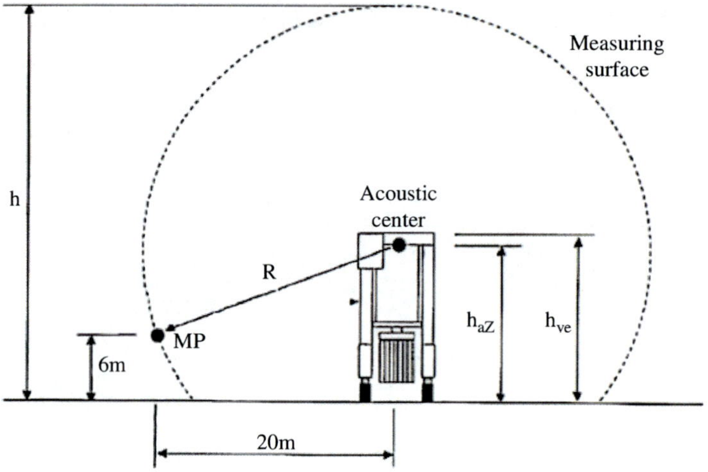

Fig. 6.30 Measurement area for determining the effective system sound power level $L_{WA,\ eff}$, [6, 7]

$$L_S = 10 \cdot \log\left(\frac{S}{S_0}\right) \tag{6.2}$$

Thereby is:

L_S	Measurement area in dB
$S = 2 \cdot \pi \cdot R \cdot h$	Measuring area in m^2
R	Radius of the sphere section in m
h	Height of the sphere section in m
$S_0 = 1\ \text{m}^2$	Reference area

The total sound pressure level can be determined from the partial sound pressure levels at 20 m distance:

$$\overline{L_{pA,20\,\text{m}}} = 10 \cdot \lg \sum_{j=1}^{n} 10^{L_{pA,20\ \text{m},j}} \quad \text{dB(A)} \tag{6.3}$$

By comparing the calculated average A-sound pressure levels at 20 m, $\overline{L_{pA,20\,\text{m}}}$ with the corresponding measured values, it is possible to localise the sound emitters that are decisive for the guarantee or target. Furthermore, the accuracy of the weak point analysis, see Chap. 5, can be checked.

The A-total sound power level of 106.9 dB(A) given in Figs. 6.24 and 6.25 was determined mathematically by energetic addition of the partial sound power levels and also agrees well here with the A-sound power level L_{WA} = 107.1 dB(A), which was determined according to (6.1) by measuring the sound pressure level for the operating condition "driving with $v_{max} = 24$ km/h" [8].

6.3.2 Noise Reduction Plans

In order to reduce the average total sound pressure level at a distance of 20 m from the centre of the device to a value of $L_{pA20\,m} \leq 65$ dB(A), the following noise reductions are necessary, as shown in investigations [8, 9]:

- The reduction of the average sound power level for L_{WA} the operating condition "**driving at maximum speed**" by 3.4 dB(A). This can be achieved, for example, by reducing the partial sound power levels of the air intake grille (inlet) by 4.7 dB(A) and the partial sound power levels of the landing gear by 4.6 dB(A). Figure 6.31 shows the corresponding noise reduction plan.
- The reduction of the average sound power level for L_{WA} the operating condition "**lifting and lowering of the hoist**" by 2.1 dB(A). This can be achieved, for example, by reducing the partial sound power levels of the air inlet grille (inlet) by 4.7 dB(A). Figure 6.32 shows the corresponding noise reduction plan.

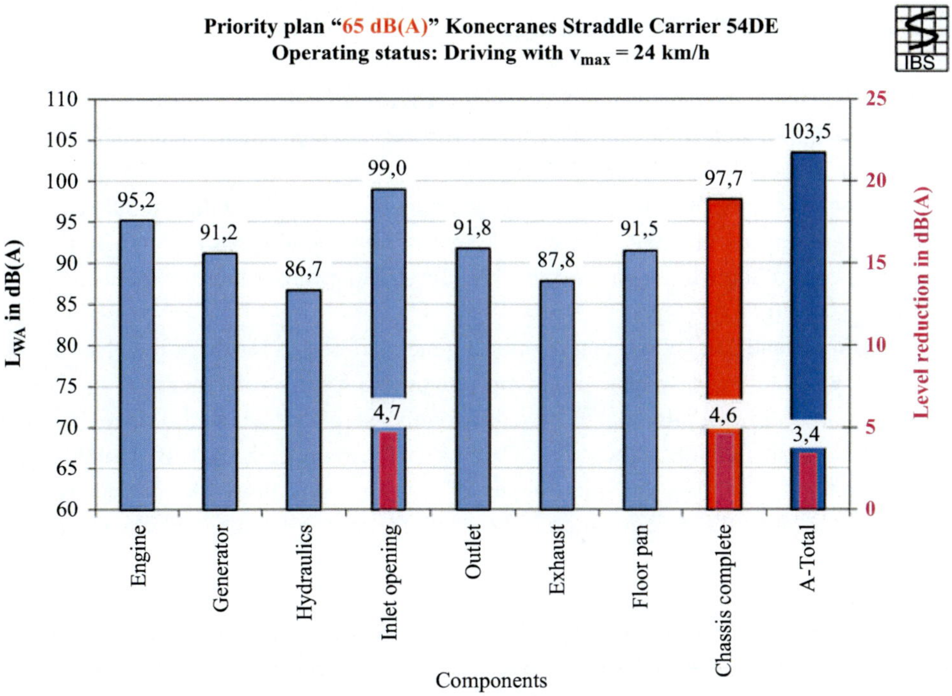

Fig. 6.31 Noise reduction or priority plan for the operating state "Driving with $v_{max} = 24$ km/h"

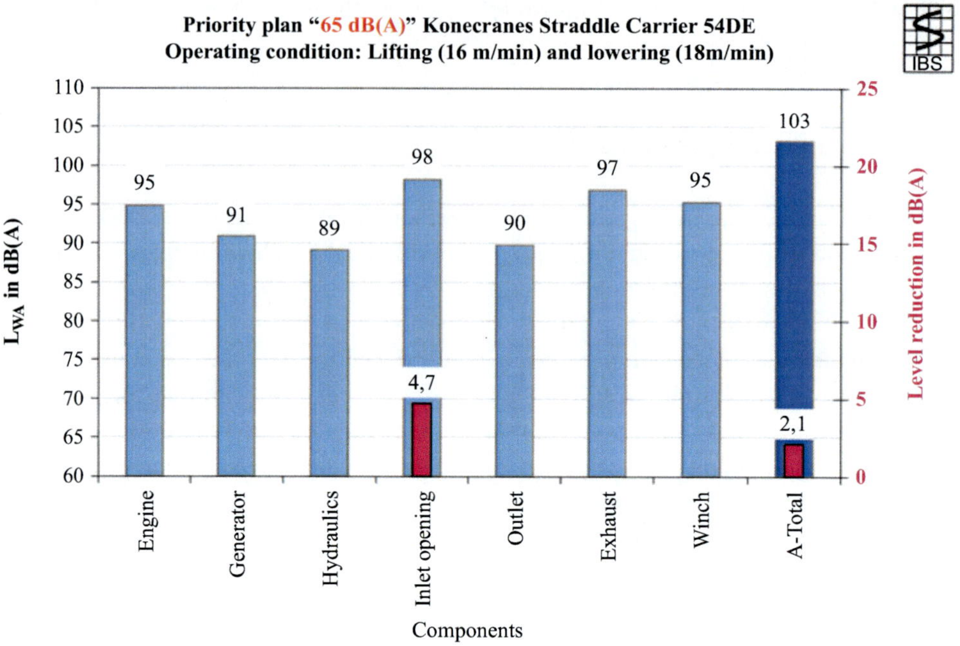

Fig. 6.32 Noise reduction or priority plan for the operating state "Lifting and lowering of the hoist"

6.3.3 Noise Reduction Measures

In order to achieve the necessary level reduction at the air inlet, IBS GmbH, in cooperation with the manufacturer of the straddle carrier, Konecranes Heavy Lifting GmbH, developed a compact intake silencer of only 120 mm.

In order to achieve the necessary level reduction of 4.6 dB(A) at the straddle carrier, the noise emission of the inside of the straddle carrier must be reduced by 3.6 dB(A) and the noise emission of the wheels on the outside of the straddle carrier must be reduced by 10 dB (A) each (see Fig. 6.33).

Measurements have shown that the noise generated on the inside of the chassis can be reduced by 3.6 dB(A) by means of appropriate covers for the bevel gearbox.

Figure 6.34 shows a principal noise reduction measure with which the noise emission of the wheels on the outside of the driving beams can be reduced by 10 dB(A) in each case.

In summary, it follows that with the aid of a design acoustic weak point analysis, it is possible to localize the decisive sources that are responsible for the overall noise development. Noise reduction plans can then be drawn up based on the target and suitable noise reduction measures can be developed.

In this case, in cooperation with Konecranes, proposed measures were developed with the aim of reducing the emission characteristic total sound pressure level to a value of $L_{pA20\,\mathrm{m}} \leq 65$ dB(A) at a distance of 20 m from the centre of the unit.

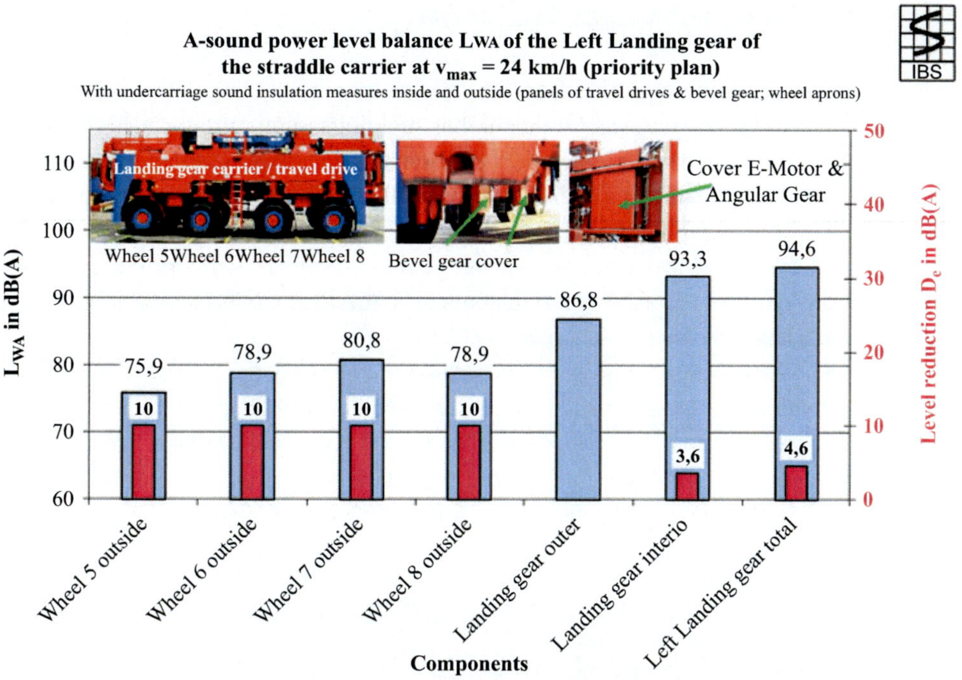

Fig. 6.33 Noise reduction or priority plan of the landing gear

Fig. 6.34 Protective plate with integrated wheel apron

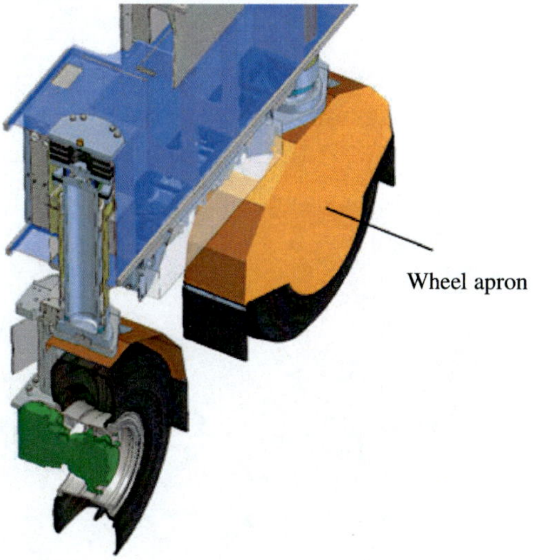

6.4 Shell and Tube Heat Exchanger

After the commissioning of a **F**lue **G**as **D**esulphurisation plant (FGD plant) of the Rauxel power plant, see Fig. 6.35, it was found out that at various immission locations outside the power plant site the permissible noise immission guide values were exceeded by 10 to 17 dB(A), depending on the immission location, both at full load and in several partial load ranges.

The FGD system is designed as a single-circuit scrubber with trickle installations without a pre-scrubber. Before entering the stack, the flue gas is heated to 78 °C in a steam-heated shell-and-tube heat exchanger.

With the help of a sound and vibration weak point analysis for the entire plant and the relevant individual components such as pumps, blowers, scrubbers and heat exchangers, the noise generation is to be described and the relevant noise sources localised. The aim of the investigation was to find possible process engineering and/or design solutions to ensure that the permissible immission guide values are complied with [10, 11].

Fig. 6.35 View of the Rauxel [11] power plant

6.4.1 Sound and Vibration Weak Point Analysis

Figure 6.36 shows the location of the FGD at the Rauxel power plant and some critical immission points.

Figure 6.37 shows the third-octave spectrum of the A-sound pressure level at the reference point $L_{pA,\ ref}$ for full-load operation. From this it can be seen that the immission level at full load operation of the unit is mainly determined by the tonal components at third octaves 250 and 315 Hz.

For the localisation of the excitation source, the main noise source of the plant was determined by extensive airborne sound, structure-borne sound and sound intensity measurements. Figure 6.38 shows the FGD and several acoustically relevant components of the plant including airborne and structure-borne sound measurement points.

The investigations have shown that the sound radiation of the GRP bend and the stack mouth are primarily responsible for the level increases at the immission locations. Figure 6.39 shows the A-part sound power level of the above-mentioned plant components.

Figure 6.40 shows the A-sound pressure level for the relevant frequency range up to 400 Hz at the place of immission "Ohmstraße" for different load conditions as a narrow-band spectrum.

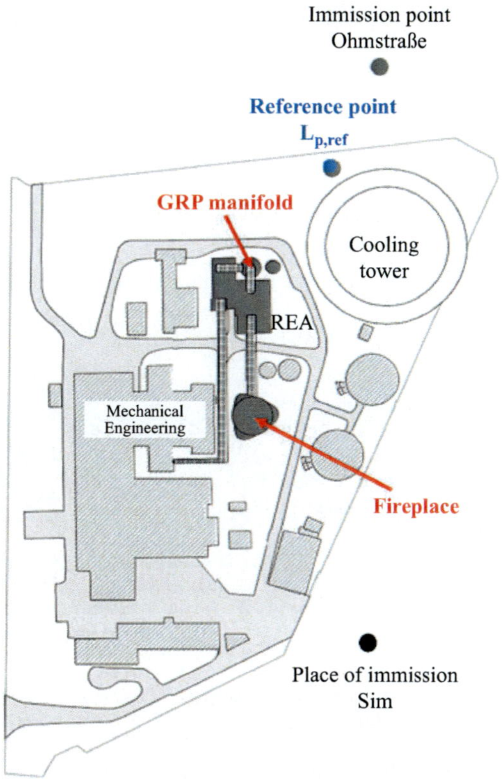

Fig. 6.36 Map of the FGD and the main immission points in and outside the Rauxel power plant

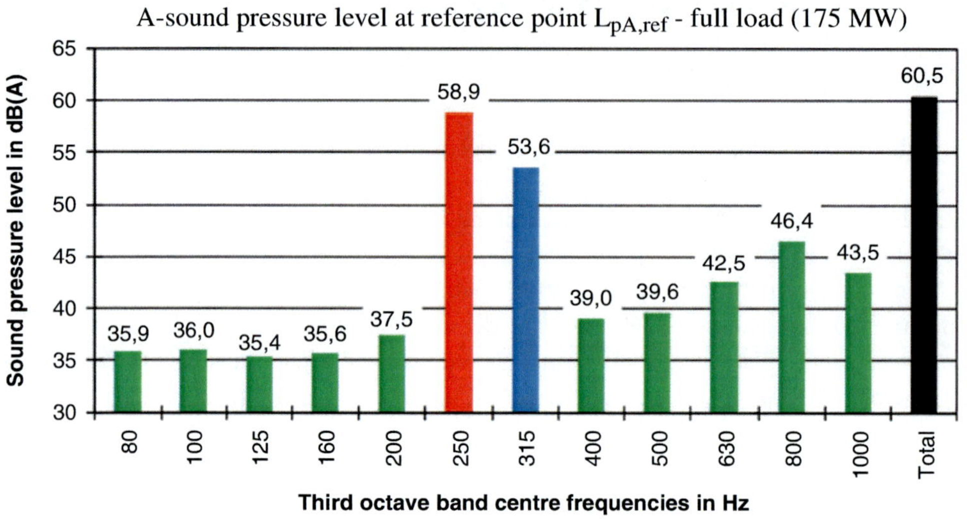

Fig. 6.37 Third-octave spectrum of the A-sound pressure level at the reference point $L_{pA,\,\mathrm{ref}}$ for full-load operation

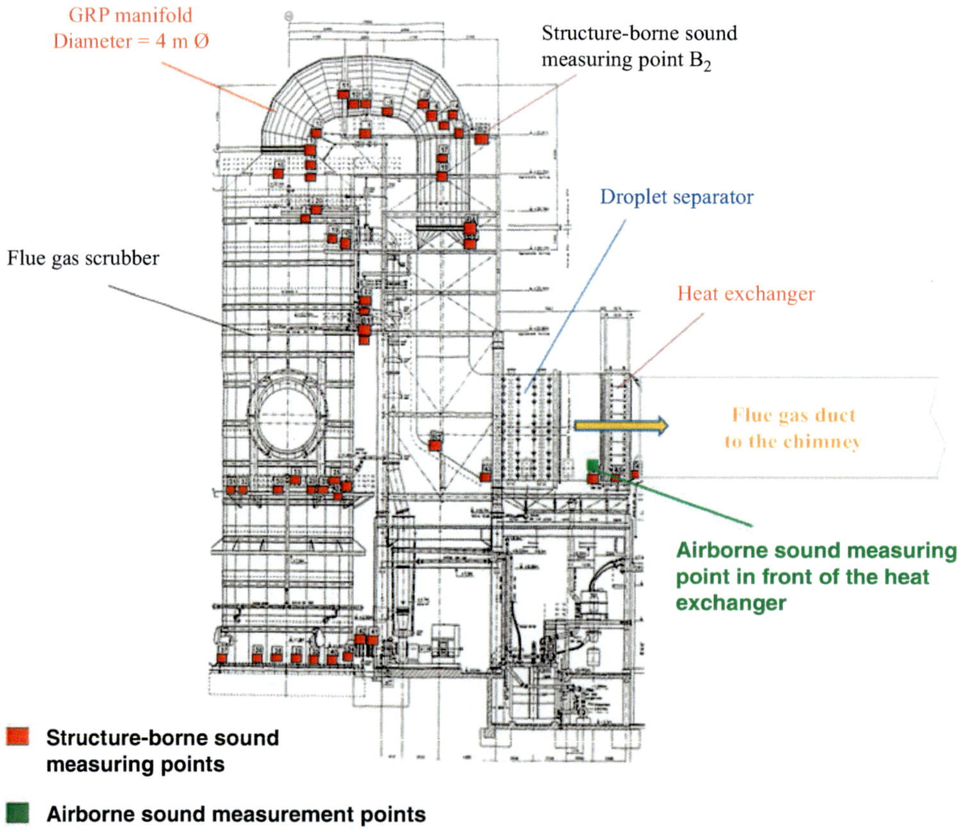

Fig. 6.38 FGD including some acoustically relevant plant components as well as the airborne and structure-borne sound measurement points

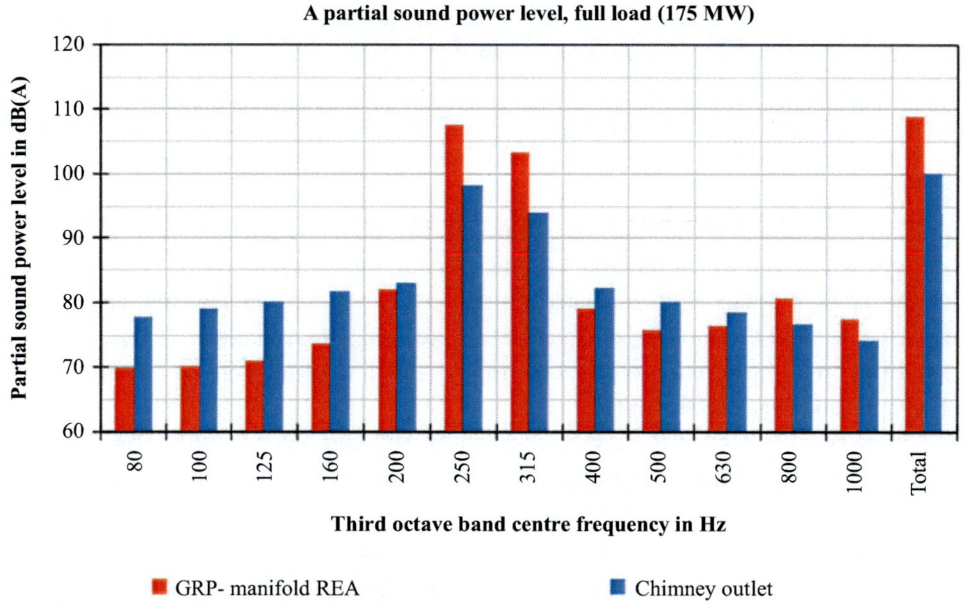

Fig. 6.39 A-partial sound power level of relevant plant components

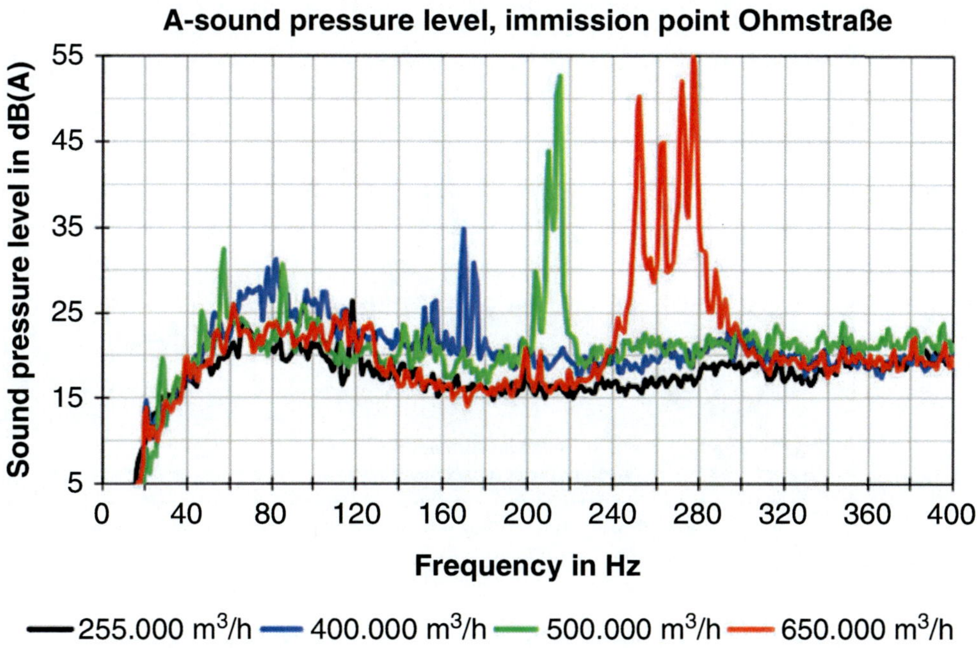

Fig. 6.40 Narrow-band spectrum of the A-sound pressure level at the "Ohmstraße" immission point at different throughputs

From this it can be clearly seen that the tonal frequency components first appear at a throughput of 400, 000 $\mathrm{N\,m^3/h}$. Further level increases were detected at 500, 000 $\mathrm{N\,m^3/h}$ and 650, 000 $\mathrm{N\,m^3/h}$ (full load operation) of the unit. It is pointed out that the tonal frequency components, which are also responsible for the level increases at the immission locations, only occur at certain throughputs, which indicates a resonance phenomenon.

The measured frequencies are not related to periodic excitation frequencies of drive units, such as induced draft fans or pumps, so that the predetermination of the excitation frequencies, similar to the rotational sound of a fan, was not possible.

The cross resonances in the duct system, which are known for shell-and-tube heat exchangers, were also not responsible for the excitation frequencies [12–15].

Further investigations proved that the tonal frequency components originate within the shell-and-tube heat exchanger [10, 11].

Figure 6.41 shows the narrowband spectra of the sound pressure level in the flue gas duct upstream of the heat exchanger and that of the structure-borne sound acceleration level at the GRP elbow.

It follows from this that the sound radiation of the GRP manifold, especially for the tonal frequency components generated in the heat exchanger, is determined by structure-borne sound radiation.

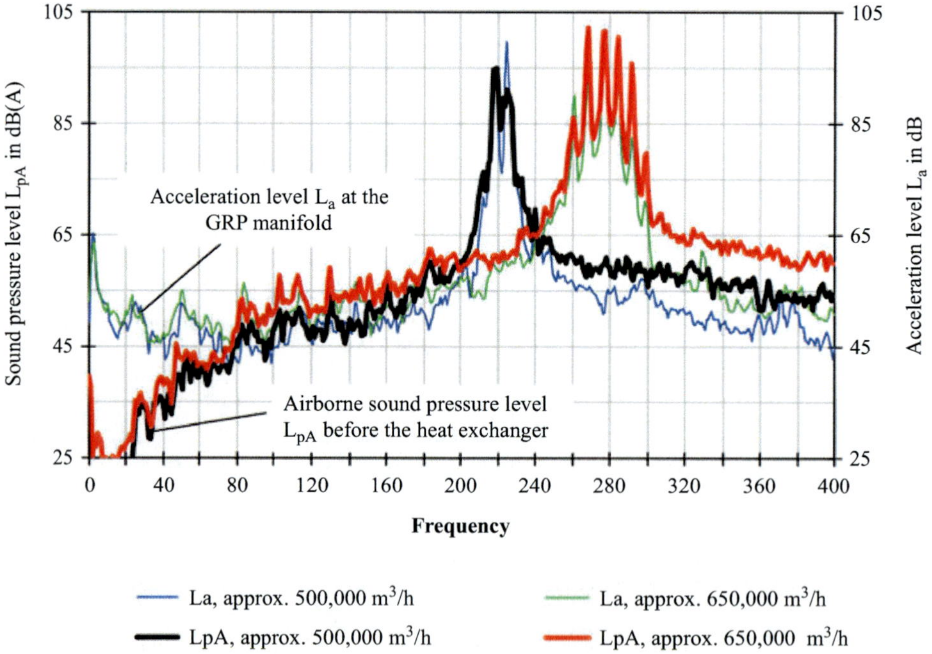

Fig. 6.41 Comparison of the sound pressure level in the flue gas duct with the structure-borne sound acceleration level at the GRP elbow under different load conditions

The tonal excitation frequencies generated in the heat exchanger are responsible for the excessive noise development of the FGD plant and the level increases at the immission locations. It is particularly critical if the excitation frequencies coincide with the natural frequencies of the plant components.

By means of natural frequency determinations, it was possible to prove, among other things, that the natural frequencies of the GRP bend lie in the range of the exciter frequencies, i.e. the level increases of the GRP bend are primarily amplified by resonance phenomena.

Since the plant was to be put into operation in the short term and at that time the mechanism of origin of the tonal components had not yet been sufficiently investigated to derive primary measures, the immission level at the critical immission point “Ohmstraße” was reduced by approx. 10 dB(A) in a first step with a secondary reduction measure. This was achieved by soundproofing the GRP bend by approx. 20 dB at 250 Hz. Since after the implementation of this measure no tonality of the noise could be detected at this immission point, the rating level was even reduced by approx. 16 dB(A) after omission of the tonality surcharge [16].

However, further immission measurements showed that even after insulation of the GRP bend the immission guide values were still exceeded, now at other immission locations that were less noticeable before, by approx. 1–6 dB(A). The immission levels were influenced by other components of the power plant, which were previously covered by the sound radiation of the GRP bend. After the sound insulation of the GRP bend, the sound radiation of the stack mouth was responsible for exceeding the immission guide values. A reduction of the sound radiation would also have been possible by installing a silencer after the heat exchanger, which, however, would have meant a further, considerable cost expenditure. In order to be able to avoid this further, costly secondary noise reduction measure, suitable primary constructional measures for subsequent installation at the existing plant for noise reduction at the chimney mouth were to be worked out by describing the sound generation and by model tests.

6.4.2 Model Tests

Since the observed phenomenon has not yet been described in this form in the literature, corresponding sound and vibration investigations were carried out on a specially built model test section at the FH Bingen, among other things as part of a diploma thesis [17]. The different throughputs or flow velocities were realised with the aid of a continuously variable blower. Similar to the FGD (duct cross-section in the area of the shell-and-tube heat exchanger: 5.87 m × 7.0 m), two tube bundles were arranged one behind the other in the duct, the dimensions of which were reduced to a scale of approx. 1: 15 (0.405 m × 0.465 m). Pipe diameter and spacing in the bundles were realized on a scale of 1: 2 compared to the original. The division ratios, the distances between the

Fig. 6.42 Test section with model shell-and-tube heat exchanger

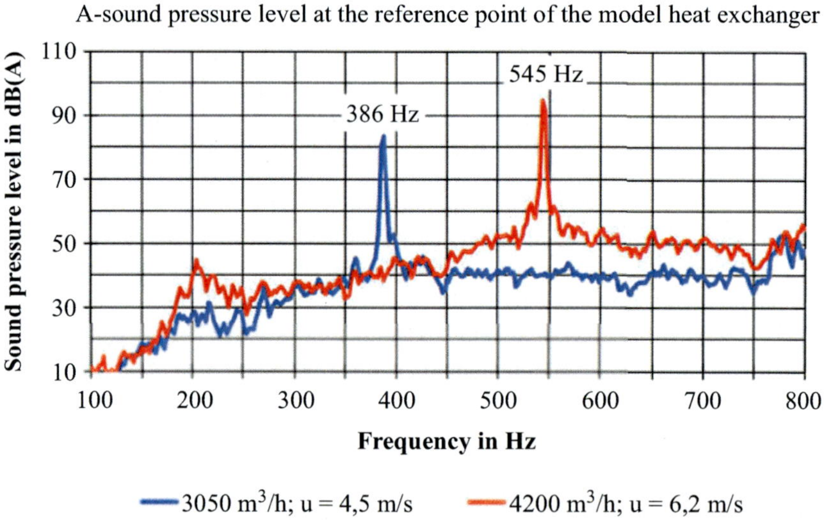

Fig. 6.43 Narrow-band spectrum of the A-sound pressure level for the model heat exchanger at a reference point, approx. 1 m outside the model duct, at various flow rates

individual pipes, were not changed compared to the original. Figure 6.42 shows the model test section.

During the model tests, it was found that, similar to the FGD system, tonal components arise at certain velocities (throughputs), which also determine the overall level. Figure 6.43

shows the narrowband spectrum of the A-sound pressure level in the model heat exchanger at a reference point, approx. 1 m outside the model duct, at various flow rates:

approx.	3050 m^3/h ($u \approx 4.5$ m/s): 386 Hz
approx.	4200 m^3/h ($u \approx 6.2$ m/s): 545 Hz

Although the frequencies do not match the frequencies observed in the FGD, the model experiment nevertheless demonstrated qualitatively the same behavior.

Through extensive literature research, the effect of various influencing parameters on the sound generation was investigated and compared with known works [12]. The investigations have shown that the sound generation mechanisms, which obviously come into play in the FGD and the model test section, cannot be explained with known works [13–15], which mainly refer to transverse resonances in the duct system.

By further model experiments in a shallow water tank at the University of Kaiserslautern (see Fig. 6.44a) it could be qualitatively proven that within a certain velocity range (Reynolds number range) stable vortex zones develop between the individual tubes of a tube bundle, the vortex frequencies of which are very probably responsible for the excitation frequencies (see Fig. 6.44b, c). The periodic vortex shedding behind the last row of tubes of a

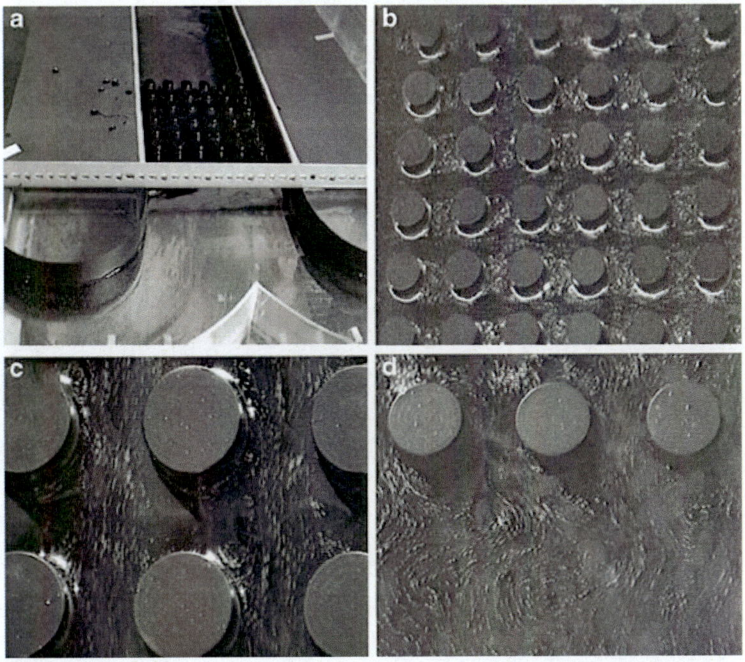

Fig. 6.44 Shallow water tank model experiment (University of Kaiserslautern); (**a**) **shallow** water basin, (**b**), (**c**) stable vortex zone between individual tubes, (**d**) Kármáin vortex street

tube bundle, Kármán's vortex street (see Fig. 6.44d), can of course also be considered as the location of sound generation, but plays only a minor role in the present case.

In summary, it can be stated that tonal excitation frequencies are generated by the flow around the pipes within a certain velocity or Reynolds number range. The coincidence of the excitation frequencies with the natural frequencies of the duct system, as was the case here both in the FGD plant and in the model test, leads to considerable airborne and structure-borne noise radiation due to resonance oscillations.

After it had been established that the stable vortex zones within the tube bundle are responsible for the sound generation, it was obvious to avoid the generation of these vortex zones by suitable design measures.

By installing adjustable baffles and deflectors (hereinafter referred to as sound or turbulence grids) in front of the respective model tube bundle (see Fig. 6.45), the flow was to be superimposed with additional turbulence and the formation of the stable vortex zones was to be avoided or at least reduced. By adjusting the angle of inclination of the turbulence grid or its distance to the tube bundle, the optimum setting with the lowest pressure loss was determined on the model [12, 16].

Figure 6.46 shows the modified model test section.

Figure 6.47 shows the results for the different angle settings. Here it can be seen that with only the installation of a turbulence grille in front of the 1st tube bundle in the flow direction the tonal component is reduced by approx. 40 dB. An increase in the angle of inclination brings only a slight improvement.

Figure 6.48 shows that the sound generation is not influenced by an arrangement of the turbulence grid behind the tube bundle. This confirms the previously made claim that the stable vortex zones within the tube bundle and not the vortex shedding behind the last tube row are responsible for the sound generation in the present case.

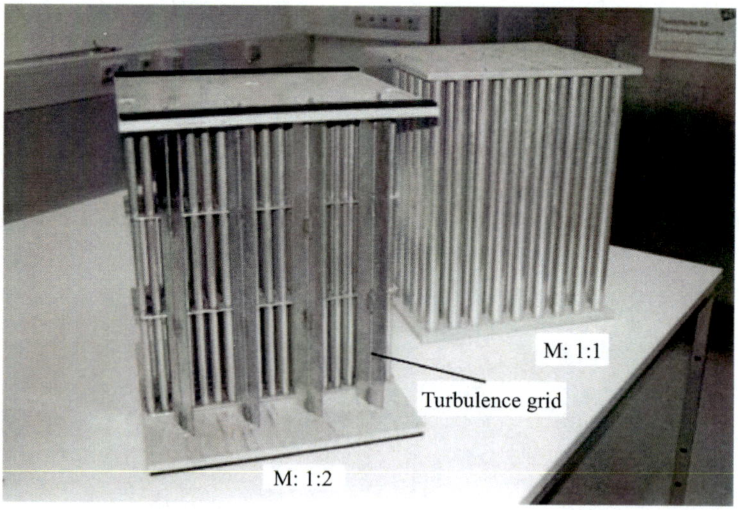

Fig. 6.45 Tube bundle model with adjustable turbulence grid

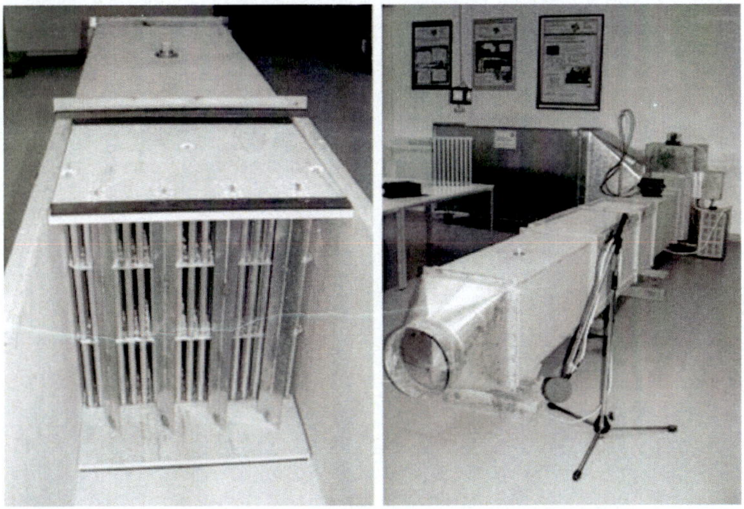

Fig. 6.46 Modified model test section with turbulence grid in front of the heat exchangers (FH Bingen)

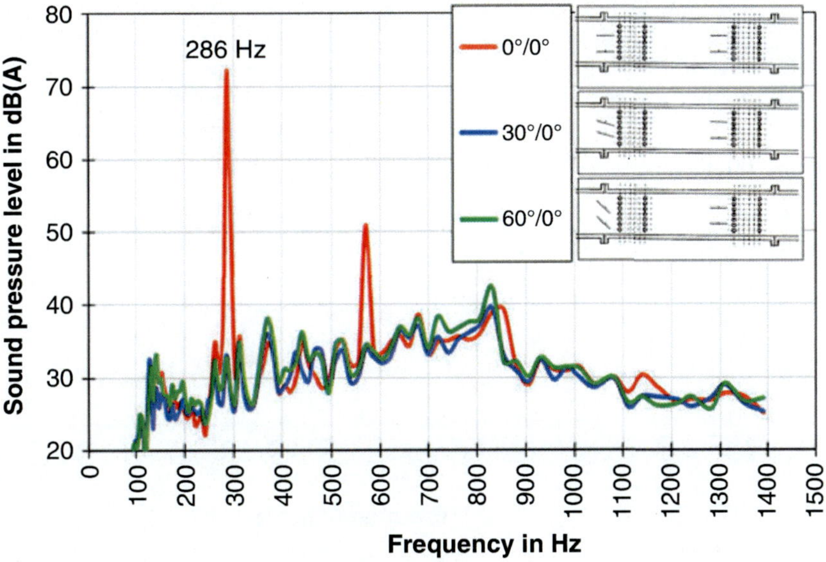

Fig. 6.47 Model shell and tube heat exchanger with turbulence grid at different baffle positions ($u = 3.7$ m/s)

In Fig. 6.49 it is shown that with increasing distance of the turbulence grid from the 1st row of pipes the effect decreases, i.e. the additional turbulence generated in this way is not sufficient to completely disturb the stable vortex zones.

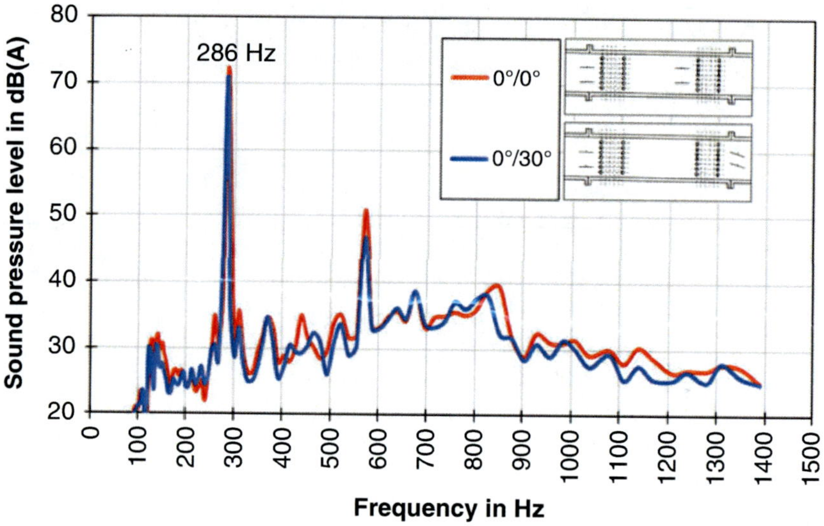

Fig. 6.48 Model shell-and-tube heat exchanger with turbulence grid at different baffle positions, effect of the turbulence grid behind the 2nd tube bundle ($u = 3.7$ m/s)

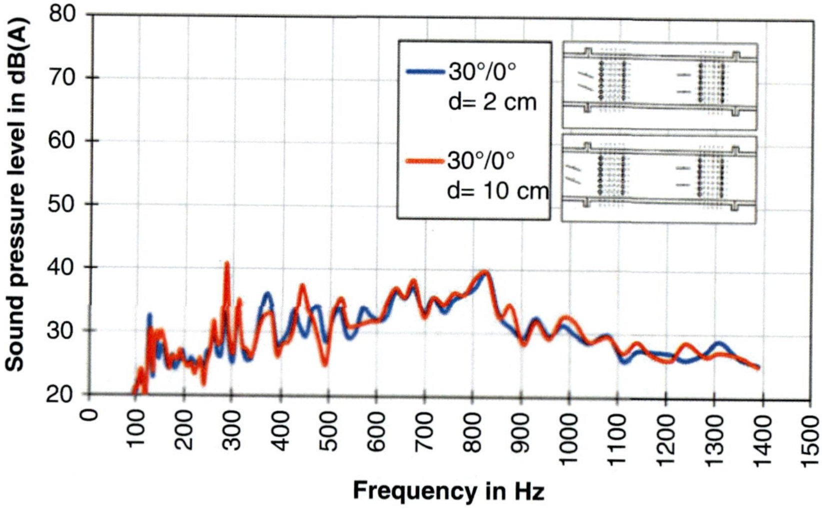

Fig. 6.49 Model shell-and-tube heat exchanger with turbulence grid in front of the 1st tube bundle at different distances of the grid to the 1st tube row ($u = 3.7$ m/s)

6.4.3 Constructional Noise Reduction Measures

Based on the results of the model tests, a sound or turbulence grille was designed and manufactured for the Rauxel power plant for installation in front of the 1st tube bundle of the heat exchanger. The angle of inclination of the individual baffles can be adjusted from

Fig. 6.50 Arrangement of the turbulence grid in front of the 1st shell-and-tube heat exchanger in the Rauxel power plant

the outside. Figure 6.50 shows a picture of the turbulence grille in front of the 1st heat exchanger tube bundle. For cost reasons, such a turbulence grid should initially only be installed and tested in front of the 1st tube bundle [11, 12, 20].

The effectiveness of the sound and turbulence grille was tested under real operating conditions by means of sound and vibration tests. The tests showed that by installing only one turbulence grille in front of the first pipe bundle, the immission guide values are complied with at all critical immission points in the entire load range. The tonal frequency components no longer occur, except for a non-critical increase at low load. The installation of a turbulence grille is obviously sufficient to prevent the build-up of the stable vortex zones.

Figure 6.51 shows the one-third octave spectra of the A-sound power levels at the stack outlet with and without turbulence grille determined by sound pressure measurements in the flue gas duct at the stack base after the heat exchanger.

As can be clearly seen from Fig. 6.51, it was possible to completely eliminate the tonal component at the time of generation for the decisive load range. The level increases in the lower frequency range can be explained primarily by increased flow noise caused by the turbulence grid. However, these have no influence on the overall noise development of the FGD system [11, 12, 20].

In summary, it can be stated that the noise generation within the FGD plant could be reduced by primary constructional measures to such an extent that the permissible immission guide values were complied with. However, this was only possible after the sound generation mechanisms had been determined or described, among other things, with the help of a weak point analysis and model investigations.

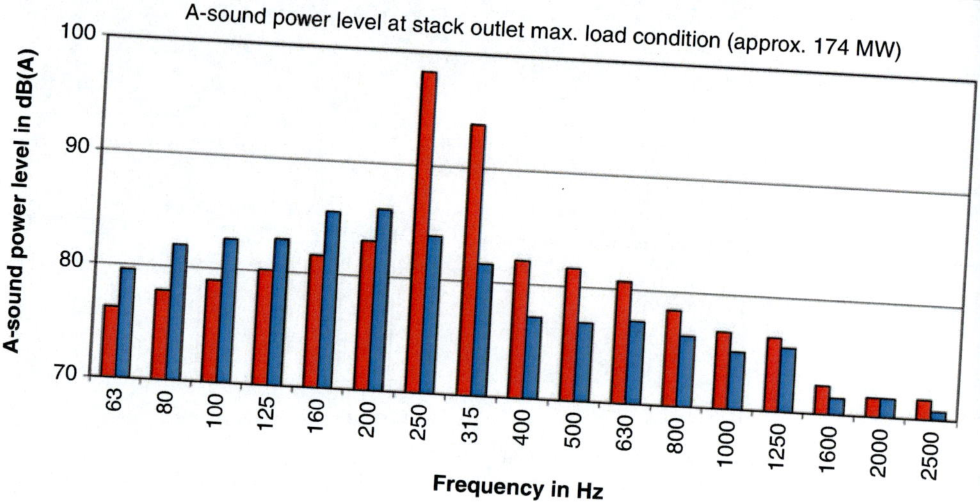

Fig. 6.51 A-sound power level at the stack outlet, with and without turbulence grid, at maximum load condition: 174 MW, 650, 000 m^3/h

In the present case, it can be assumed that it would very likely have been possible to dispense with the secondary noise protection measure on the GRP bend as a sound-emitting plant component if the noise causes had first been avoided or reduced by primary measures. However, it was necessary to reduce the noise immission in the short term in order to obtain an operating license for the power plant.

It should also be noted that, as a positive side effect, the heat output of the FGD plant was also increased. This is most likely achieved by better heat exchange between flue gas and superheated tube bundles, since the build-up of the stable vortex zones is prevented by the superposition of the turbulence caused by the turbulence grid.

6.5 Acoustic Optimisation of a Vane Gearbox

In an open-cast mine in the Central German lignite mining area, a bucket wheel excavator SRs2000 from MAN TAKRAF Fördertechnik GmbH is to be equipped with a new bucket wheel drive as part of the modernisation of an overburden conveyor line. The bucket wheel drive is designed for the following operating data:

- Drive power of 800 kW
- Translation from $n_1/n_2 = 990/3.3\ \text{min}^{-1}$

In addition to the fundamental objective of significantly increasing the reliability and availability of the equipment technology, one focus of the entire modernization program was to significantly reduce the noise immission on the local communities in the immediate vicinity of the open pit mine. Since the bucket wheel drive in the original design was one of the main sources of noise, an extremely low noise emission value was specified for the new drive to be developed.

To meet this requirement, the entire development and design process was geared to optimizing the acoustics of the transmission right from the start. The objective was to comply with the specified emission values through primary noise reduction measures alone. Secondary measures, such as the complete enclosure of the drive, are not only far more expensive to implement, but also cause considerable problems, such as poor heat dissipation, difficult accessibility and high assembly and disassembly costs for maintenance work. In addition, the increase in mass at the bucket wheel head of the excavator associated with the enclosure requires a careful check of the statics and stability of the equipment.

Primary measures are intended to minimize both the noise generation by appropriate modification of the gear teeth and the noise radiation of the gearbox by appropriate design of the housing. Measures for the acoustic optimisation of the gearbox housing are described below [18, 19].

6.5.1 Weak Point Analysis

The noise development of the gear unit is almost exclusively determined by the structure-borne noise radiation of the housing. In the case of large gearboxes, the overall noise development is decisively influenced by structure-borne noise radiation from the individual plate arrays, especially when they are excited to resonance vibrations.

The frequency content of gear noise is essentially determined by the tooth meshing frequencies and their harmonics. If, as in the present case, the speed is kept constant, the design analysis can be limited to these excitation frequencies. Table 6.3 lists the tooth meshing frequencies f_0 and their harmonics (multiples of the tooth meshing frequencies).

Since, on the one hand, the low frequencies are perceived much more weakly – among other things because of the A-weighting – and, on the other hand, the plates have a lower degree of radiation than at high frequencies, in the present case and with these dimensions one can limit oneself without hesitation to frequencies above 100 Hz. That means only the bevel gear and the planetary set are considered with their excitation frequencies.

Table 6.3 Tooth meshing frequencies in Hz, gearbox 800 kW

Multiple	f_0	f_1	f_2	f_3
Bevel gear stage	181	363	544	726
Planetary set	62	125	187	249
Reversing pinion	12	24	35	47
Drive stage	6	12	19	25

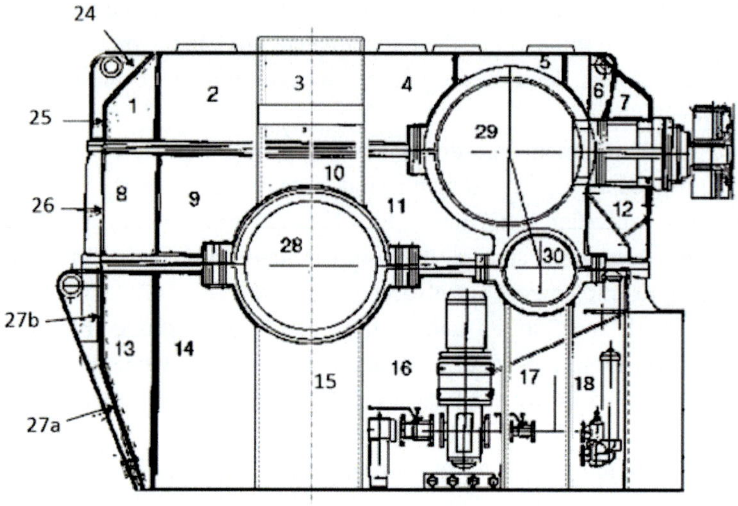

Fig. 6.52 Division of the gearbox surface into plate fields

For the acoustic vulnerability analysis based on design drawings, the enclosure surface was divided into panel bays (see Fig. 6.52).

For each bay, the natural frequencies in the assembled state were calculated mathematically. The natural frequencies can be calculated using estimation formulas [21] or the FFM method. The accuracy of the calculated natural frequencies depends on the modeling, the restraint conditions and the damping. In addition to the natural frequencies and natural modes, the amount of damping is particularly decisive for the occurrence of resonant vibrations. If no damping values are available, damping measurements on comparable constructions may be necessary.

For designs similar to the present gearbox, plate arrays are critical with regard to noise generation if the excitation frequencies (tooth natural frequencies and their harmonics) coincide with the natural frequencies in the range of $\pm 10\%$ of the excitation frequencies. This assumes that, on the one hand, comparable damping as in the case of the gearbox is present and, on the other hand, only the first natural frequencies (up to approx. 4th harmonic) are excited.

In this way, the plates of the gearbox housing, which could significantly influence the noise development of the gearbox through resonance vibrations, were localized.

Table 6.4 lists some critical plate fields whose natural frequencies are close to excitation frequencies.

Table 6.4 Critical plate fields

Plate field	Natural frequencies	Exciter frequencies
No.	Hz	Hz
3	360	363
8	581	544
11	747	726
14	171	182, 187
15	183	182, 187
18	543	545
20	571	545
22	185	181, 187
23	506	545
29	129 186	125 181, 187

6.5.2 Noise Reduction Measures

Noise reduction measures are intended to achieve the contractually required A total sound power level of $L_{WA} = 108.9$ dB(A) (engine + gearbox). An exact quantification of the noise reduction to be expected by optimizing the gearbox housing is not possible in the planning phase. Based on rough estimates and empirical values on similar gearboxes, it can be expected that the set target can be achieved by avoiding all resonance vibrations and increasing the damping. At least the acoustic weak points can be avoided, since a subsequent improvement is not easily possible. The actual noise reduction achieved is to be verified by measurements after completion.

Detuning of the Plate Fields

In order to avoid resonances, all critical plate fields of the gear unit should be detuned by stiffening before production or assembly. This increases the natural frequencies. Care was taken to ensure that the natural frequencies of the stiffened plate fields do not lie in the range of the other excitation frequencies. The stiffeners are to be attached in the oscillation antinodes of the fundamental oscillations. They must not touch the edges – i.e. the clamping points. This is to prevent additional vibration energy from being transferred to the plates.

Increase in Damping

To ensure that the sound radiation of the stiffeners does not increase the noise development of the gearbox, the stiffeners were damped by filling them with dried quartz sand, grain size approx. 1 mm. To increase the overall damping, the cavities of all stiffener boxes were also filled with the same quartz sand.

After the implementation of these measures, the noise development of the new gearbox was experimentally tested on a stress test bench. Figure 6.53 shows the test bench setup and the new gearbox including stiffeners [19].

Figure 6.54 shows the gearbox with the stiffeners and the plate-related measured sound intensities. The numbers are the measured intensity levels of the individual plates in dB(A).

The natural frequencies and damping measured by impact tests have shown that almost all plates are not excited to resonance vibrations. Resonant vibrations at higher frequencies were still detected in circular plate No. 29 (measuring point 69, Fig. 6.54). This is most likely due to the fact that the arranged stiffeners are located in the vibration node of the higher eigenmodes of plate [19, 22].

After installation of the acoustically optimized gearbox, its A-sound power level was also measured under real operating conditions on the bucket wheel excavator SRs2000. Figure 6.55 shows the acoustically optimised gearbox on the SRs2000 bucket wheel excavator. A measurement carried out by MAN TAKRAF showed an A-sound power level at full load of $L_{WA} = 107.8$ dB(A). This meant that the contractually agreed limit values were complied with. The noise level of the acoustically optimised gearbox is

Test bench setup: transmission front and drive side

Test bench set-up: rear side of gearbox and braking device

Fig. 6.53 Stress test bench and the new gearbox

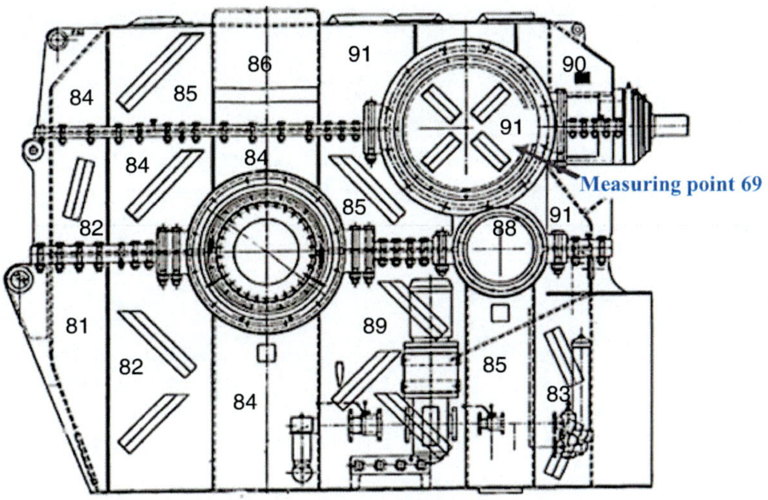

Fig. 6.54 Measured A-sound intensity levels at the gearbox with measures (stiffening angle and sand damping)

Fig. 6.55 Acoustically optimized gearbox in installed condition

approx. 6 dB(A) below the specification of VDI 2159 [23] (Fig. 6.56). Further measures were no longer necessary.

The present application example clearly shows that it is possible to localize acoustic weak points of a construction in the planning phase and to work out suitable constructive measures. Comparable measures on an already built gearbox are not readily possible and are associated with considerably greater expense. In such cases, often only passive measures are possible to reduce noise, e.g. enclosure by means of a sound enclosure.

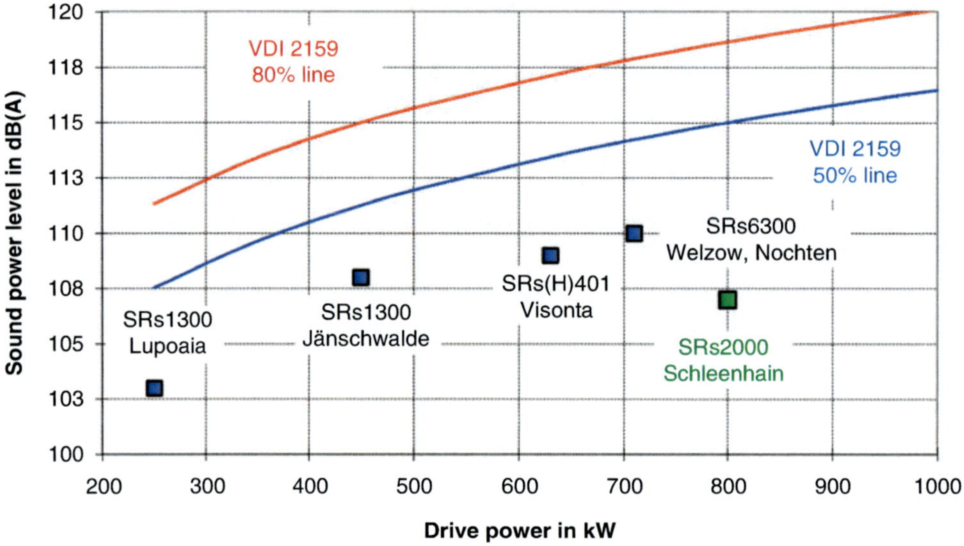

Fig. 6.56 Noise development of the SRs2000 paddlewheel gearbox compared to the state of the art [19]

6.6 Vibration Reduction

Component vibrations not only lead to excessive noise generation, they are also responsible for increased wear and not infrequently for vibration-related material damage. Here, the resonance-excited components are particularly critical. For the sake of completeness, reference is made below to two application examples in which the focus is exclusively on vibration reduction. Here, only the essential correlations and results are reproduced. With regard to the investigations carried out and detailed results, reference is made to the corresponding publications [24, 25].

6.6.1 Cause of Vibration-Induced Damage to the Slideway Supports of a Tablet Press

In this example, the slideway supports broke off during operation of the tablet press. This was caused by excessive vibration. Figure 6.57 shows the tablet press under investigation.

With the help of detailed FEM calculations, which are usually carried out in the planning phase and in the development stage, many weak points of a machine can be identified at an early stage, so that the design can be optimized as far as possible in advance. However, due to the complexity of the designs of modern processing machines and the large number of possible additional components within a machine, it is often not possible, or only possible with a disproportionately large amount of effort, to locate all the

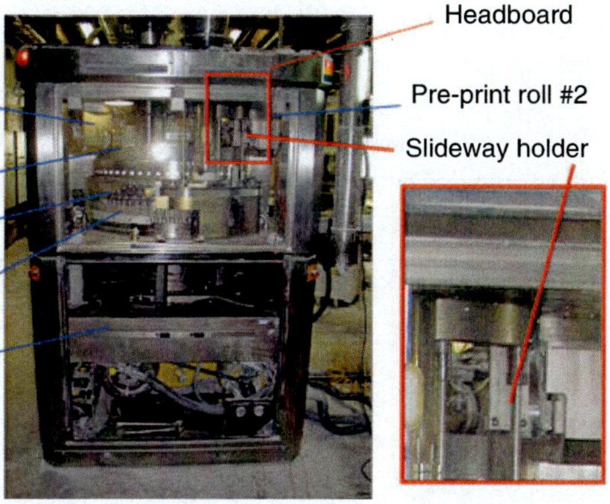

Fig. 6.57 Tablet press KTS 1000

components in question that could possibly be excited to resonance vibrations. It is correspondingly difficult to estimate the stress on add-on components which, for example, experience high vibration stresses due to interaction with components vibrating in resonance.

The investigations have shown that the slideway support was torn off due to excessive vibration stress at a frequency of 35.5 Hz (punch frequency), although the component itself has no natural frequency in this frequency range. The 1st natural frequency of the slideway support is 48 Hz and therefore does not resonate.

The vibration stress is caused by the resonance vibrations of the press head section with a 1st natural frequency of approx. 34 Hz. Due to these resonance vibrations, the slideway support is excited to extremely high vibrations of up to approx. 110 mm/s.

Figure 6.58 shows an example of the results of the FFT and order analyses at measuring point 1 (press head section) during a speed ramp-up of the press from standstill to maximum speed (n = 41 1/min 300,000 Tab/h) in the form of so-called contour plots (Campbell diagram). In the upper diagram the frequency and in the lower diagram the order is plotted against the punch frequency of the press. This clearly shows that the press is excited to very high vibrations at the 1st order (punch frequency) at approx. 35 Hz due to resonance.

Since the resonance vibrations of the press head section also stress all other press components with excessive vibrations in addition to the slideway supports, it does not make sense to develop measures only for the slideway supports. In the present case, it was recommended to avoid the resonance vibration of the press head section by constructive measures.

Since the press is designed for a maximum number of tablets of 300,000 Tabs/h, the highest punch frequency occurring during operation is: $f_{0\max} \approx 42$ Hz. In order to avoid a resonance condition, the 1st natural frequency of the press upper part should be approx.

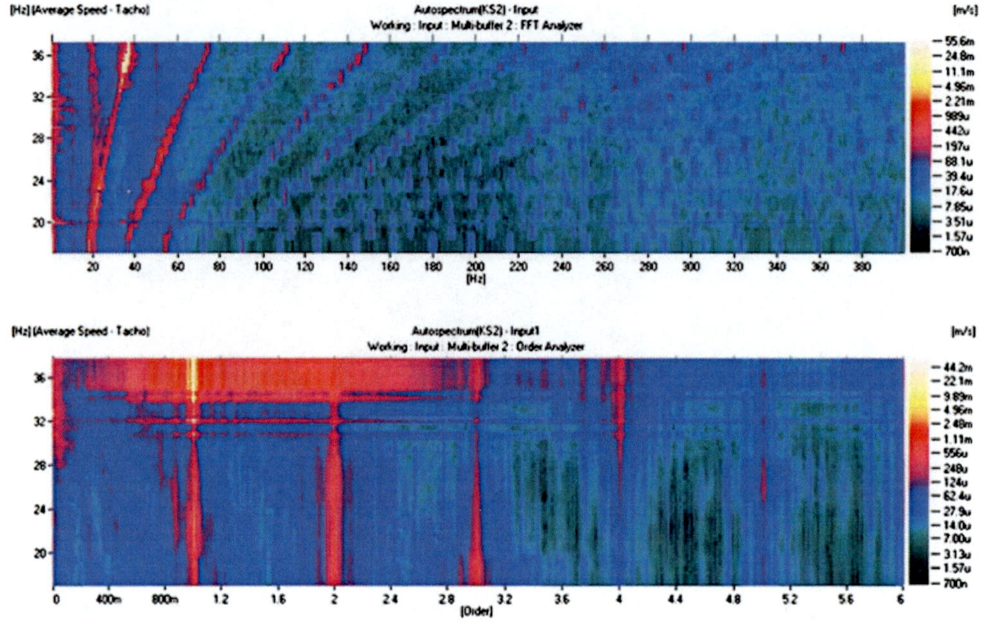

Fig. 6.58 Order and FFT analysis on MP 2 (press head section)

10 to 15 Hz higher than the highest punch frequency occurring in operation, taking into account the existing system damping. Therefore, it was recommended, among other things, to increase the 1st natural frequency of the press head to 50 Hz. This can be achieved by increasing the stiffness, e.g. by increasing the column diameter.

For the quantitative design, an estimation formula for the calculation of the 1st natural frequency of the press upper part was determined based on theoretical considerations and model considerations. These and other investigations and results are compiled in [24].

6.6.2 Vibration Reduction in Pipelines

At a new vacuum plant of MiRO Mineralölraffinerie Oberrhein GmbH & Co KG, vibrations occurred at the furnace outlet pipes after commissioning, which were to be regarded as critical for the fatigue strength of the pipelines.

Within the scope of a vibration analysis, the vibration behaviour of the pipelines in the critical frequency range was therefore immediately investigated and appropriate measures to control the vibration problem were developed.

Figure 6.59 shows the vacuum system and Fig. 6.60 the spatial measuring point plan of the furnace outlet line F-102 B. Table 6.5 shows the measured maximum values of the vibration velocities.

Fig. 6.59 Vacuum system

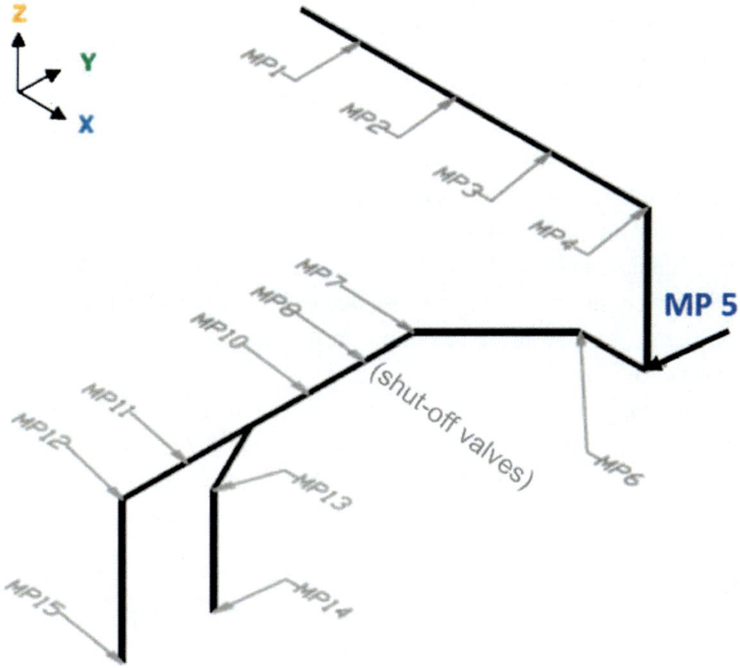

Fig. 6.60 Measuring point diagram of the furnace outlet line F-102 B

Table 6.5 Maximum values of the vibration velocities at the furnace outlet pipe F-102 B

	$\lvert v_{max} \rvert$		
	X	Y	Z
	mm/s	mm/s	mm/s
MP 1	21.0	22.0	27.3
MP 2	12.4	20.4	15.4
MP3	10.0	20.8	15.3
MP4	18.5	28.9	33.4
MP 5	75.6	70.5	23.9
MP 6	84.6	49.2	22.1
MP 7	72.4	45.2	36.8
MP 8	58.6	36.8	33.6
MP 10	45.8	24.1	15.5
MP 11	54.8	31.4	17.5
MP 12	50.3	20.0	19.1
MP 13	23.8	29.1	28.7
MP 14	28.6	21.7	19.8
MP 15	34.0	34.8	26.3

The maximum values of the vibration velocities at several locations are higher than 50 mm/s ($v_{max} > 50$ mm/s), therefore the vibrations, with respect to the fatigue strength of the pipelines, are to be classified as critical [26]. Since the risk of pipeline damage due to vibration cannot be excluded, mitigation measures were necessary.

Figure 6.61 shows an example of the time and frequency characteristics of the vibration velocities for measurement point 5 (MP5) of the furnace outlet pipe (F-102B).

With regard to the vibration excitation, it should be mentioned that the piping system is excited to vibrate by stochastic shock-like pulses caused by the two-phase flow within the piping. During the shock period, the piping system responds with a very low frequency ($f < 1$ Hz) forced oscillation (initial oscillation!). During the decay phase, the piping system exhibits resonant oscillations (residual oscillations!), i.e. excitation in the natural frequency range of the piping system.

Since the plant could not be shut down, the operating vibration modes of the pipelines in three spatial directions x, y and z were determined by measurement for the critical frequencies. Figure 6.62 shows an example of the measured operating vibration mode in the *x-direction* at 3.5 Hz.

Since forced vibrations are always measured during operation, i.e. the measured frequencies are essentially only excitation frequencies, it is reasonable to assume, based on the excitation mechanisms described above, that the measured frequencies and vibration modes are natural frequencies and natural modes, especially in the case of amplitude overshoots. For an exact clarification of this question, measurements at standstill of the plant would have been necessary.

Order No: 04.9.701, Client: MiRO Mineralo elraffinerie Oberrhein GmbH

Vibration measurements within the scope of the project Repl. Vacuum 4 at the furnace outlet lines F-102A/ F-102B and support of the development of measures for control.

Furnace outlet pipe **F -102B:**

Triaxial measurement of the vibration velocity at **MP5**

MP5_B/K:2/X-direction

r	Tim s	v mm/s
1	33.647	75.6

MP5_B/K:1/Y-direction

r	Tim s	v mm/s
1	71.771	–70.5

MP5_A/K:1/Z-direction

r	Tim s	v mm/s
1	47.957	–23.9

MP5_B/K:2/X-direction

r	Frequency Hz	v mm/s
1	3.4	6.9

MP5_B/K:1/Y-direction

r	Frequency Hz	v mm/s
1	11.0	4.8

MP5_A/K:1/Z-direction

r	Frequency Hz	v mm/s
1	3.6	1.6

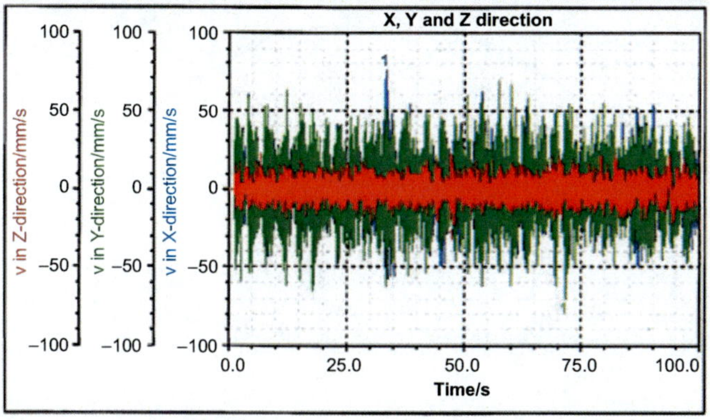

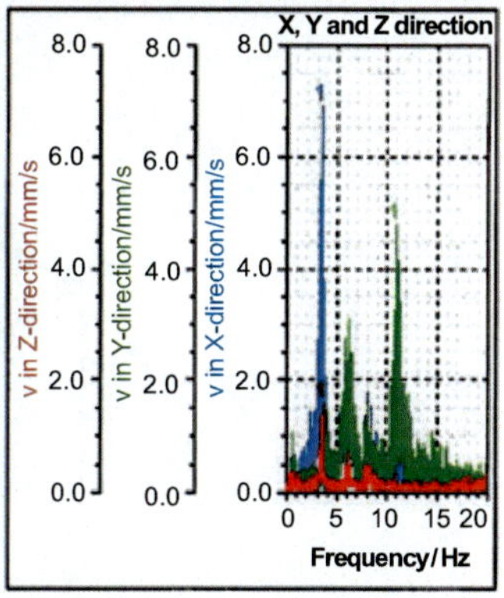

Fig. 6.61 Triaxially measured vibration velocities at MP 5 (F-102 B)

Fig. 6.62 Measured operating vibration waveform in *x-direction* at 3.5 Hz (F-102 B)

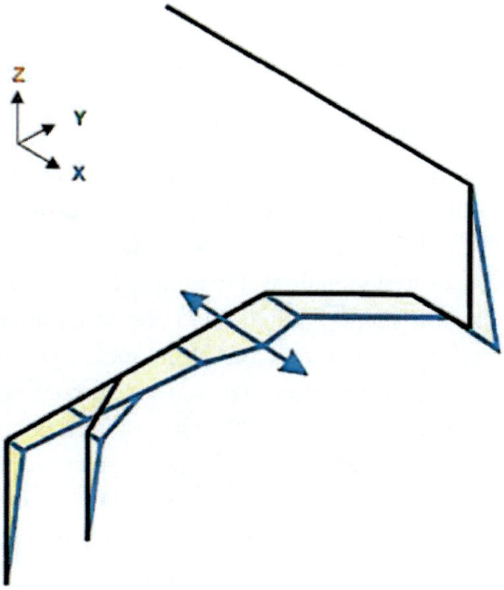

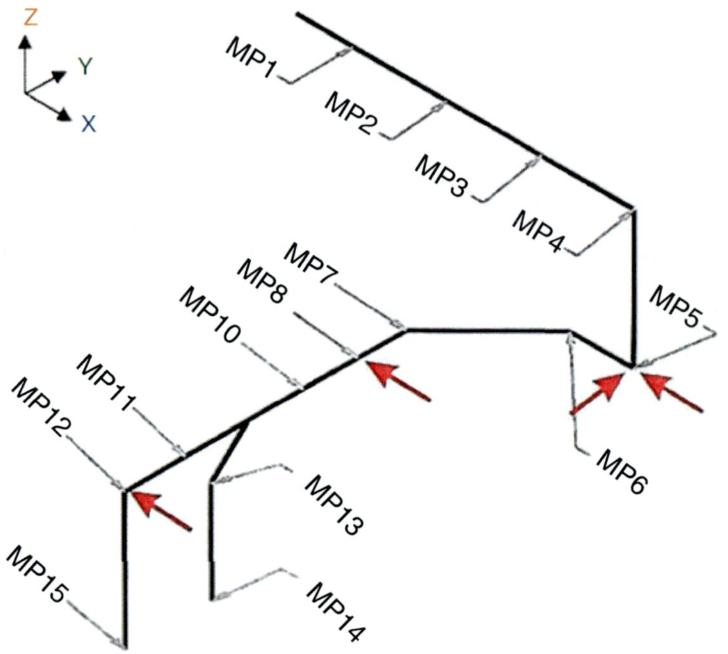

Fig. 6.63 Planned locations where the furnace outlet pipe F-102 B is to be braked

With the help of the operational vibration shapes, the pipe locations for hydraulic vibration brakes were determined. In Fig. 6.63, the locations where the furnace outlet pipe F-102B is to be braked using hydraulic vibration brakes are indicated by arrows, 3 times in the *x-direction* and once in the *y-direction*.

The characteristic values required for the design of the vibration brakes, such as the moving mass and its self-damping at critical frequencies, could not be determined exactly by measurement, since on the one hand the system could not be switched off and on the other hand the operating vibrations of the pipelines were too high for external excitation (e.g. by means of an impulse hammer). Therefore, based on the empirical values and the measured vibrations, the co-moved mass m_{dyn}, the loads F_{dyn} to be dynamically supported at the support points, and the maximum dynamic reaction displacements w_{max} were estimated. On the basis of the measured and estimated dynamic characteristic values and taking into account the cold-hot movements of the pipeline system, the 2 1/2″ hydraulic vibration brakes of PSS GmbH, Neunkirchen, Germany (nominal load capacity 45 kN) were selected for the vibration reduction of the pipeline.

In order to install the planned vibration brakes at the intended locations, brackets decoupled from the piping system had to be built beforehand. To ensure that the effectiveness of the vibration brakes is not reduced, the stiffness of the attachment points of the steel structure (brackets) must be considerably higher than that of the pipelines.

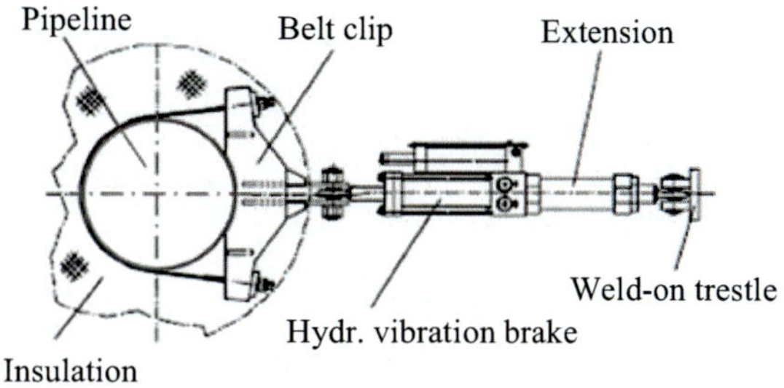

Fig. 6.64 Hydraulic vibration brake from PSS GmbH and exemplary coupling by means of a belt clamp on the pipeline

Figure 6.64 shows the hydraulic vibration brake of PSS GmbH as well as an example of the coupling of a vibration brake to the pipeline by means of a belt clip.

After the implementation of the measures, their effectiveness was checked by measurement. With the help of the implemented measures, the vibrations were reduced to a level from which it can be assumed from experience that there is no risk of damage [27]. Further investigations and detailed results are compiled in [25].

6.7 Field-Tested Application Examples

The following selection of examples is intended to stimulate ideas among users for the implementation of product-related noise reduction measures. In contrast to the examples presented in Sects. 6.1, 6.2, 6.3, 6.4, 6.5 and 6.6, only the task and the noise reduction measures applied are presented here.

As already mentioned, the following application examples were mainly developed between 1970 and 1990 [28–37]. Copying relatively old publications explains the poor quality of the following pictures.

On a CD, in the appendix of the VDI guideline 3720, sheet 1, further application examples are compiled [36].

6.7.1 Sound Generation

The examples given here aim at influencing the sound generation by changing the force-time curve.

Example 6.1 (Cutting Shape of Punch in a Press)
By changing the cutting tool (punch head), the force-time curve was changed or stretched. This significantly reduced the high-frequency excitation forces and achieved a sound level reduction of approx. 7.5 dB(A), Fig. 6.65.

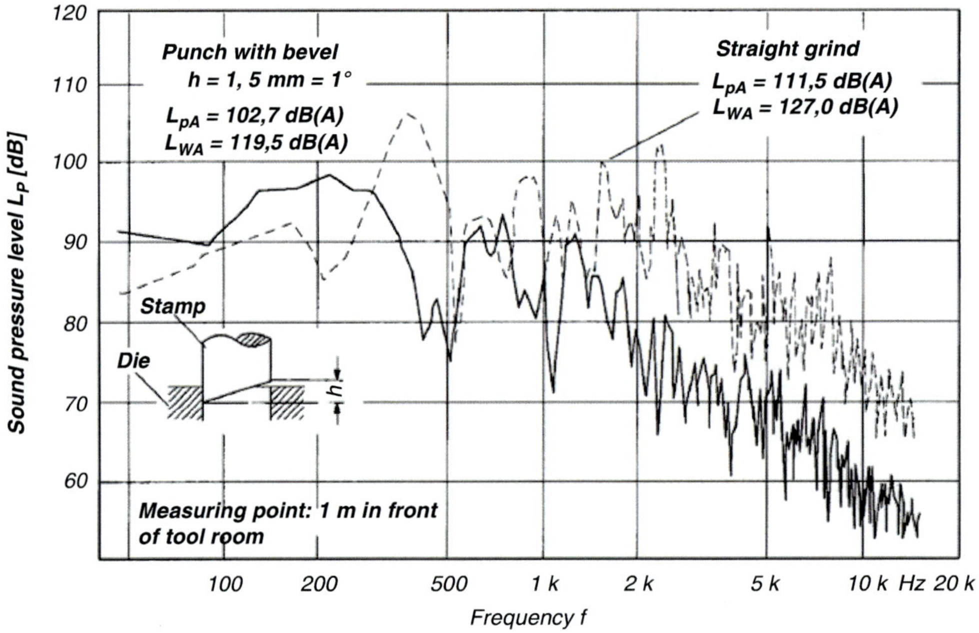

Fig. 6.65 Airborne sound spectra when cutting 4 mm thick sheet metal for punches with oblique and straight bevels

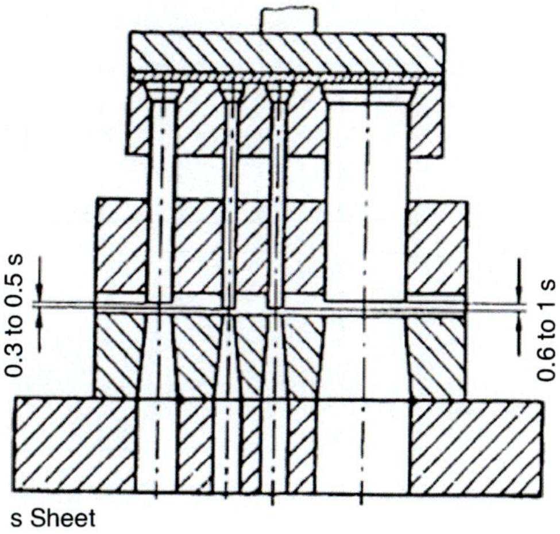

Fig. 6.66 Multiple cutting of a tool by an offset arrangement of the cutting punches

Example 6.2 (Punch Height of a Punching Tool)
By staggering the cutting punches, the cutting process can be stretched in time, Fig. 6.66. Here, the punches should be arranged in such a way that shortly before the first punch finishes the cut, the following punch engages.

By superimposing the individual time histories, a stretched force-time history is obtained.

Example 6.3 (Toggle Press)
With conventional cutting, noise is primarily generated at the end of the cutting process when the workpiece is torn off. By means of a stop damper installed in the press frame, the abrupt retraction of the press frame is prevented or extended in time. As a result, the A-sound pressure level could be reduced by approx. 16 dB(A). Figure 6.67 shows the corresponding force-time curves and the octave spectrum of the measured sound pressure levels, with and without stop damper.

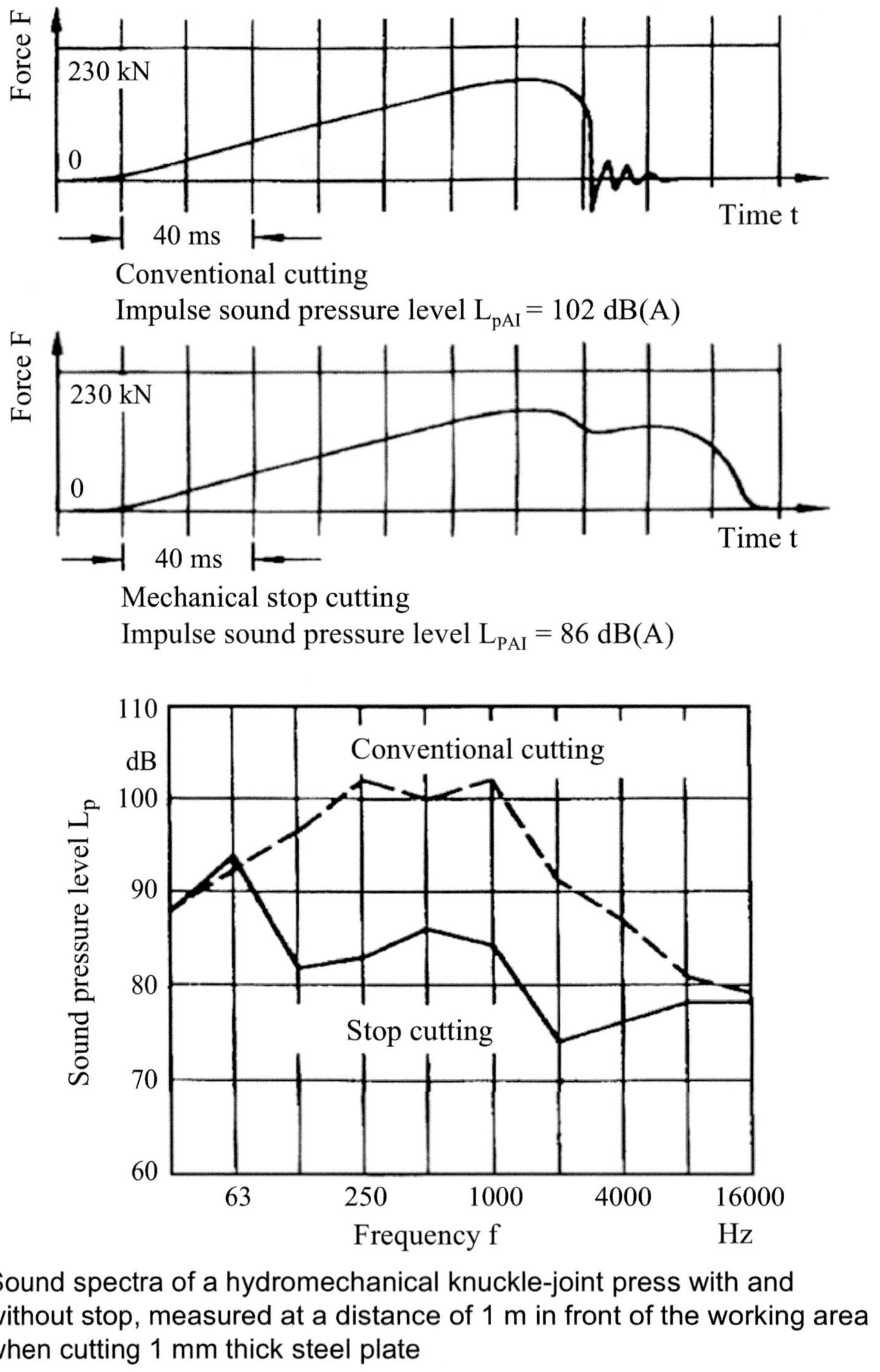

Fig. 6.67 Force-time curves and measured octave spectrum of the sound pressure level

Example 6.4 (Pincer Feed)

The force-time curve of metallic impact of a clamp on the "stop" provided for positioning was stretched with the aid of two rubber elements, thus a noise reduction of approx. 7 dB (A) was achieved, see Fig. 6.68.

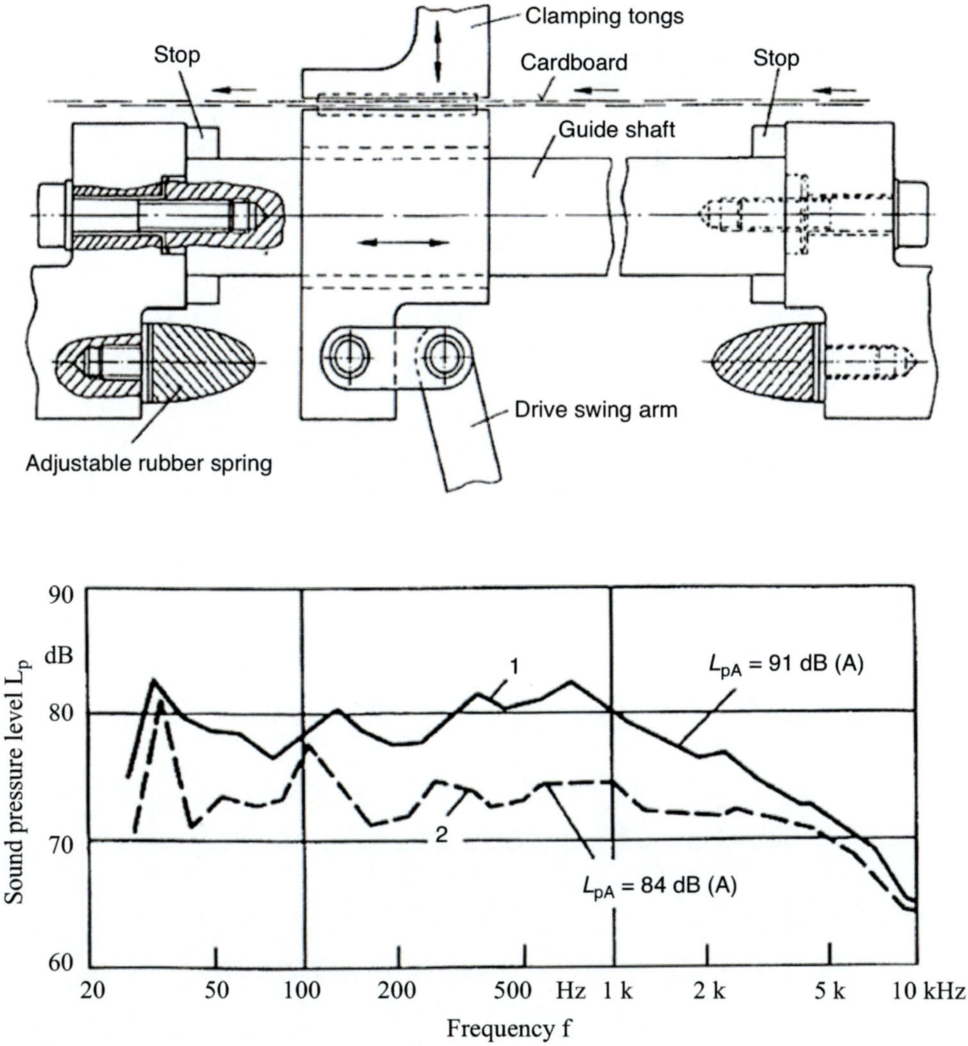

Sound spectra before (curve 1) and after (curve 2) the implementation of reduction measures at the stops of the eccentric-controlled pincer feed of a machine for the production of letterpress inserts, measured at the workplace of this machine.

Fig. 6.68 Noise reduction of a pincer feed by stretching the force-time curve

Example 6.5 (Bottle Transport System)
The reduction of the noise development of a bottle transport system was realised by redesigning the bottle entrainment element. This means that the bottles can no longer fall, but are transported downwards via a sliding surface, see Fig. 6.69. A noise reduction of approx. 4 dB(A) was achieved by stretching the force-time curve.

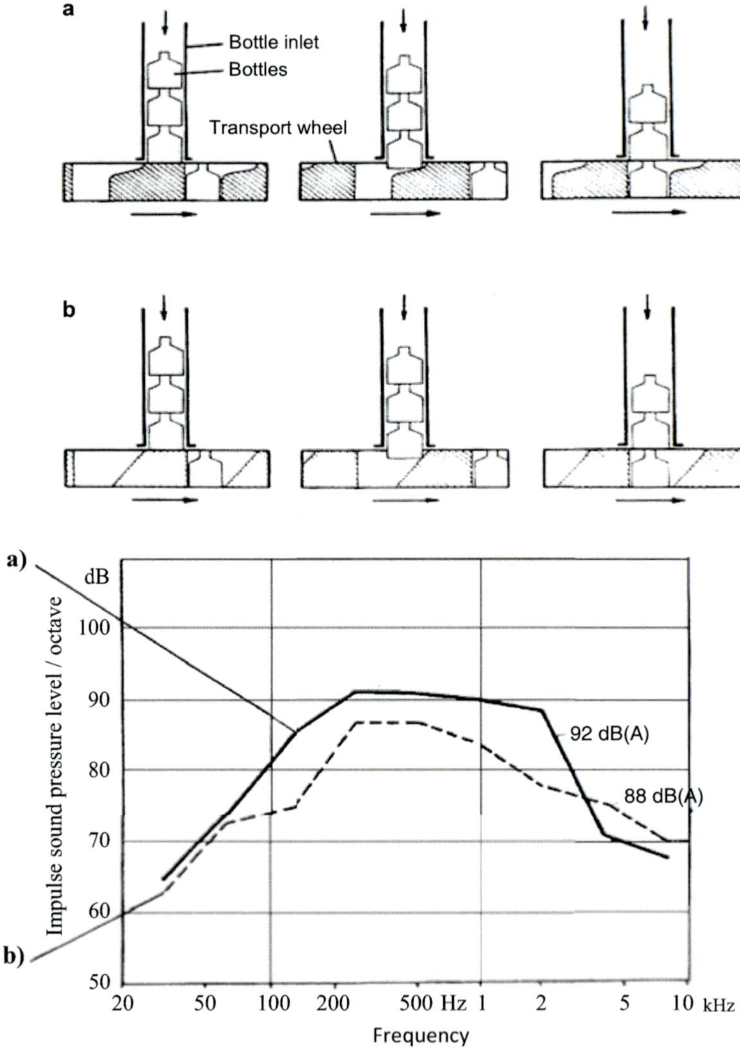

Fig. 6.69 Noise reduction of a bottle transport system; reduction of the airborne sound level at a distance of 1 m from the transport system by avoiding impact excitation

Example 6.6 (Noise Reduction on a Collecting Trough for Pipes)
The very high noise level of a collecting trough was caused by the beating of the pipes, which were collected in the trough for further transport after completion. The noise reduction was achieved with the help of a specially manufactured device. Here, the pipes are guided into the trough via a chute and braked at the end with the aid of a braking element. The pipes are then slowly deposited in the trough from a significantly lower height, see Fig. 6.70. By stretching the force-time curve, a noise reduction of approx. 24 dB (A) could be achieved here.

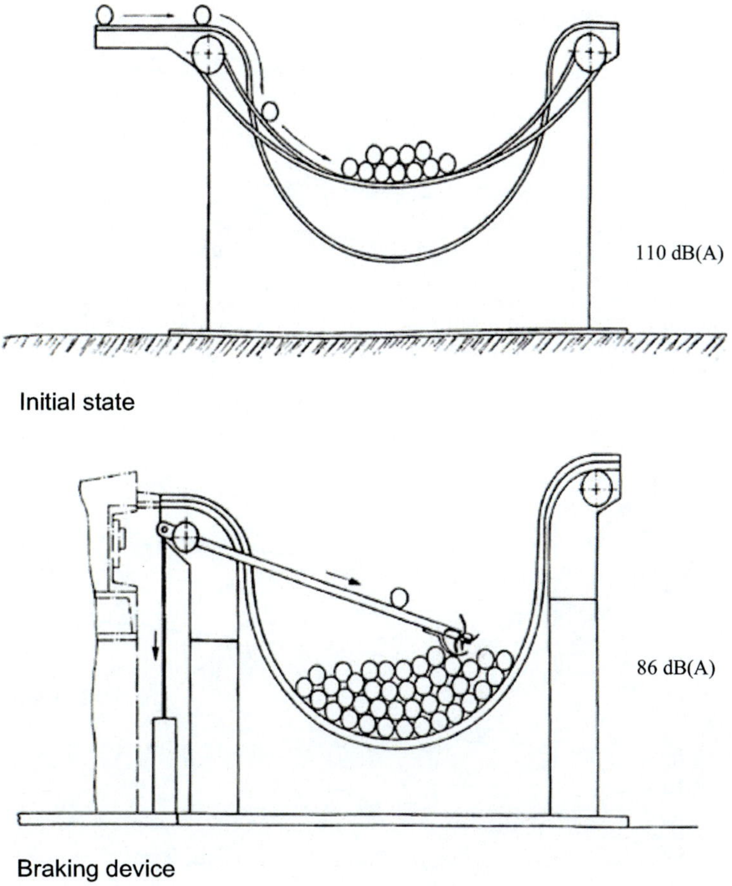

Fig. 6.70 Noise reduction on a collecting trough by installing a braking device

Example 6.7 (Rope Drill)

In this example, during rope manufacture in a stranding machine, the wires are fed through a so-called stranding head and wound helically around a core wire. The noise caused by the stranding heads hitting hard against each other was softened by means of rubber buffers in the end positions. The stranding heads were additionally guided over plastic plates and partially encapsulated, see Fig. 6.71. Here, too, the force-time curve was primarily stretched and an overall noise reduction of approx. 14 dB(A) was achieved.

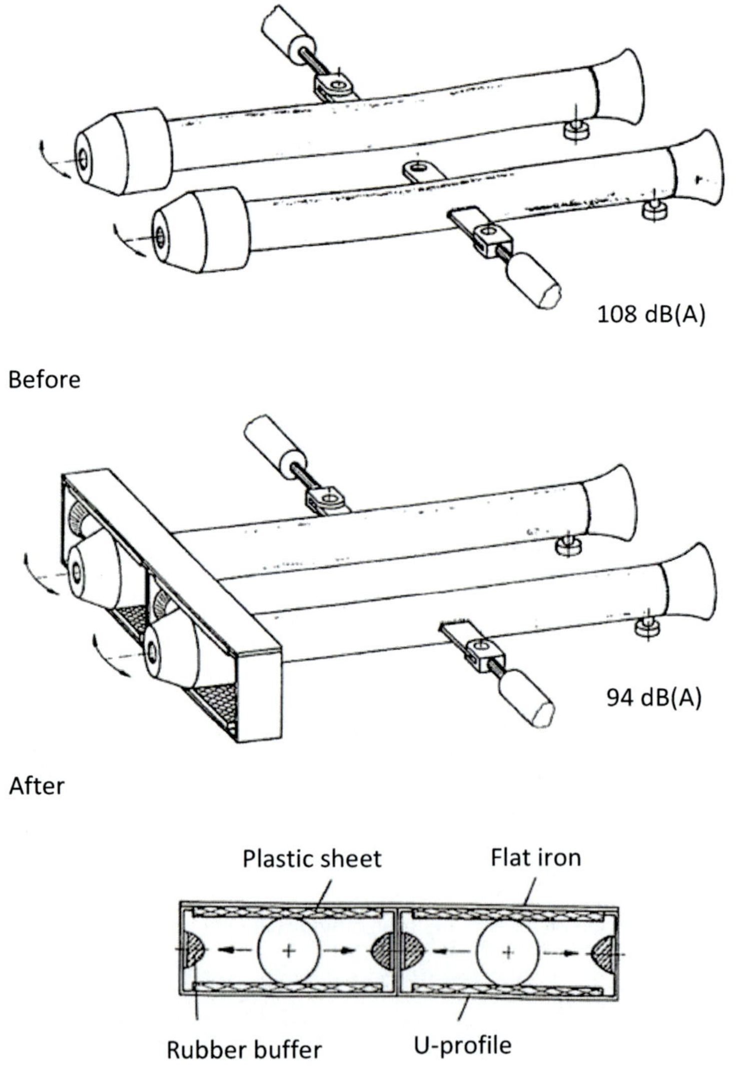

Fig. 6.71 Noise reduction of a rope drill

6.7.2 Sound and Vibration Transmission

If the possibilities of noise reduction by influencing the sound generation are exhausted or are no longer reasonable for technical and/or financial reasons, measures to influence the sound and vibration transmission should be considered for further noise reduction. As described in Sect. 3.2, either isolation with a high connection impedance or elastic decoupling with a low connection impedance can be considered, depending on whether the connected components that are to be designed with low noise are force- or speed-excited.

Figure 6.72 shows the principle arrangement for elastic coupling with an increase in the input impedance of the coupling point. The input impedance is increased, for example, by force-locking an additional mass to the structure. In the case of elastic coupling, e.g. with the aid of a spring element, the coupling point must have a much higher dynamic mass than the mass of the machine or excitation source acting at the coupling point. For this reason, the coupling points of spring elements, i.e. the foundation, are often designed with additional masses.

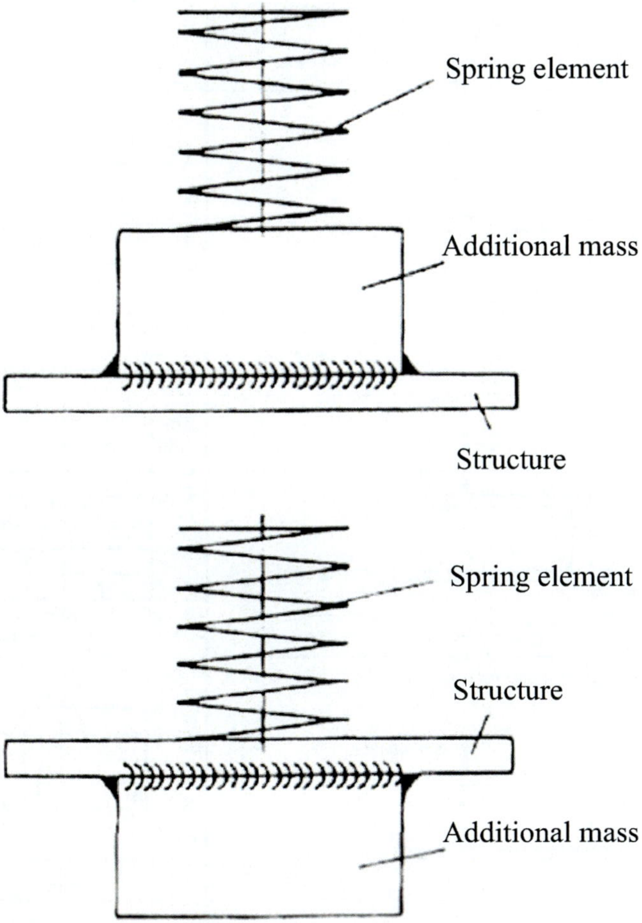

Fig. 6.72 Principle arrangement for elastic coupling with increase of the input impedance at the coupling point

Example 6.8 (Increasing the Mechanical Input Impedance)
Using the example of a cantilever beam excited by the constant force *F*, the acoustic effect of various measures was demonstrated by measuring the sound pressure level at a distance of 1 m in a model study, see Fig. 6.73. Noise reductions of approx. 8 to 18 dB(A) were achieved depending on the measure.

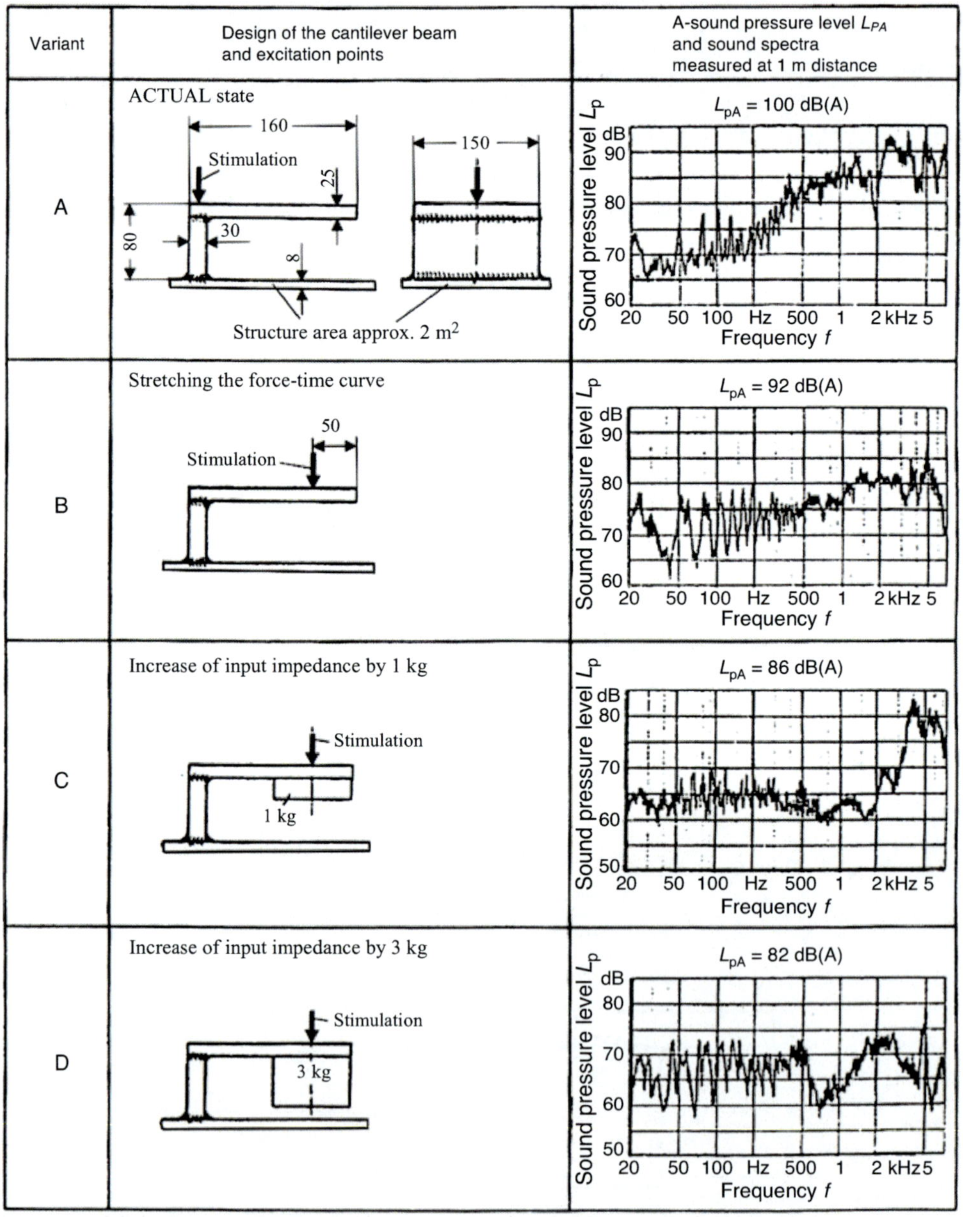

Fig. 6.73 Noise generation of a cantilever beam

Example 6.9 (Noise Reduction of a Slide)
The noise generated during the transfer of workpieces produced by an automatic lathe is mainly caused when the parts hit the chute. The noise is reduced by redesigning the chute. For this purpose, the input impedance in the area of the impact of the workpieces was realized by an 80 mm steel block, which is elastically connected to the chute by means of rubber elements. In addition, the slide surface was damped, see Fig. 6.74. This resulted in a noise reduction of approx. 11 dB(A).

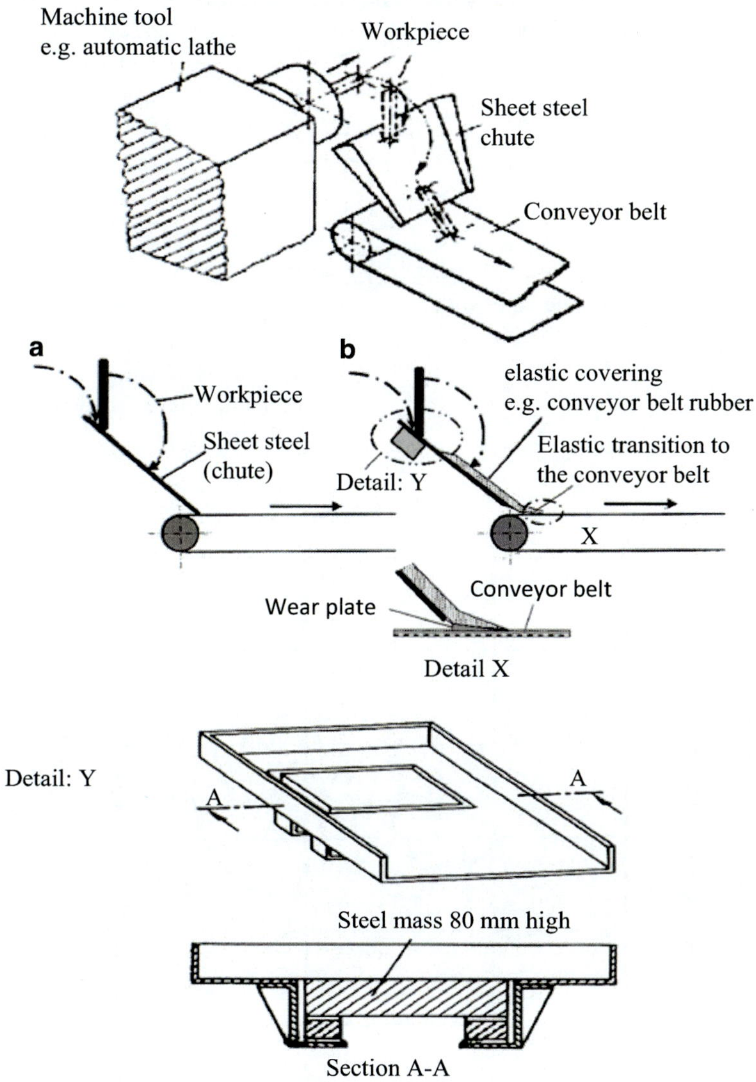

Fig. 6.74 Schematic representation of the noise reduction of the workpiece transfer before (design A) and after (design B) the measures

Example 6.10 (Coupling of Machine Guards)
Unless they are load-bearing, the machine guards should not have a rigid connection to the machine frame if possible. The coupling points should be designed as points. Line and surface couplings should be avoided. Figure 6.75 shows various coupling options for

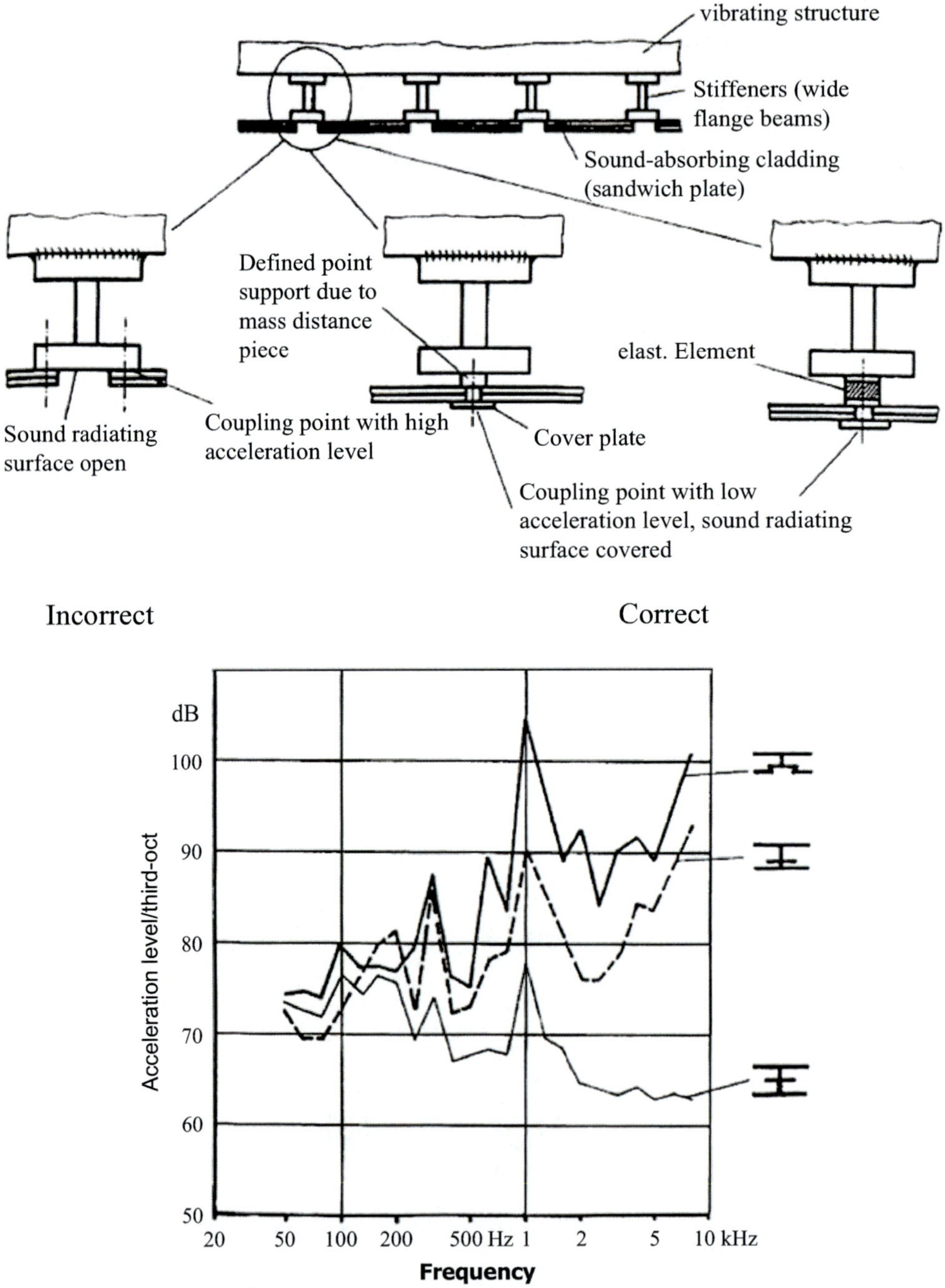

Fig. 6.75 Various couplings of machine enclosures

machine protective cladding. The effect of the individual connections was checked by measuring the structure-borne sound accelerations after the coupling. In principle, it is possible to design the protective cladding as a sound-insulating hood.

Example 6.11 (Pipeline Feed-Through)
The pipeline bushings must always be designed to be elastically decoupled. In the case of rigid coupling, especially if the pipelines are connected by coupling to other excitation sources such as a pump, the pipeline vibrations are transmitted to the connected structures, machine housings, walls, etc., and the noise development is significantly influenced as a result. Figure 6.76 schematically shows the elastic feed-through of pipelines. The barrier masses are intended to reduce the transmission of vibrations above the pipeline.

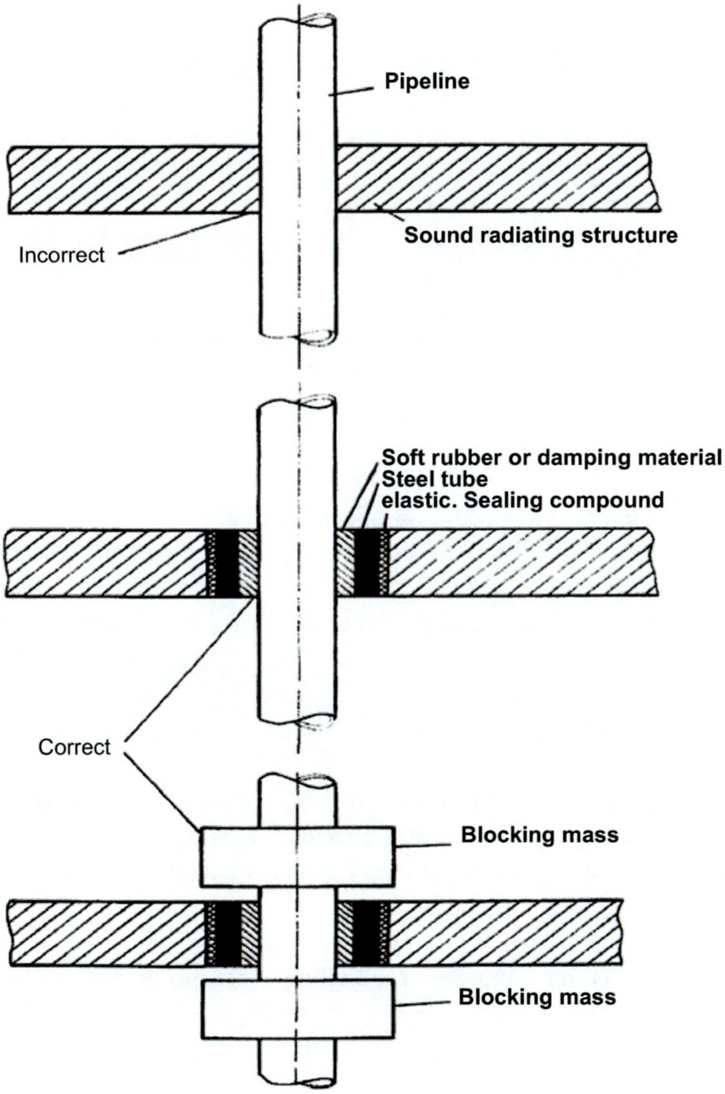

Fig. 6.76 Structure-borne sound insulation of pipe penetrations

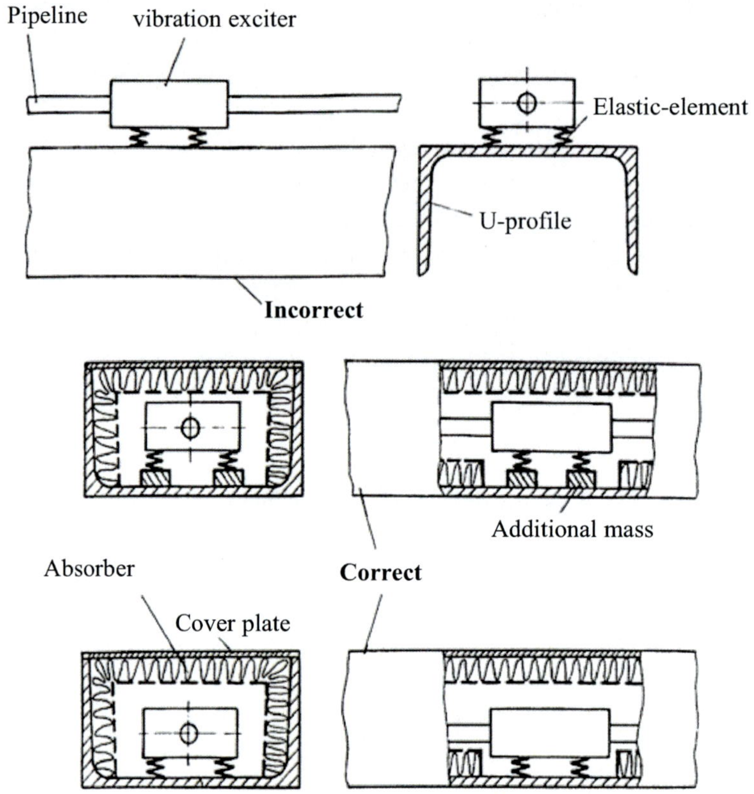

Fig. 6.77 Schematic diagram of the optimum coupling of an excitation source including enclosure of the sound-emitting pipelines

Example 6.12 (Coupling of Excitation Sources)
The coupling of excitation sources to structures, e.g. pumps, and the pipelines connected to them should, if possible, be elastic, and the sound-emitting surfaces should be encapsulated. Figure 6.77 schematically shows the correct coupling of an excitation source.

6.7.3 Sound Radiation

Noise reduction by influencing sound radiation should only be considered if the possibilities for influencing sound generation and sound transmission have been exhausted. In practice, however, measures that influence sound radiation, e.g. encapsulation, are often applied first for reasons of cost and/or time.

Example 6.13 (Collecting Container for Falling Workpieces, Fig. 6.78)

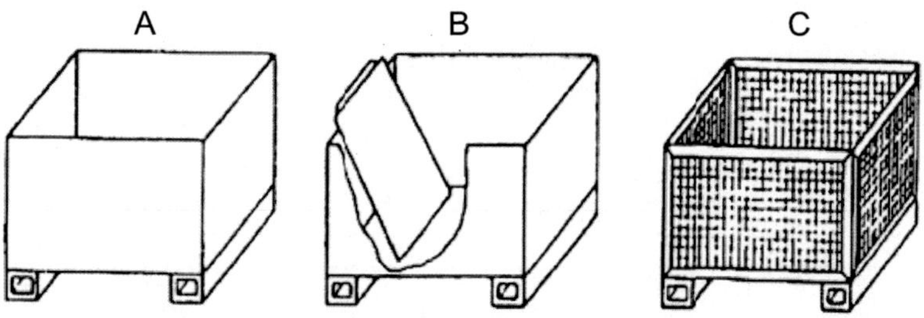

Variant A: Solid sheet metal container (initial state)
Noise generation due to sound radiation of the container plate

Variant B: Solid sheet metal container with plastic baffle board
Noise reduction through reduction of impact forces

Variant C: Wire container
Noise reduction by influencing the radiation level

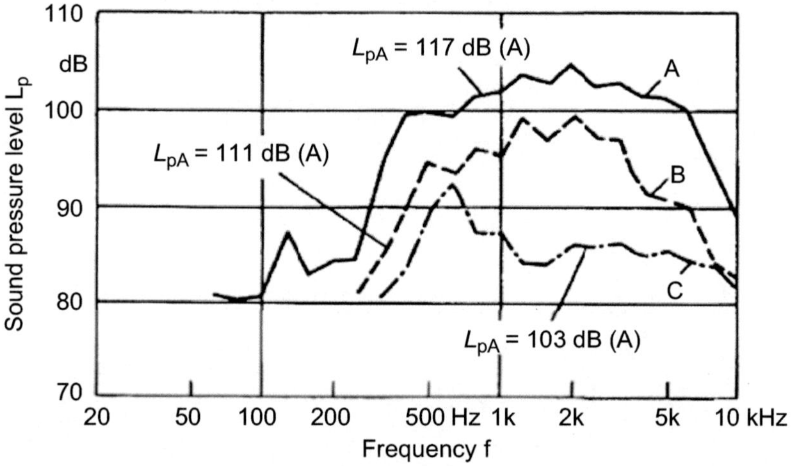

Sound spectra of collecting containers for falling workpieces, measured at a distance of 1 m during the fall of screws weighing approx. 0.5 kg from a height of approx. 1 m into empty containers corresponding to the design variants A to C shown in the figure above

Fig. 6.78 Noise reduction of a collecting tank

Example 6.14 (Structure-Borne Sound Insulated Partial Enclosure)
The noise development of a machine group, which is rigidly mounted on a framework, is decisively determined by the middle machine. The reduction of the total noise development was realized with the help of a partial encapsulation. The enclosure must be set up on the frame in such a way that it is insulated against structure-borne noise, since the structure-

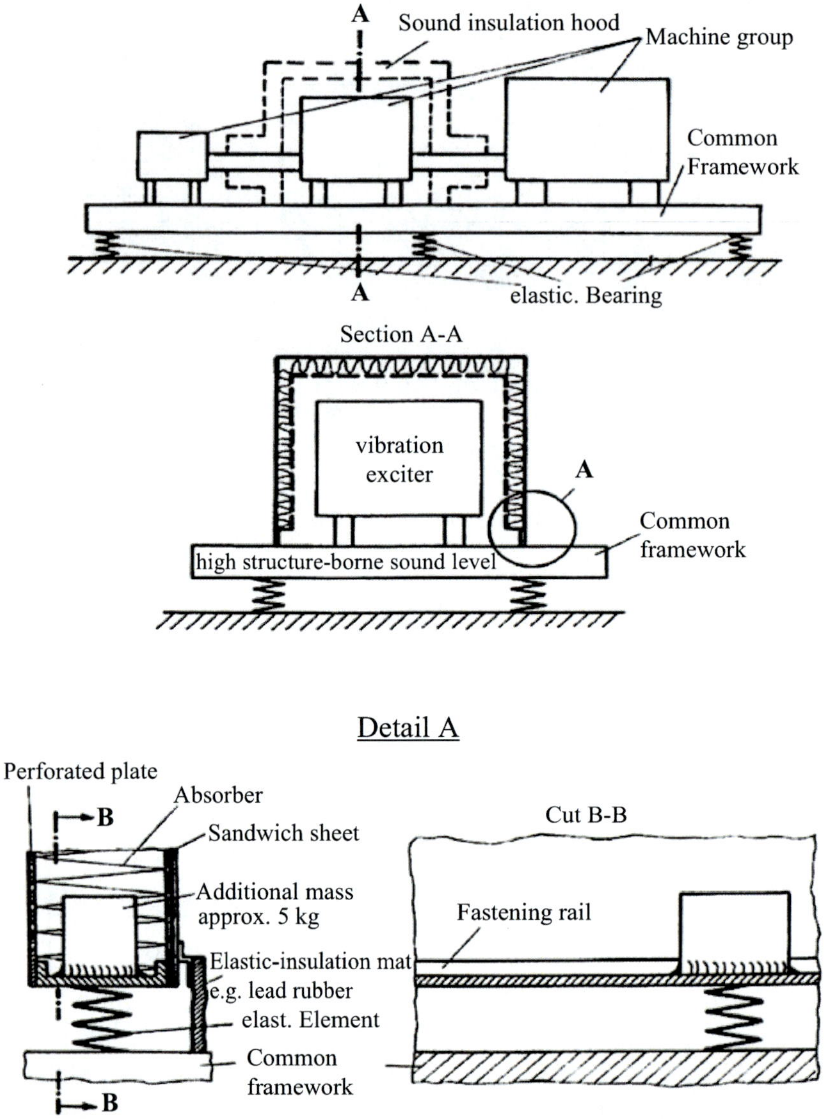

Fig. 6.79 Partial encapsulation of a noise source on a framework with structure-borne sound insulation

borne noise vibrations of the framework can cause the enclosure to vibrate. The air gap between the enclosure and the frame resulting from the structure-borne sound insulation was sealed, see Fig. 6.79.

6.7.4 Interaction of Different Mechanisms of Action

In practice, noise reduction of machines is often realized by the interaction of different mechanisms in sound generation, transmission and radiation.

Example 6.15 (Noise Reduction of a Foundry Cleaning Drum)
In a fettling drum, castings are deburred by hitting them against each other and against the drum wall. The very high noise development was reduced by approx. 27 dB(A) by coating the cleaning drum wall, see Fig. 6.80. The high level reduction was achieved, among other

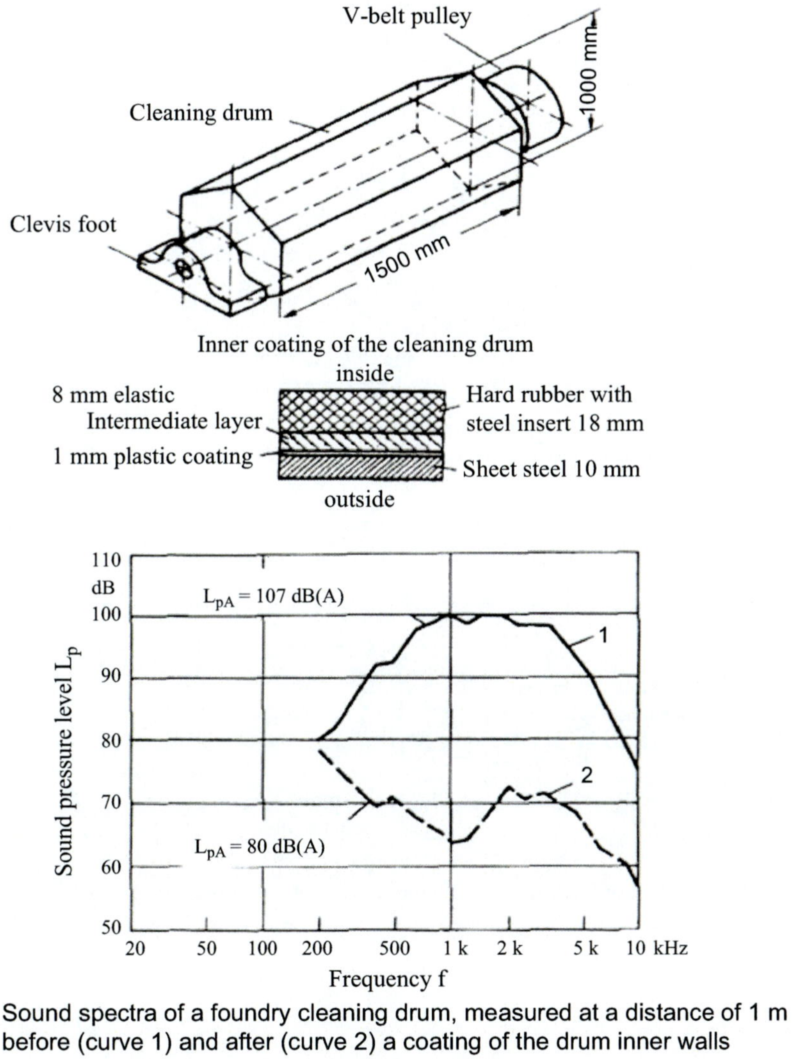

Fig. 6.80 Noise reduction of a cleaning drum

things, by the resilience of the inside of the cleaning drum, the stretching of the force-time curve, the influencing of the sound generation and the increase in the insulation of the wall, i.e. by influencing the sound transmission and radiation.

Example 6.16 (Noise Reduction of an Automatic Spot Welding Machine)
The noise reduction was achieved by encapsulating the pneumatic cylinders and damping the machine housing, see Fig. 6.81. Due to structure-borne noise vibrations of the machine housing, the coupling of the enclosure had to be designed elastically. After the encapsulation, the noise development was decisively influenced by the structure-borne sound radiation of the machine housing. For this purpose, the machine housing was damped. The effect of the damping was increased by strip damping using 1 mm sheet metal strips. These measures resulted in an overall level reduction of 6 dB(A).

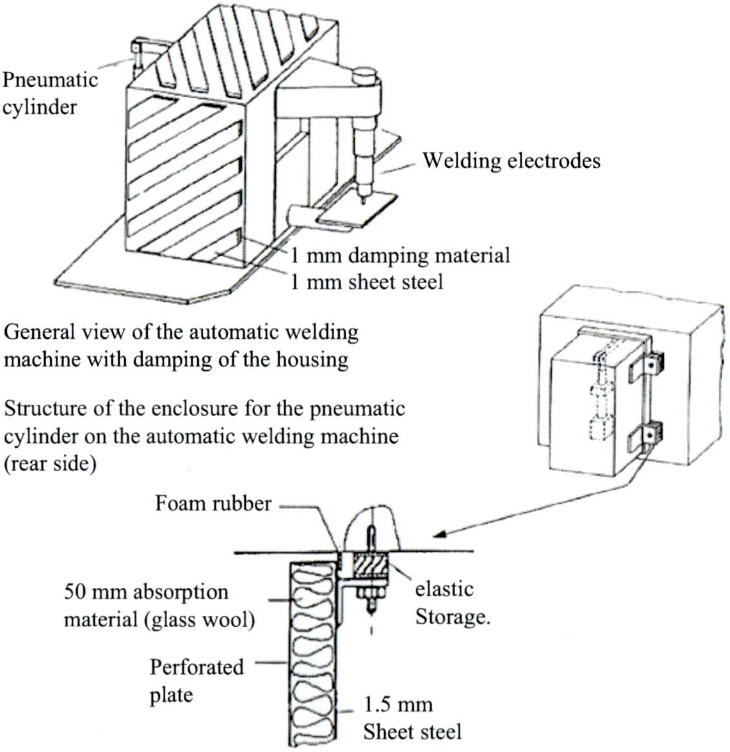

Fig. 6.81 Noise reduction measures of an automatic spot welding machine

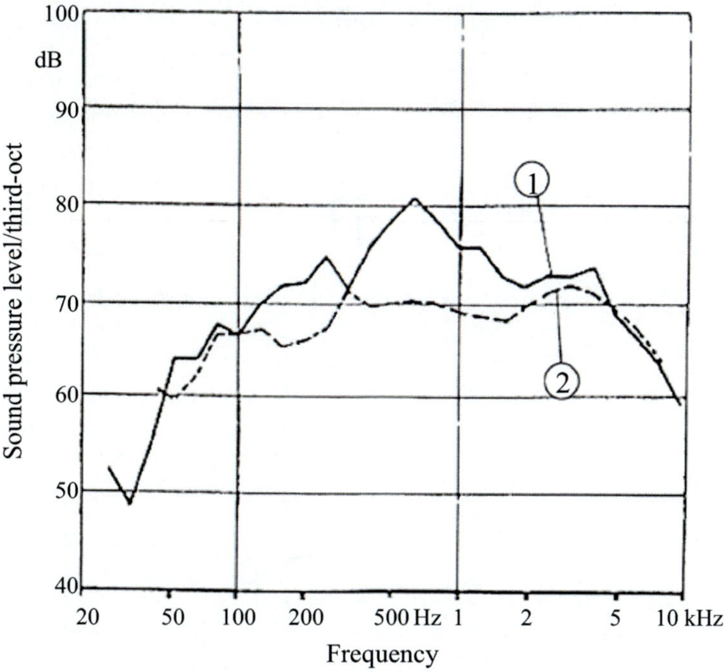

Sound pressure level tertiary spectrum L_{Terz} and A sound pressure level in front of the automatic welder at the ear of the operator

(1) ——— measured level without noise reduction measures: **87 dB(A)**

(2) - - - measuredlevel with noise reduction measures: **81 dB(A)**

Fig. 6.82 Reduction of sound emission by decoupling and damping of sound-emitting components of an automatic spot welding machine

Figure 6.82 shows the sound pressure levels at the ear of the operator, Example 6.16.

Example 6.17 (Noise Reduction Through Optimum Use of Gearboxes)
Before applying or trying out measures for noise reduction, it makes sense to check whether alternative techniques, processes and/or workflows can be considered for the solution of the task. In the following example, it was shown that it would make more sense to use a planetary gearbox as an alternative instead of encapsulating the noisy hydraulic gearbox, see Fig. 6.83.

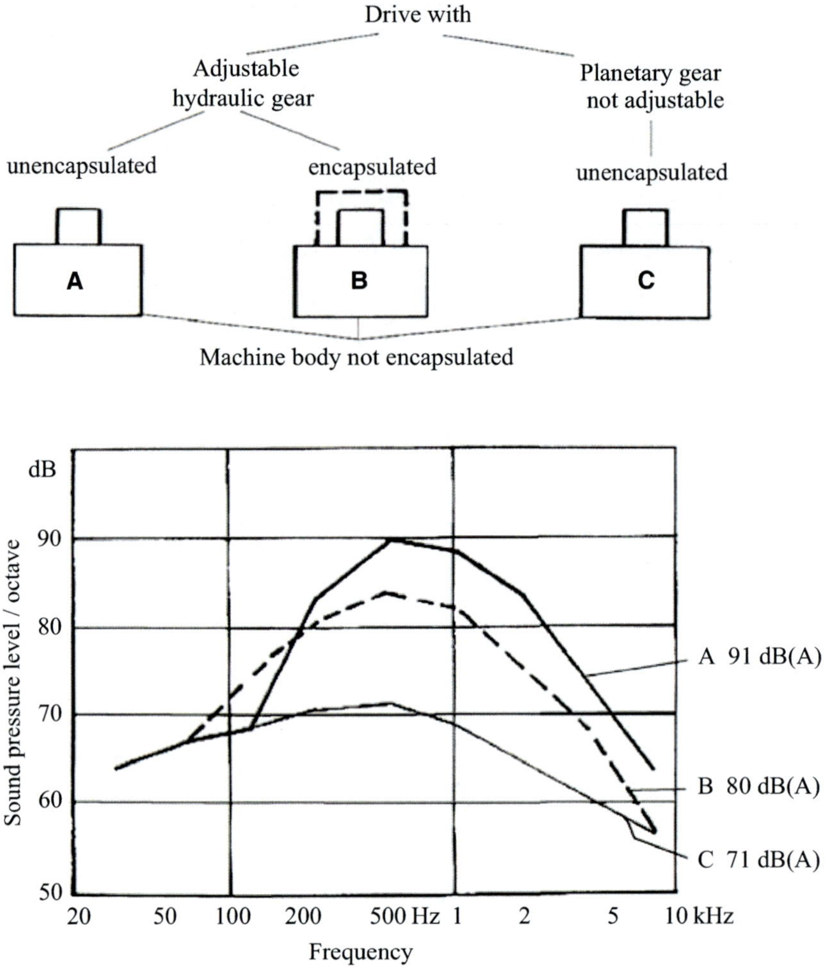

Fig. 6.83 Influence of the gearbox on noise development, comparison of primary and secondary measures

Example 6.18 (Noise Reduction, Interaction of Airborne and Structure-Borne Sound Radiation)
The overall noise development of a machine is often determined by the sound radiation of several parts and/or components. The noise reduction of the individual components then determines the overall noise development of other components that were previously uncritical. The following example schematically illustrates the above-described noise reduction problem using various design variants of the machine protective cladding, Fig. 6.84.

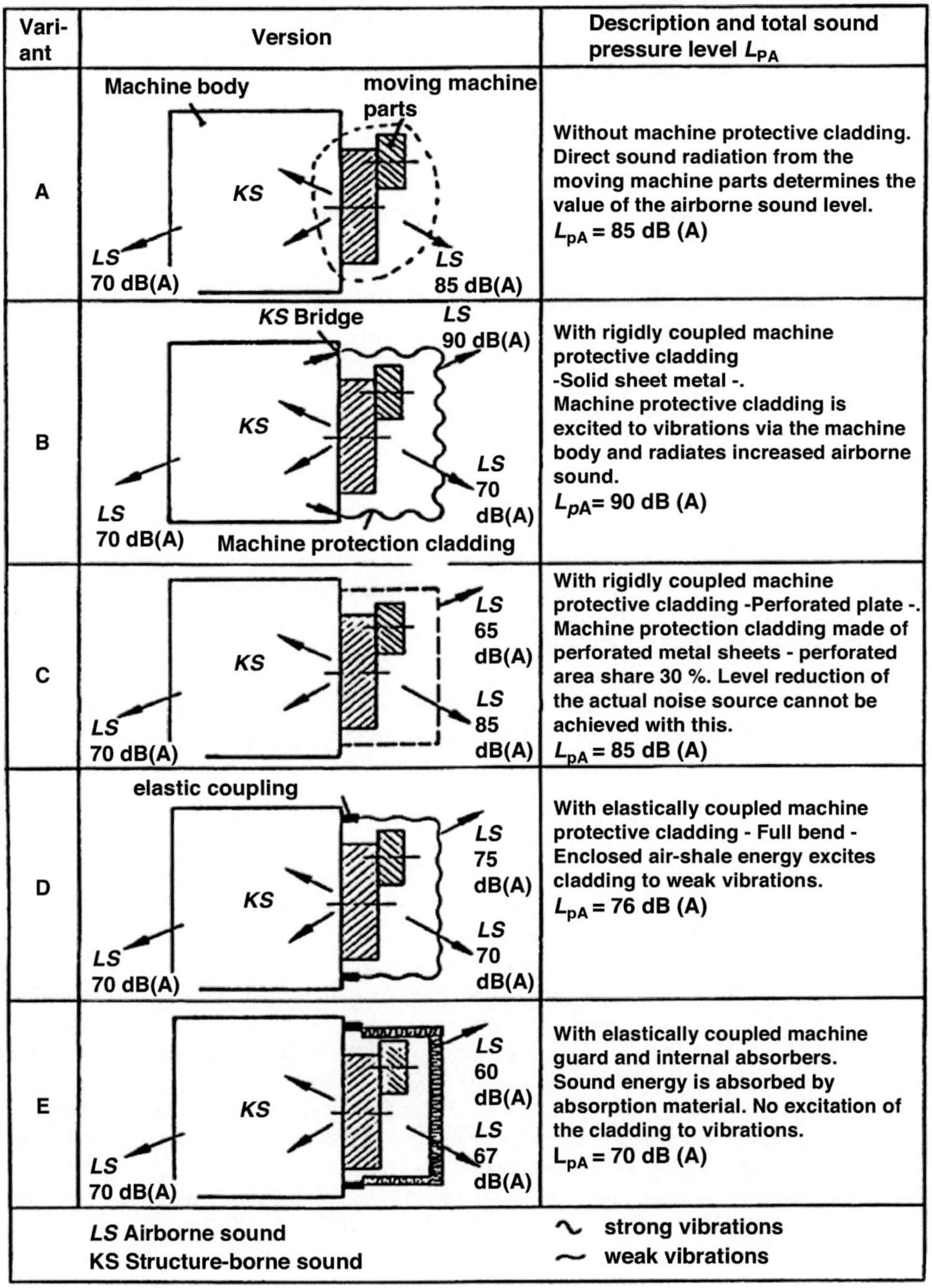

Fig. 6.84 Schematic representation of various design variants for machine protective cladding

Note

According to design variants E, machine structure-borne noise (KS) is also decisive!

References

1. Bartsch, H.-W., Sinambari, Gh.R.: Akustische Schwachstellenanalyse eines Antriebsfundaments. HDT-Fachveranstaltung E-H035-09-002-0, 2000
2. IBS GmbH, Frankenthal: Akustische Optimierung von Antriebsfundamenten. Nicht veröffentlichter Untersuchungsbericht Nr. 99.1.318e. Auftraggeber: RWE-Power (früher Rheinbraun AG), Köln (2000)
3. Sinambari, Gh.R., Thorn, U., Tschöp, E.: "Konstruktionsakustik", Seminarveranstaltung. IBS-Seminar, St. Martin/Pfalz, 2016
4. Sinambari, Gh.R., Felk, G., Thorn, U.: Konstruktionsakustische Schwachstellenanalyse an einer Verpackungsmaschine, VDI-Berichte Nr. 2052 (2008)
5. IBS-GmbH: Konstruktionsakustische Schwachstellenanalyse an einem Kartonierer vom Typ C 150 der Fa. Uhlmann Pac-Systeme GmbH & Co. KG, unveröff. Untersuchungsbericht Nr. 04.1.752-7 (2005)
6. Schalldruckpegel EUROGATE in 20 m Entfernung zum Mittelpunkt des Gerätes. Hauseigene Messvorschrift. EUROGATE Container Terminal Bremerhaven GmbH
7. EUROGATE Hausnorm Schall 01: Geräuschmessung an Hafenumschlaggeräten, Teil 1 – Luftschallemission von Van Carrier. Hauseigene Messvorschrift. EUROGATE Container Terminal Bremerhaven GmbH
8. Thorn, U., Wachsmuth, J.: Konstruktionsakustische Schwachstellenanalyse eines Straddle Carriers Typ Konecranes 54 DE und Erarbeiten von prinzipiellen Lärmminderungsmaßnahmen. VDI-Berichte 2118 (2010)
9. IBS GmbH: Akustische Messungen an einem Straddle Carrier Typ Konecranes 54 DE auf dem EUROGATE Container Terminal Hamburg. Unveröff. Untersuchungsbericht Nr. 08.1.206-1 (2008)
10. IBS GmbH: Schalltechnische Ursachenanalyse und Ausarbeitung von Lärmminderungsmaßnahmen an der Rauchgasreinigung VKR-KW Rauxel. Nicht veröffentlichter Untersuchungsbericht (1995)
11. Reuter, F.D., Sinambari, Gh.R.: Noise emission caused by a tube-nest heat exchanger behind a flue gas desulphurisation plant (FGD): Detecting and eliminating the causes. POWER-GEN Europe '99 Conference, Frankfurt, 1–3 June 1999
12. Sinambari, Gh.R., Thorn, U.: Strömungsinduzierte Schallentstehung in Rohrbündel-Wärmetauschern. Z. Lärmbekämpfung **47**(1) (2000)
13. Bühlmann, E.T., Oengören, A., Ziada, S.: Strömungserregte akustische Querresonanzen von Rohrbündeln in Wärmetauschern. Minderung von Rohrleitungsschwingungen. VDI-Berichte **748**, 43–68. VDI-Verlag, Düsseldorf (1989)
14. Fitzpatrick, J.A.: A design guide proposal for avoidance of acoustic resonances in inline heat exchangers. ASME J. Vib. Acoust. Stress Reliab. Des. **105**, 296–300 (1986)
15. Ziada, S., Oengören, A., Bühlmann, E.T.: On acoustical resonance in tube arrays, Part II: Damping criteria. J. Fluids Struct. **3** (1989)

16. IBS GmbH: Ausarbeitung und Überprüfung von konstruktiven Lärmminderungsmaßnahmen zur Reduzierung der tonalen Luftschall-Komponenten am DaGaVo des Kraftwerks Rauxel. Nicht veröffentlichter Untersuchungsbericht (1998)
17. Max, S.: Geräuschentwicklung am Rohrbündel-Wärmetauscher des Kraftwerks Rauxel – Ursachenanalyse anhand von Modellversuchen. Diplomarbeit, FH Bingen (1996)
18. Sinambari, Gh.R., Gnilke, M., Kunz, F.: Primäre Lärmminderung durch Maßnahmen des Maschinenhaus. VDI-KUT Jahrbuch (1999/2000)
19. Gnilke, M., Sinambari, G.R.: Akustische Optimierung eines Schaufelradgetriebes. HDT-Fachveranstaltung, Essen (2001)
20. Sinambari, Gh.R., Kunz, F.: Primäre Lärmminderung durch akustische Schwachstellenanalyse. VDI Berichte Nr. 1491 (1999)
21. Sinambari, Gh.R., Sentpali, S.: Ingenieurakustik, 6. Aufl. Springer Vieweg, Wiesbaden (2020)
22. Weber, A.: Eigenfrequenzanalyse an versteiften Rechteckplatten. Diplomarbeit, FH Bingen (2011)
23. VDI 2159: Characteristic values of emissions of technical sound sources; gearing noise (1985-7)
24. Sinambari, G.R., Carstens, J., Thorn, U.: Ursache schwingungsbedingter Schäden an den Gleitbahnhalterungen einer Tablettenpresse. VDI-Berichte Nr. **1887** (2005)
25. Sinambari, G.R., Thorn, U.: Schwingungsminderung bei Rohrleitungen. 3R Int. **44**(6) (2005)
26. DIN 4150-3: Vibrations in buildings - Part 3: Effects on structures (2016-12)
27. VDI 3842: Vibrations in piping systems (2004)
28. Grotheer, W.: Lärmminderung bei der Blechverarbeitung mit Pressen. VDI-Bericht 134 (1969)
29. Schmidt, K.-P.: Lärmarm konstruieren – Beispiele für die Praxis. Forschungsbericht Nr. 129. BAU, Wirtschaftsverlag NW (1974)
30. Schmidt, K.-P.: Arbeitswissenschaftliche Erkenntnisse 4/81 – Lärmminderung – Zangenvorschub 2. BAU, Wirtschaftsverlag NW (1981)
31. Doege, E.; Seidel, H.-J.: Der heutige Stand der Schneidtechnik unter Berücksichtigung der verfahrensseitigen Möglichkeiten zur Lärmminderung. VDI-Berichte 527 (1984)
32. Schmidt, K.-P.: Lärmminderung am Arbeitsplatz III; Änderung der Eingangsimpedanz als Maßnahme zur Lärmminderung. Forschungsbericht 169. BAU, Wirtschaftsverlag NW (1985)
33. Schmidt, K.-P.: Lärmminderung am Arbeitsplatz IV, ILK Institut für lärmarme Konstruktion. Schriftenreihe der Bundesanstalt für Arbeitsschutz (1987)
34. Matischka, G.: Geräuschemission ausgewählter mechanischer Pressen zum Schneiden und Maßnahmen zur Lärmminderung. Forschungsbericht 351. BAU, Wirtschaftsverlag (1983)
35. Robeck, R.; Schmidt, K.-P.: Arbeitswissenschaftliche Erkenntnisse 9/79 – Lärmminderung – Putztrommel. BAU, Wirtschaftsverlag (1979)
36. VDI-3720-1, Design of low-noise machinery and equipment - construction tasks and methodology (2014)
37. Dietz, P.; Gummersbach, F.: Fb 883, Lärmarm konstruieren XVIII – Systematische Zusammenstellung maschinenakustischer Konstruktionsbeispiele. Wirtschaftsverlag NW Verlag für neue Wissenschaft, Bremerhaven (2001)

Index

G. R. Sinambari, *Design Acoustics*,
https://doi.org/10.1007/978-3-658-40183-2

L

M

N

O